Dewatering, Desalting, and Distillation in Petroleum Refining

This book presents a detailed and practical description of various processes – dewatering, desalting, and distillation – that prepare refinery feedstocks for different conversion processes they will go through. Relevant process data are provided, and process operations are fully described. This accessible guide is written for managers, professionals, and technicians as well as graduate students transitioning into the refining industry.

Key Features:

- Describes feedstock evaluation and the effects of elemental, chemical, and fractional composition.
- Details the equipment and components and possible impacts due to composition.
- Explores the process options and parameters involved in dewatering, desalting, and distillation.
- Considers next-generation processes and developments.

Petroleum Refining Technology Series

Series Editor:
James G. Speight

This series of books is designed to address the current processes used by the refining industry and take the reader through various steps that are necessary for crude oil evaluation and refining. Technological advancements and processing innovations are highlighted in each of the volumes.

Refinery Feedstocks
James G. Speight

Dewatering, Desalting, and Distillation in Petroleum Refining
James G. Speight

Dewatering, Desalting, and Distillation in Petroleum Refining

James G. Speight

CRC Press
Taylor & Francis Group
Boca Raton London New York

CRC Press is an imprint of the
Taylor & Francis Group, an **informa** business

Cover image: © Shutterstock

First edition published 2023
by CRC Press
6000 Broken Sound Parkway NW, Suite 300, Boca Raton, FL 33487-2742

and by CRC Press
4 Park Square, Milton Park, Abingdon, Oxon, OX14 4RN

CRC Press is an imprint of Taylor & Francis Group, LLC

© 2023 James G. Speight

Library of Congress Cataloging-in-Publication Data
Names: Speight, James G., author.
Title: Dewatering, desalting, and distillation in petroleum refining / James G. Speight.
Description: First edition. | Boca Raton : CRC Press, 2022. | Series:
Dewatering, desalting, and distillation in petroleum refining |
Includes bibliographical references and index.
Identifiers: LCCN 2022028320 | Subjects: LCSH: Petroleum—Refining—Desalting. | Petroleum—Dewatering. |
Distillation, Fractional. Classification: LCC TP690.45 .S6283 2022 | DDC 622/.3382—dc23/eng/20220808
LC record available at https://lccn.loc.gov/2022028320

ISBN: 9781032027340 (hbk)
ISBN: 9781032027388 (pbk)
ISBN: 9781003184959 (ebk)

DOI: 10.1201/9781003184959

Typeset in Times
by codeMantra

Contents

Contents

Preface

As the second book of the refinery series, this book presents a description of various dewatering, desalting, and distillation processes that prepare the refinery feedstocks for various conversion processes that lead to the production of specification grade products.

Crude oil introduced to a refinery for any form of processing contains many undesirable impurities, such as sand, inorganic salts, drilling mud, polymer, and corrosion byproduct. The purpose of crude oil dewatering and desalting is to remove these undesirable impurities, especially salts and water, from the crude oil prior to distillation. During this processing, steps are necessary to remove the oleophobic impurities and typically they occur in two ways which are (1) field separation and (2) dewatering and desalting. Field separation after the crude oil is produced from the reservoir is the first step in which gases are removed and the water and dirt that accompany crude oil coming from the ground and are located in the field near the site of the wells.

The first processes (desalting and dewatering) are focused on the cleanup of the feedstock, particularly the removal of the troublesome brine constituents. This is followed by distillation to remove the volatile constituents with the concurrent production of a residuum that can be used as a cracking (coking) feedstock or as a precursor to asphalt. Other methods of feedstock treatment that involve the concept of volatility are also included here even though some of the methods (such as stripping and rerunning) might also be used for product purification. However, the distillation step has recently been recognized as a viable step that produces additional valuable high-boiling fractions for the bitumen.

The book is designed to address the processes involved in the cleanup (dewatering and desalting) of crude oil and various aspects of distillation leading to the production of bulk products that are sent for further refining to produce saleable products.

Chapter 1 presents an overview of various types of feedstocks that are currently accepted and used by refineries and emphasizes on the dewatering, desalting, and distillation processes within a refinery. This chapter also presents description of the nature of the dewatering, desalting, and distillation processes, and the behavior of these feedstocks in dewatering and desalting operations in these processes needs to be better understood.

Chapter 2 presents the process options that can be used for the dewatering and desalting processes that prepare the feedstock for introduction to the distillation section of the refinery and then to a variety of conversion processes which are often, incorrectly, referred to as refining proper. This chapter also includes process options for heavy feedstocks and the potential for corrosion and fouling of refinery equipment if the dewatering–desalting processes are not performed with the maximum efficiency.

Chapter 3 introduces the concept of incompatibility (and instability) of feedstocks in the refinery and how this event might be affected by dewatering-desalting and distillation processes. In addition, some of the more prominent methods are used for determining the potential for incompatibility and instability of crude oil feedstocks and crude oil products. The choice of the method is subject to the composition and properties of the feedstock and the process parameters as well as desired product.

Chapter 4 presents a description of the distillation processes that are applied to a feedstock in a refinery which remove the volatile constituents with the concurrent production of a residuum that can be used as a cracking (coking) feedstock or as a precursor to asphalt. Current methods of bitumen processing involve direct use of the bitumen as feedstock for delayed or fluid coking.

Chapter 5 deals with the ancillary distillation processes. Because of the presence of the lower boiling and higher boiling constituents within each side-stream distillation fraction, there is a need to reprocess the side-stream fractions using any one (or more of a variety of processes to ensure that the requirements of the distillation fractions are met.

Chapter 6 discusses the disadvantages and consequences of not having a robust and efficient dewatering–desalting system in place, and it is necessary to point out the types of corrosion and fouling that can occur and the consequences that follow when corrosion occurs. Thus, the focus of this chapter is to present a more detailed presentation of the corrosion and fouling that can occur in crude oil refineries to emphasize the need for feedstock cleanup as soon as the feedstocks (that are used to produce the feedstock blend) enter the refinery and the consequences of inefficient dewatering and desalting processes.

The reader will be guided through various steps that are necessary for the pretreatment and distillation of crude oil in each chapter. This book updates the reader's knowledge, provides more information, and offers processing options that could be used in the processes of the 21st century.

This book will meet the needs of engineers and scientists at all levels from academia to the refinery by understanding the evolutionary changes that have occurred till date; it will also help them in understanding the initial refining processes and preparing them for new changes and industry's evolution.

The target audience includes engineers, scientists, and students who are interested in learning more about petroleum processing and where the industry is headed in the next 50 years. The book will also be useful for non-technical readers, with the aid of the extensive glossary.

James G. Speight,
Laramie, Wyoming, USA.

Author

James G. Speight has a B.Sc. and Ph.D. in Chemistry; he also holds a D.Sc. in Geological Sciences and a Ph.D. in Petroleum Engineering. He has more than 50 years of experience in areas associated with (1) the properties, recovery, and refining of conventional petroleum, heavy oil, and tar sand bitumen; (2) the properties and refining of natural gas; and (3) the properties and refining of biomass, biofuels, biogas, and the generation of bioenergy. His work has also focused on environmental effects, environmental remediation, and safety issues associated with the production and use of fuels and biofuels. He is the author (and co-author) of more than 95 books related to petroleum science, petroleum engineering, biomass and biofuels, and environmental sciences.

Although he has always worked in private industry which focused on contract-based work, he has served as a Visiting Professor in the College of Science, University of Mosul (Iraq) and has also been a Visiting Professor in Chemical Engineering at the Technical University of Denmark and the University of Trinidad and Tobago as well as held adjunct positions at various universities. He has also reviewed more than 25 theses as a thesis examiner.

As a result of his work, he has been honored with the following awards:

- Diploma of Honor, United States National Petroleum Engineering Society. *For Outstanding Contributions to the Petroleum Industry*, 1995.
- Gold Medal of the Russian Academy of Sciences. *For Outstanding Work in the Area of Petroleum Science*, 1996.
- Einstein Medal of the Russian Academy of Sciences. *In recognition of Outstanding Contributions and Service in the field of Geologic Sciences*, 2001.
- Gold Medal – Scientists without Frontiers, Russian Academy of Sciences. *In recognition of His Continuous Encouragement of Scientists to Work Together across International Borders*, 2005.
- Gold Medal – Giants of Science and Engineering, Russian Academy of Sciences. *In recognition of Continued Excellence in Science and Engineering*, 2006.
- Methanex Distinguished Professor, University of Trinidad and Tobago. *In Recognition of Excellence in Research*, 2007.
- In 2018, he received the American Excellence Award for Excellence in Client Solutions from the United States Institute of Trade and Commerce, Washington, DC.

1 Refinery Operations

1.1 INTRODUCTION

Products from crude oil feedstocks, such as various crude oils (including heavy crude oil) and various viscous feedstocks (such as extra heavy oil and tar sand bitumen), and unconventional sources, such as liquids from coal (coal liquids), oil shale (shale oil), and biomass (bio-oil), are important aspects of current and future energy sources. In addition, it is projected with some degree of speculation that the continued use fossil resources at current rates will have serious and irreversible consequences for the global climate, although other natural effects are also in play (Speight, 2020c). Whatever the rationale and however the numbers are manipulated, the supply of crude oil, which is the basic feedstock for refineries and for the petrochemicals industry, is finite, and in this context, the dominant position of crude oil will become unsustainable as supply and demand issues erode the economic advantage of crude oil over other alternative feedstocks.

In fact, whatever the type of feedstock, refinery feedstocks come from different geographical locations that affect the properties of the oil. In the natural state, unrefined state, crude oil, heavy

crude oil, extra heavy crude oil, and tar sand bitumen range in density and consistency from very thin, lightweight, and volatile fluid liquids to an extremely viscous, semi-solid material (Parkash, 2003; Gary et al., 2007; Speight, 2014, 2017; Hsu and Robinson, 2017) all of which require a variety of processes to provide the necessary saleable products. There is also a wide gradation in the color of the feedstocks, which range from a light, golden yellow (conventional crude oil) to the black bitumen that is isolated from tar sand formations (tar sand bitumen).

The crude oil that is received directly from the wellbore in the natural state is often termed raw crude oil before it is treated in a separation plant (at the production site prior to transportation as well as at the refinery) and usually contains non-hydrocarbon contaminants in the natural state crude oil. A similar rationale also applies to heavy crude oil, extra heavy crude oil, and tar sand bitumen which, along with conventional crude oil, are not homogeneous materials and the physical characteristics differ depending on where the feedstock was produced.

After recovery of the crude oil from the reservoir, there are necessary steps to remove the oleophobic impurities and it typically occurs in two ways which are (1) field separation and (2) dewatering and desalting at the refinery. Field separation after the crude oil is produced from the reservoir is the first step in which gases are removed and the water and dirt that accompany crude oil coming from the ground and is located in the field near the site of the wells. The crude oil is treated to meet the specification for the crude oil to enter a pipeline (or transportation) system and the consequences of omitting these field processes will become evident through pipeline fouling (i.e. the deposition of solids) and corrosion.

However, the crude oil that is introduced to a refinery for any form of processing will still contain undesirable impurities, such as sand, inorganic salts, drilling mud, polymer, and corrosion byproducts. The purpose of crude oil dewatering and desalting is to remove these undesirable impurities, especially salts and water, from the crude oil prior to distillation. Failure to do so can have an adverse effect on the refinery operations which will suffer from incompatibility and instability that can lead to fouling and corrosion of refinery equipment (Chapters 3 and 6).

Thus, crude oil, heavy crude oil, extra heavy crude oil, and tar sand bitumen are not (within the individual categories) uniform materials, and the chemical and physical (fractional) composition of each of these refinery feedstocks can vary not only with the location and age of the reservoir or deposit but also with the depth of the individual well within the reservoir or deposit. On a molecular basis, the three feedstocks are complex mixtures containing (depending upon the feedstock)

DOI: 10.1201/9781003184959-1

hydrocarbons with varying amounts of hydrocarbonaceous constituents that contain sulfur, oxygen, and nitrogen as well as constituents containing metallic constituents, particularly those containing vanadium nickel, iron, and copper. The hydrocarbon content may be as high as 97% w/w, for example, in a light crude oil or less than 50% w/w in heavy crude oil, extra heavy crude oil, and tar sand bitumen (Parkash, 2003; Gary et al., 2007; Speight, 2014, 2017; Hsu and Robinson, 2017).

Furthermore, whatever the source of the feedstock for a refinery, there is always the need for pretreatment of which dewatering and desalting will need to be performed to reduce any water and mineral salts from the feedstock. The most recent challenges are to refine more heavy crude oil, extra heavy crude oil, and tar sand bitumen as well as foamy oils, high-acid crude oils, and opportunity crude oils (Speight, 2014, 2015).

By the way of clarification (since more details of these feedstock are presented later in this chapter), foamy crude oil is in the form of an oil-continuous foam that contains dispersed gas bubbles produced at the wellhead from heavy crude oil reservoirs under solution gas drive, while high-acid crude oils are crude oils that contain considerable proportions of naphthenic acid derivatives in which (collectively) all of the organic acids are present in the crude oil. Finally, opportunity crude oils are either new crude oils with unknown or poorly understood properties relating to processing and are more difficult to process due to high levels of solids (and other contaminants) that are produced from the well with the oil and may also have high levels of acidic constituents as well as high viscosity. The opportunity crude oils often contain constituents that are incompatible with other crude oil feedstocks and, as such, be liable to cause excessive equipment fouling (and/or corrosion) when processed either separately or in a blend.

In addition, constituents that are found in crude oil, heavy crude oil, extra heavy crude oil, and tar sand bitumen can have major effects on the dewatering, desalting, and distillation processes and, although descriptions of these feedstocks have been presented elsewhere (Speight, 2013a, 2014, 2017, 2021a), for completeness of the current text, brief descriptions of conventional crude oil, heavy crude oil, extra heavy crude oil, and tar sand bitumen are also presented here. In fact, the issues that arise when processing heavy crude oil, extra heavy crude oil, and tar sand bitumen can also be equated to the chemical character and the amount of complex, higher boiling constituents in the feedstock. Refining these materials is not just a matter of applying know-how derived from refining conventional crude oils but it requires knowledge of the chemical structure and chemical behavior of these more complex constituents, which also include the metal-containing porphyrin derivatives.

The purpose of this chapter is to present a brief overview of the types of feedstocks that are currently accepted and used by refineries and to place an emphasis on the dewatering, desalting, and distillation processes within a refinery. This chapter also presents descriptions of the nature of the dewatering, desalting, and distillation processes, so that the behavior of these feedstocks in these process operations can be better understood.

1.2 REFINERY CONFIGURATION

Before describing various types of feedstocks sent to refineries, it is advisable to understand various types of configuration that are used in the refining industry. However, it must be recognized that whatever the configuration, the feedstock must go through the dewatering and desalting operations.

A crude oil refinery is an industrial processing plant that is composed of a collection of integrated unit processes (Figure 1.1) (Parkash, 2003; Gary et al., 2007; Speight, 2014, 2017; Hsu and Robinson, 2017).

The feedstock that is used by the typical refinery – once a conventional crude oil from a single reservoir – is, in the modern refinery, typically a blend of two or more crude oils from different reservoirs that may also include a viscous feedstock (or two) in compatible amounts insofar as blending (mixing) does not cause the formation of a separate phase (Chapter 3). The separate phase may be a combination of the higher molecular weight and/or the more polar constituents of the viscous

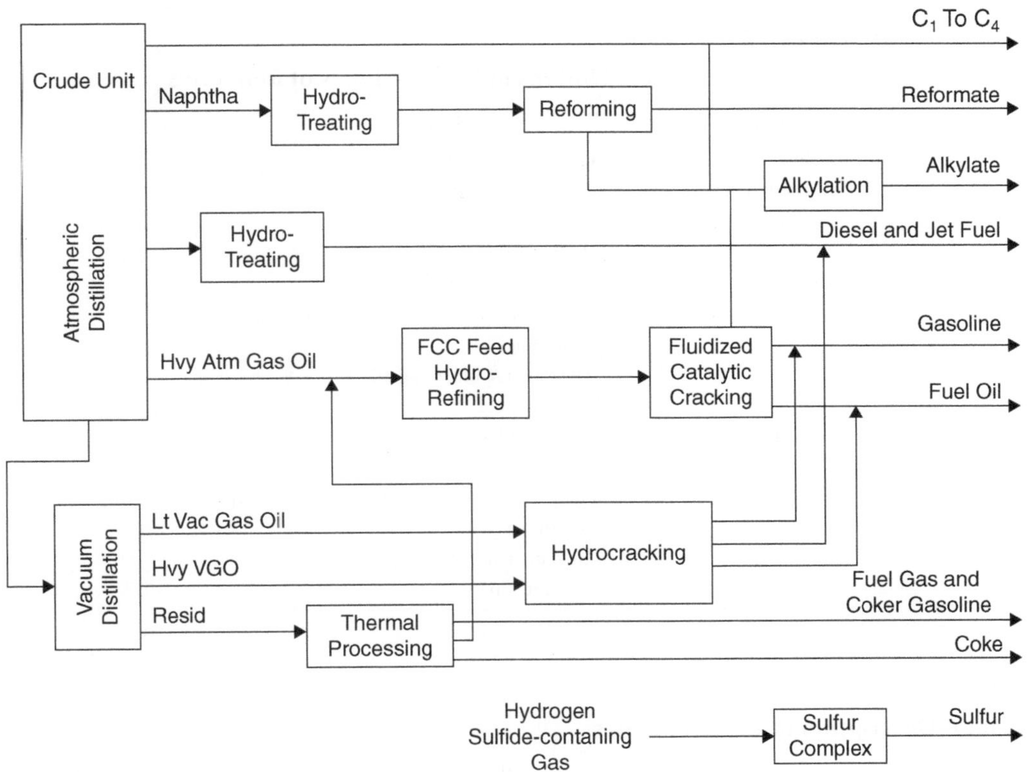

FIGURE 1.1 Schematic overview of a refinery.

feedstock (i.e. the asphaltene constituents and some associated resin constituents) that are incompatible with the other constituents of the blend. Thus, caution is advised when choosing crude oils as blend stock and a series of test methods are available to identify the potential for incompatibility (Chapter 3).

In the early to middle of the 20th century, refineries were originally designed to operate within a narrow range of feedstock properties and to produce a relatively fixed slate of products. Since the "oil shocks" of the 1970s, the flexibility of the refinery needs to be increased in order to adapt to a system that requires processing of the unconventional (heavier) feedstocks. Several possible paths may be used by refiners to increase their flexibility within the existing refineries. Example of the flexibility is the severity of the operating parameters of a variety of process units by varying the range of feedstock inputs used by the refinery.

In addition, there has been the need to modify the existing processes and accommodate an influx of new processes to accommodate the need for refinery flexibility, but it is limited by the constraint of strict complementarity of the new units with the rest of the existing plant and involves a higher risk than the previous ones. It is not surprising that many refiners decide to modify the existing processes. As a result, refineries have generally evolved into three types of which examples are (1) the topping refinery, (2) the hydroskimming refinery, and (3) the conversion refinery, which include the deep conversion refinery (Table 1.1).

Each type of refinery has its own distinctive character and process train (Parkash, 2003; Gary et al., 2007; Speight, 2017; Hsu and Robinson, 2017). However, the configuration of various refinery options may vary from refinery to refinery or there may be two or more refinery configurations on the same refinery site and it is the choice (any of the refiners) of the active process units that give the name to the so-called refinery type.

TABLE 1.1

Examples of Process Options in Various Types of Refineries

Refinery Type

Topping	Crude distillation
Hydroskimming	Crude distillation
	Reforming
	Hydrotreating
Conversion	Crude distillation
	Fluid catalytic cracking and/or hydrocracking
	Reforming
	Alkylation & other upgrading
	Hydrotreating
Deep Conversion	Crude distillation
	Coking
	Fluid catalytic cracking and/or hydrocracking
	Reforming
	Alkylation and other upgrading
	Hydrotreating

TABLE 1.2

General Description of Refinery Types

Refinery Type	Processes	Other Name	Complexity	Complexity
Topping	Distillation	Skimming	Low	1
Hydroskimming	Distillation	Hydroskimming	Moderate	2–5
	Reforming			
	Hydrotreating			
Conversion	Distillation	Cracking	High	6–7
	Fluid catalytic cracking			
	Hydrocracking			
	Reforming			
	Alkylation			
	Hydrotreating			
Deep conversion	Distillation	Coking	Very high	10
	Coking			
	Fluid catalytic cracking			
	Hydrocracking			
	Reforming			
	Alkylation			
	Hydrotreating			

1.2.1 THE TOPPING REFINERY

The topping refinery (sometimes referred to as a modular mini refinery) is designed to prepare feedstocks for petrochemical manufacture or for the production of industrial fuels in remote oil-production areas (Speight, 2014, 2017). The topping refinery consists of tankage, a distillation unit, recovery facilities for gases and low-boiling hydrocarbon derivatives, and the necessary utility systems (steam, power, and water treatment plants) (Table 1.2). The actual topping unit is an atmospheric distillation unit that produces a variety of fractions (Chapter 4) and is a technically simple method to enhance the value of crude oil near the source of production.

Feedstock	Distillation	Primary Products	Final Products
		Gases	
		Naphtha	Light naphtha
			Heavy naphtha
		Kerosene	
		Gas oil	Light gas oil
			Heavy gas oil
		Residuum	

This type of refinery configuration is best utilized in remote locations where naphtha and kerosene are the desired products. The refinery feedstock may be the lowest cost feedstock because the transportation costs are minimized. A topping refinery with a viscous heavy crude and low API gravity will produce more fuel oil and less naphtha and kerosene. On the other hand, a crude oil with a high API gravity will produce more low-boiling distillates in the form of naphtha and kerosene and less fuel. Additionally, the sulfur content of the refinery feedstock determines refinery cost as low sulfur crudes may not require hydrotreaters.

1.2.2 THE HYDROSKIMMING REFINERY

A hydroskimming refinery is a refinery equipped with an atmospheric distillation unit, a reforming unit, and a hydrotreating unit. The configuration gets its name from having a reformer as the only real conversion unit.

A hydroskimming refinery is therefore more complex than a topping refinery (which just separates the crude into its constituent petroleum products by atmospheric distillation and produces naphtha and kerosene).

Feedstock	Distillation	Products	Hydrotreater/Reformer Products
		Gases	Petrochemical feedstocks
		Naphtha	Gasoline
		Kerosene	Diesel fuel
		Residuum	

The hydroskimming refinery has the ability to switch between conventional (low-density, high API gravity, conventional) feedstocks and viscous feedstocks which means that processing a viscous crude oil may require the presence of a larger naphtha hydrotreater, a larger naphtha reformer, and a larger kerosene hydrotreater.

1.2.3 THE CONVERSION REFINERY

The conversion refinery (Figure 1.1) is the most versatile configuration of the three and is typically based on coking technology, and the refinery based on catalytic cracking is an example (Speight, 2014, 2017). Conversion process units are used to convert one hydrocarbon stream into another by changing the molecular size and structure of the constituents. The objective is to shift the yield of the refinery away from less valuable products (dictated by market demand) and toward more valuable products, such as naphtha and kerosene which are precursors of gasoline and diesel fuel.

Most refineries therefore add vacuum distillation and catalytic cracking, which adds one more level of complexity by reducing fuel oil by conversion to low-boiling distillates and middle distillates. The conversion refinery incorporates all the basic units found in both the topping and hydroskimming refineries, but it also features gas oil conversion plants such as catalytic cracking and hydrocracking units, olefin conversion plants such as alkylation or polymerization units, and frequently, coking units for sharply reducing or eliminating the production of residual fuels.

Another named option, which is similar to the conversion refinery, is the arbitrarily named deep conversion refinery (often simply referred to as a coking refinery) which is, as the name implies, a special class of conversion refinery that is configured to convert all of the feedstock to lower boiling products, including gases. The deep conversion refinery includes not only catalytic cracking and/or hydrocracking to convert gas oil fractions but also one or more coking units. Coking units convert the highest molecular weight constituents of the feedstock (the residuum) by conversion to lower boiling product streams that serve as additional feedstocks to other conversion processes (such as catalytic cracking) and to upgrading processes (such as catalytic reforming) that produce the more valuable low-boiling products. A deep conversion refinery with sufficient coking capacity can convert essentially all of the residua produced from the crude feedstock(s) to lower boiling products (Parkash, 2003; Gary et al., 2007; Speight, 2014a, 2017; Hsu and Robinson, 2017).

The yields and quality of refined products produced by any particular refinery depend on the mixture of components used in the feedstock blend as well as the configuration of the refinery facilities. Light/sweet (low sulfur) crude oil produces higher yields of higher value low-boiling products such as naphtha, kerosene, and gas oil. Viscous sour (high sulfur) feedstocks are generally less expensive and produce higher yields of lower value higher boiling products (such as vacuum gas oil (VGO) and residua) that must be converted into lower saleable boiling products (Speight, 2103, 2014).

Thus, the configuration of refineries may vary from refinery to refinery and may be more oriented toward the production of naphtha for gasoline manufacture by the use of reforming and/or catalytic cracking processes, whereas the configuration of other refineries may be more oriented toward the production of distillates such as kerosene, low-boiling atmospheric gas oil, high-boiling atmospheric gas oil, and VGO in order to convert these distillates to saleable products.

The selection of refinery feedstocks must take into account compatibility in order to have a successful refinery operation. The constantly changing crude oil slate composed of conventional and unconventional crudes as well as mixtures of these crudes provides an opportunity for refiners to maximize the potential of the refinery. However, caution is advised when choosing conventional feedstocks and unconventional feedstocks to be blended as a composite feedstock for the refinery because without the application of the appropriate test methods there is the potential for incompatibility within the blend (Chapter 3) (Sayles and Routt, 2011; Speight, 2014, 2017). However, maximizing the potential of a refinery does come with a risk of equipment and/or fouling or the failure to achieve desired objectives. The use of crude oil compatibility prediction processes can be used to predict the performance of unconventional crudes and crude oil blends in a specific refinery before the blend is processed (Chapter 3).

It is certain, however, that whatever the configuration of the refinery, the refinery feedstocks will need pretreatment (dewatering, desalting, and in the case of blends, compatibility tests) prior to introduction into the distillation unit. If not, serious corrosion can occur thereby shutting down the unit and the refinery until repairs are completed.

1.3 REFINERY FEEDSTOCKS

In the simplest definition, a refinery feedstock is the crude oil that is destined for processing in the refinery and is produced from a reservoir or a deposit (in fact from any geological formation) by means of one or more wells drilled into the formation. By this means, the crude oil is transformed into one or more components and/or finished products (Parkash, 2003; Gary et al., 2007; Speight,

2014, 2017; Hsu and Robinson, 2017). In the modern refinery, this has changed due to the acceptance of the refinery of two or more crude oil as a blend.

In recent years, the average quality of crude oil has become deteriorated and continues to do so as more heavy crude oil, extra heavy crude oil, and tar sand bitumen are being sent to refineries (Speight, 2013a, 2014, 2017, 2020a). This has caused the nature of crude oil refining change considerably. Indeed, the declining reserves of lighter (low-density, high API) crude oil have resulted in an increasing need to develop options to desulfurize and upgrade the viscous (high-density, low API) feedstocks, specifically the viscous feedstocks. This has resulted in a variety of process options that specialize in maintaining the feedstock flow through the reactor. Therefore, refineries need to be constantly adapted and upgraded to remain viable and responsive to ever-changing patterns of crude supply and product market demands. As a result, increasingly complex process options have been introduced into refineries to produce higher yields of lower boiling products (such as naphtha and kerosene) from the viscous feedstocks.

Because of the need for a thorough understanding of crude oil and the associated technologies, it is essential that the definitions and the terminology of crude oil science and technology be given prime consideration. This will aid in a better understanding of crude oil, its constituents, and its various fractions. Of various forms of terminology that have been used not all have survived, but the more commonly used are illustrated here. Particularly troublesome, and more confusing, are those terms that are applied to the more viscous materials, for example, the use of the terms extra heavy crude oil, bitumen, asphalt, and pitch.

Briefly, in the context of this book, extra heavy crude oil is a viscous material that is near-solid or solid at ambient temperature but has some mobility in the deposit because of the high temperature of the deposit (i.e. the temperature of the deposit abides with the pour point of the material). Bitumen is the viscous near-solid or solid material that occurs in tar sand formations and is not recoverable by any of the conventional recovery methods. Asphalt is not a naturally occurring material but is the product of a refinery and is a mixture of aggregates (inorganic materials, such as crushed rock, sand, gravel, or slags), binder, and filler that is used for constructing and maintaining roads, parking areas, railway tracks, ports, airport runways, bicycle lanes, sidewalks, and also play and sport areas. Pitch is the organic residue that is near-solid or solid residue from the thermal decomposition of coal and which remains after volatile tars have been removed by distillation (Speight, 2013a, 2014).

Each type of feedstocks (as enumerated below) has sufficiently different properties that require consideration before the feedstock is subjected to the dewatering/desalting operations either prior to transportation or when the feedstock reached the refinery. In addition, blending two or more of these feedstocks (especially without the use of a pre-blending testing protocol) could lead to the serious consequences of incompatibility of the blend (Chapters 3 and 6) and the depletion of a separate phase (typical a solid phase) within the dewatering/desalting equipment or in the lead up to, and in, the distillation tower. Either event could lead to shutdown of the equipment and possible shutdown of the refinery.

Thus, the properties of each of these feedstocks can affect the dewatering and desalting operations. Contribution of the constituents to the viscosity of the feedstock determines the temperature at which the dewatering and desalting operations will be most efficient. In addition, the presence of polar constituents can also affect the operations.

1.3.1 CONVENTIONAL CRUDE OIL

The terms crude oil and the equivalent term petroleum cover a wide assortment of materials consisting of mixtures of hydrocarbon derivatives and other compounds containing variable amounts of sulfur, nitrogen, and oxygen, which may vary widely in volatility, specific gravity, and viscosity. Metal-containing constituents, notably those compounds that contain vanadium and nickel and which usually occur in the asphaltene fraction, usually occur in the more viscous crude oils in amounts up to several thousand parts per million and can have serious consequences during the processing of these feedstocks (Speight, 2014, 2019a). Because crude oil is a mixture of widely varying constituents and proportions, its physical properties also vary widely and the color varies from colorless to black. Thus, the definition of crude oil has been varied, unsystematic, diverse, and often archaic. Furthermore, the terminology of crude oil is a product of many years of growth. Thus, the long-established use of an expression, however inadequate it may be, is altered with difficulty, and a new term, however precise, is at best adopted only slowly.

Since there is a wide variation in the properties of crude oil, the proportions in which different constituents occur vary with origin (Speight, 2014, 2015, 2017). Thus, some crude oils have higher proportions of the lower boiling components and others (such as extra heavy crude oil and tar sand bitumen) have higher proportions of higher boiling components (i.e. resin constituents and asphaltene constituents).

Conventional crude oil (often referred to simply as crude oil) is a mixture of gaseous, liquid, and solid hydrocarbon derivatives (such as waxes) that occurs typically in the liquid state in sedimentary rock deposits throughout the world and also contains small quantities of nitrogen-, oxygen-, and sulfur-containing compounds as well as trace amounts of metallic constituents (Speight, 2014, 2021a).

The constituents of crude oil, in addition to hydrocarbon derivatives, may also include compounds of sulfur, nitrogen, oxygen, and metals and other elements. Some crude oils have higher proportions of the lower boiling components and others (such as heavy oil and bitumen) have higher proportions of higher boiling components (asphaltic components and residuum). In fact, the molecular boundaries of crude oil cover a wide range of boiling points and carbon numbers which dictate the options to be used in a refinery (Parkash, 2003; Gary et al., 2007; Speight, 2014, 2017; Hsu and Robinson, 2017). However, the actual boundaries of such a *crude oil map* can only be arbitrarily defined in terms of boiling point and carbon number. In fact, crude oil is so diverse that materials from different sources exhibit different boundary limits, and for this reason alone, it is not surprising that crude oil has been difficult to *map* in a precise manner.

Of particular interest in terms of the content of metals, metal content of crude oil is then naturally occurring in porphyrin derivatives which are macrocyclic organic compounds, composed of four modified pyrrole subunits interconnected at their α carbon (alpha-carbon) atoms via methine (–CH=) bridges. The parent of molecule of the porphyrin derivatives is porphine (Figure 1.2).

These derivatives usually occur in the non-basic portion of the nitrogen-containing concentrate and a large number of different porphyrin compounds exist in nature or have been synthesized. Most of these compounds have substituents other than hydrogen on many of the ring carbons. For example, in a porphyrin molecule, the metal (such as nickel or vanadium and sometimes iron) is

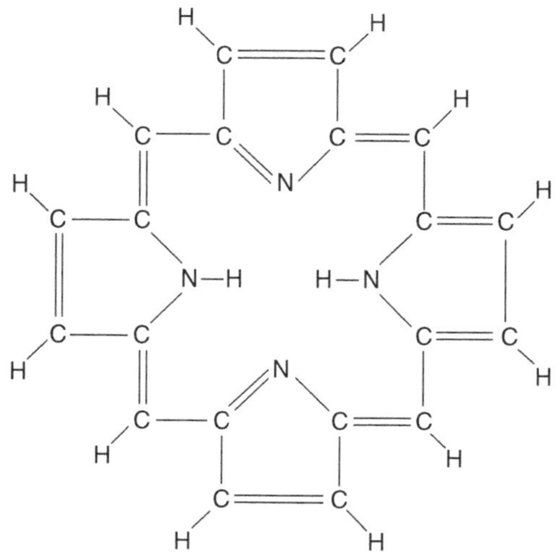

FIGURE 1.2 Porphine.

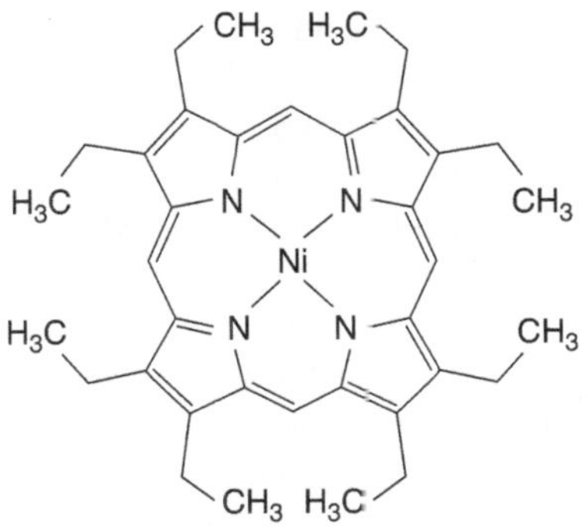

FIGURE 1.3 Nickel porphyrin.

bonded in the center of the molecule between the four nitrogen atoms with alkyl groups replacing some of the hydrogen atoms on the outer edges of the molecule (Figure 1.3).

Almost all crude oil, heavy oil, and bitumen contain detectable amounts of vanadium and nickel porphyrin derivatives. More mature, lighter crude oils usually contain only small amounts of these compounds. Heavy crude oils may contain large amounts of vanadium and nickel porphyrin derivatives. For example, the concentration of vanadium is known for some crude oils to be in excess of 1000 ppm and a substantial amount of the vanadium in these crude oils is chelated with porphyrins. In high-sulfur crude oil of marine origin, vanadium porphyrin derivatives are more abundant than nickel porphyrin derivatives. Low sulfur crude oils of lacustrine origin usually contain more nickel porphyrin derivatives than vanadium porphyrin derivatives.

Tight oil (tight crude oil) is another type of crude oil that occurs in tight (low-permeability) sandstone formations as well as various shale formations that have an effective permeability of less than 1 milliDarcy (<1 mD). The tight crude oil is typically a light (low-density) highly volatile crude oil. The production of oil also yields a significant amount of volatile gases (including propane and butane) and low-boiling liquids. Because of the presence of low-boiling hydrocarbon derivatives, low-boiling naphtha (light naphtha) can become extremely explosive even at relatively low ambient temperatures. Some of these gases may be burned off (flared) at the field wellhead – provided local and federal standards are met for the flaring process – but others remain in the liquid products extracted from the well (Speight, 2014).

Typically, crude oil from tight formations is characterized by low asphaltene content, low sulfur content, and a significant molecular weight distribution of the paraffinic wax content (Speight, 2014, 2015). Paraffin carbon chains on the order of C10 to C60 have been identified in the tight oil.

Also, there are several other types of crude oil that need consideration. These are (1) foamy oil, (2) high-acid crude oil, and (3) opportunity crude oil.

1.3.2 Foamy Oil

Foamy crude oil is oil-continuous foam that contains dispersed gas bubbles produced at the wellhead from heavy crude oil reservoirs under solution gas drive. The nature of the gas dispersions in oil distinguishes foamy oil behavior from conventional heavy crude oil. The gas that comes out of solution in the reservoir does not coalesce into large gas bubbles or into a continuous flowing gas phase. Instead, it remains as small bubbles entrained in the crude oil, keeping the effective oil viscosity low while providing expansive energy that helps drive the oil toward the producing well. Foamy oil accounts for unusually high production in heavy crude oil reservoirs under solution gas drive.

During primary production of heavy crude oil from solution gas drive reservoirs, the oil is pushed into the production wells by energy supplied by the dissolved gas. As fluid is withdrawn from the production wells, the pressure in the reservoir declines and the gas that was dissolved in the oil at high pressure starts to come out of solution (foamy oil). As pressure declines further with continued removal of fluids from the production wells, more gas is released from solution and the gas already released expands in volume. The expanding gas, which at this point is in the form of isolated bubbles, pushes the oil out of the pores and provides energy for the flow of oil into the production well. This process is very efficient until the isolated gas bubbles link up and the gas itself starts flowing into the production well. Once the gas flow starts, the oil has to compete with the gas for available flow energy. Thus, in some heavy crude oil reservoirs, due to the properties of the oil and the sand and also due to the production methods, the released gas forms foam with the oil and remains subdivided in the form of dispersed bubbles much longer.

Thus, foamy oil is formed in solution gas drive reservoirs when gas is released from solution with a decline in reservoir pressure. It has been noted that the oil at the wellhead of these heavy oil reservoirs resembles the form of foam, hence the term foamy oil. The gas initially exists in the form of small bubbles within individual pores in the rock. As time passes and pressure continues to decline, the bubbles grow to fill the pores. With further declines in pressure, the bubbles created in different locations become large enough to coalesce into a continuous gas phase. Once the gas phase becomes continuous (i.e. when gas saturation exceeds the critical level) – the minimum saturation at which a continuous gas phase exists in porous media – traditional two-phase (oil and gas) flow with classical relative permeability occurs. As a result, the production of gas–oil ratio (GOR) increases rapidly after the critical gas saturation has been exceeded.

Foamy oil is oil-continuous foam that contains dispersed gas bubbles produced at the wellhead from heavy crude oil reservoirs under solution gas drive. The nature of the gas dispersions in oil distinguishes foamy oil behavior from conventional heavy crude oil. The gas that comes out of solution in the reservoir does not coalesce into large gas bubbles or into a continuous flowing gas phase. Instead, it remains as small bubbles entrained in the crude oil, keeping the effective oil viscosity low while providing expansive energy that helps drive the oil toward the producing well. Foamy oil accounts for unusually high production in heavy crude oil reservoirs under solution gas drive.

However, it has been observed that many heavy crude oil reservoirs in Alberta and Saskatchewan exhibit foamy oil behavior which is accompanied by sand production, leading to anomalously high oil recovery and lower GOR. These observations suggest that the foamy oil flow might be physically linked to sand production. It is apparent that some additional factors, which remain to be discovered, are involved in making the foamy solution gas possible at field rates of decline. One possible mechanism is the synergistic influence of sand influx into the production wells. Allowing 1%–3% w/w sand to enter the wellbore with the fluids can result in propagation of a front of sharp pressure gradients away from the wellbore. These sharp pressure gradients occur at the advancing edge of solution gas drive. It is still unknown how far from the wellbore the dilated zone can propagate.

As a word of caution, the flow of foamy oil during recovery operations may be accompanied by sand production along with the oil and gas – the presence of sand at the wellhead leads to sand dilation and the presence of high porosity, high permeability zones (wormholes) in the reservoir.

1.3.3   High-Acid Crude Oil

High-acid crude oil is a crude oil that contains a high amount of acidic species. Typically, the acid number (total acid number, TAN) is used to distinguish the acidity of crude oil. Acid crude oil means that the acid number is >0.5 mg KOH/g, and high TAN crude means that the acid number is >1.0 mg KOH/g (Speight), For a crude oil with a high acid number, there are two main types that can be identified on the basis of the sulfur content: (1) a high acid number but low sulfur heavy crude oil and (2) a high acid number but high-sulfur heavy crude oil. More generally, the high-acid crude oils are crude oils that contain considerable proportions of naphthenic acid derivatives.

The term naphthenic acid, as commonly used in the crude oil industry, refers collectively to all of the organic acids present in the crude oil (Kane and Cayard, 2002; Shalaby, 2005). The name was originally derived from the early discovery of monobasic carboxylic acids in crude oil but there is a variety of organic acids to be present in crude oil. These include fatty acids as low in molecular weight as formic and acetic as well as saturated and unsaturated acids based on single and multiple five- and six-membered rings. The general chemical formula of naphthenic acids is $R(CH_2)_n$ COOH, where R is one or more cyclopentane rings and n is typically greater than 12, although the naphthenic acid fraction is now known to have complex compositional heterogeneity and range of molecular weight (Baugh et al., 2005). The amounts of the naphthenic acids present in crude oils vary from one crude to another, and variations of the constituents of the naphthenic acid fraction include variations in molecular weight, boiling point, and ring structure which can influence both their fraction characteristics and chemical reactivity.

By the original definition, a naphthenic acid is a monobasic carboxyl group attached to a saturated cycloaliphatic structure. However, it has been a convention accepted in the oil industry that all organic acids in crude oil are called naphthenic acids. Naphthenic acids in crude oils are now known to be mixtures of low to high molecular weight acids and the naphthenic acid fraction also contains other acidic species. The naphthenic acids content in crude oils is expressed as the TAN, which is measured in units of milligrams of potassium hydroxide required to neutralize a gram of oil.

Regardless of the source, the acids present in the oil (Chapter 1) cause much corrosion in the refinery equipment (Chapter 6). The most common current measures of the corrosive potential of a crude oil are the neutralization number or TAN. These are total acidity measurements determined by base titration. Commercial experience reveals that while such tests may be sufficient for providing an indication of whether any given crude may be corrosive, the tests are poor quantitative indicators of the severity of corrosion.

Current methods for the determination of the acid content of hydrocarbon compositions are well established (ASTM D664) which includes potentiometric titration in non-aqueous conditions to clearly define end points as detected by changes in millivolts readings versus volume of titrant used. A color indicator method (ASTM D974) is also available.

1.3.4 Opportunity Crude Oil

There is also the need for a refinery to be configured to accommodate opportunity crude oils and/or high-acid crude oils which, for many purposes, are often included with heavy feedstocks (Speight, 2014). Opportunity crude oils are either new crude oils with unknown or poorly understood properties relating to processing issues or the existing crude oils with well-known properties and processing concerns. Opportunity crude oils are often, but not always, heavy crude oils but in either case are more difficult to process due to high levels of solids (and other contaminants) produced with the oil, high levels of acidity, and high viscosity. These crude oils may also be incompatible with other oils in the refinery feedstock blend and cause excessive equipment fouling when processed either in a blend or separately (Speight, 2015). There is also the need for a refinery to be configured to accommodate opportunity crude oils and/or high-acid crude oils which, for many purposes, are often included with heavy feedstocks.

In addition to taking preventative measure for the refinery to process these feedstocks without serious deleterious effects on the equipment, refiners need to develop programs for detailed and immediate feedstock evaluation so that they can understand the qualities of a crude oil very quickly and it can be valued appropriately and management of the crude processing can be planned meticulously (Speight, 2014). For example, the compatibility of opportunity crudes with other opportunity crudes and with conventional crude oil and heavy crude oil is a very important property to consider when making decisions regarding which crude to purchase. Blending crudes that are incompatible can lead to extensive fouling and processing difficulties due to unstable asphaltene constituents (Speight, 2014, 2015). These problems can quickly reduce the benefits of purchasing the opportunity crude in the first place. For example, extensive fouling in the crude preheat train may occur resulting in decreased energy efficiency, increased emissions of carbon dioxide, and increased frequency at which heat exchangers need to be cleaned. In a worst-case scenario, crude throughput may be reduced leading to significant financial losses.

Opportunity crude oils, while offering initial pricing advantages, may have composition problems which can cause severe problems at the refinery, harming infrastructure, yield, and profitability. Before refining, there is the need for comprehensive evaluations of opportunity crudes, giving the potential buyer and seller the needed data to make informed decisions regarding fair pricing and the suitability of a particular opportunity crude oil for a refinery. This will assist the refiner to manage the ever-changing crude oil quality input to a refinery – including quality and quantity requirements and situations, crude oil variations, contractual specifications, and risks associated with such opportunity crudes.

Opportunity crude oils (also called challenging crudes) are generally characterized by a variety of properties undesirable to a refiner, such as high TAN, high sulfur, nitrogen and the content of aromatic derivatives, as well as high viscosity.

The main characteristics typical of opportunity crudes include (1) high TAN >1.0 mg KOH/g sample, (2) high sulfur content, >1% w/w, and (3) low API gravity or specific gravity, <26° API or >0.9 g/mL. In addition to these typical properties, they may also present processing challenges due to high levels of water, salt, metals, solids, asphaltene incompatibility, high pour point, or high conductivity. Depending on their characteristics, these crudes impact multiple units in the refinery.

Opportunity crudes are those that are sold at a discount (relative to benchmark crude oil). The discount is most often due to the increased processing costs or risks associated with the crude. Because of the economic advantages, many refiners are looking increasingly at processing higher levels of opportunity crude oils in their crude slates. There are many potential problems associated with processing opportunity crudes, and to properly manage the potential risk of bringing in a new crude, the refiner should consider the impact that crude will have on reliability and operations due to (1) corrosion from naphthenic acid derivatives in the high-temperature regions, (2) desalter malfunction, (3) increased fouling arising from the deposition of solids, (4) diesel cetane reduction, (5) corrosion in the overhead regions from increased low molecular weight organic acids, and (6) issues related to instability of the product(s).

Although these crudes can cause corrosion and fouling problems in a downstream facility, they offer attractive discounts in crude prices. Effective methods and technologies to upgrade and process opportunity crudes can resolve these problems and provide attractive margins to the refiner. Defining crude characteristics and understanding how unit operations might be impacted will help the refiner explore various control strategies and profit on the opportunity of processing these challenging crudes.

Thus, processing opportunity crude can play an important role in refinery profitability, but the risks are high because these crude oils usually come laden with contaminants which cause high maintenance costs and equipment losses due to excessive corrosion. In particular, overhead corrosion is caused by the mineral salts, magnesium, calcium, and sodium chloride which are hydrolyzed to produce volatile hydrochloric acid, causing a highly corrosive condition in the overhead exchangers. Therefore, these salts present a significant contamination in opportunity crude oils. Other contaminants in opportunity crude oils which are shown to accelerate the hydrolysis reactions are inorganic clays and organic acids.

As refiners target improved operating margins, one of the more likely methods to accomplish this is through the processing of opportunity crudes, sometimes called challenging crudes. Opportunity crudes may command a discounted price due to known processing issues, or they may be new crudes with unknown or poorly understood properties and processing challenges. The economics of processing opportunity crudes is often so attractive that refineries are updating their strategy for purchasing these difficult crudes. The experience gained from treating over 50 high-acid crude units over the last 20 years is used to manage the risks of processing new opportunity crudes. When developing a strategy for processing opportunity crudes, it is necessary to consider the total impact on the refinery – whether the effect is positive or negative. This paper shows an effective way to analyze opportunity crudes for potential negative impacts on the process. Risk managing techniques for corrosivity studies, desalter emulsion stability, fouling prediction, and stability issues are reported. Listed are laboratory and field evaluations utilizing online monitoring systems, corrosion probes, and corrosion coupons. Proper monitoring strategy is critical to successfully manage the risk of processing opportunity crudes. Finally, the use of high-temperature corrosion inhibitors was successfully evaluated as a means to mitigate naphthenic acid corrosion.

Some of these problems can be predicted, but others are too dependent on unit operations to predict but can be controlled. The solution to managing the risk against these problems is to perform a thorough risk assessment on the unit expected to run the opportunity crude. For a refiner, a successful strategy for processing opportunity crudes involves identifying their processing issues and assessing the inherent risks associated with running a particular crude or crude blend. With this knowledge, the refiner can anticipate the possible processing problems and implement cost-effective mitigation measures.

1.3.5 TIGHT OIL

Tight oil is a light (low-density, low-viscosity) crude oil that is sometimes erroneously referred to as shale oil because the oil is confined in the pore spaces of impermeable shale formations.

By way of definition, shale oil is the liquid product produced by the decomposition of the kerogen component of oil shale. Oil shale is a kerogen-rich crude oil source rock that was not buried under the correct maturation conditions to experience the temperatures required to generate oil and gas (Speight, 2014).

Tight formations (such as shale formations) are heterogeneous and vary widely over relatively short distances. Tight oil reservoirs subjected to fracking can be divided into four different groups which are (1) Type I reservoirs, which have little matrix porosity and permeability – leading to fractures dominating both storage capacity and fluid-flow pathways; (2) Type II reservoirs, which have low matrix porosity and permeability, but here the matrix provides storage capacity while fractures provide fluid-flow paths; (3) Type III reservoirs, which are microporous reservoirs with high matrix porosity but low matrix permeability, thus giving induced fractures dominance in fluid-flow paths; and (4) Type IV reservoirs, which are macroporous reservoirs with high matric porosity and permeability, thus the matrix provides both storage capacity and flow paths while fractures only enhance permeability.

Even in a single horizontal drill hole, the amount recovered may vary within a field or even between adjacent wells. This makes evaluation of plays and decisions regarding the profitability of wells on a particular lease difficult. Production of oil from tight formations requires a gas cap representing at least 15%–20% natural gas in the reservoir pore space to drive the oil toward the borehole; tight reservoirs which contain only oil cannot be economically produced but such reserves may be limited (Wachtmeister et al., 2017).

1.3.6 HEAVY CRUDE OIL

Although often separated as a different type of crude oil, heavy crude oil is considered to be (based on the method of recovery and properties in the reservoir) a member of the crude oil family.

Heavy crude oil is a type of crude oil that is different from the conventional crude oil insofar as it is much more difficult to recover from the subsurface reservoir. These materials have a much higher viscosity (and lower API gravity) than conventional crude oil, and primary recovery of these crude oil types usually requires thermal stimulation of the reservoir. When crude oil occurs in a reservoir that allows the crude material to be recovered by pumping operations as a free-flowing dark to light-colored liquid, it is often referred to as conventional crude oil. Heavy (high-density) crude oil is more difficult to recover from the subsurface reservoir than conventional (low-density) crude oil. The definition of heavy crude oil is usually based on the API gravity or viscosity, and the definition

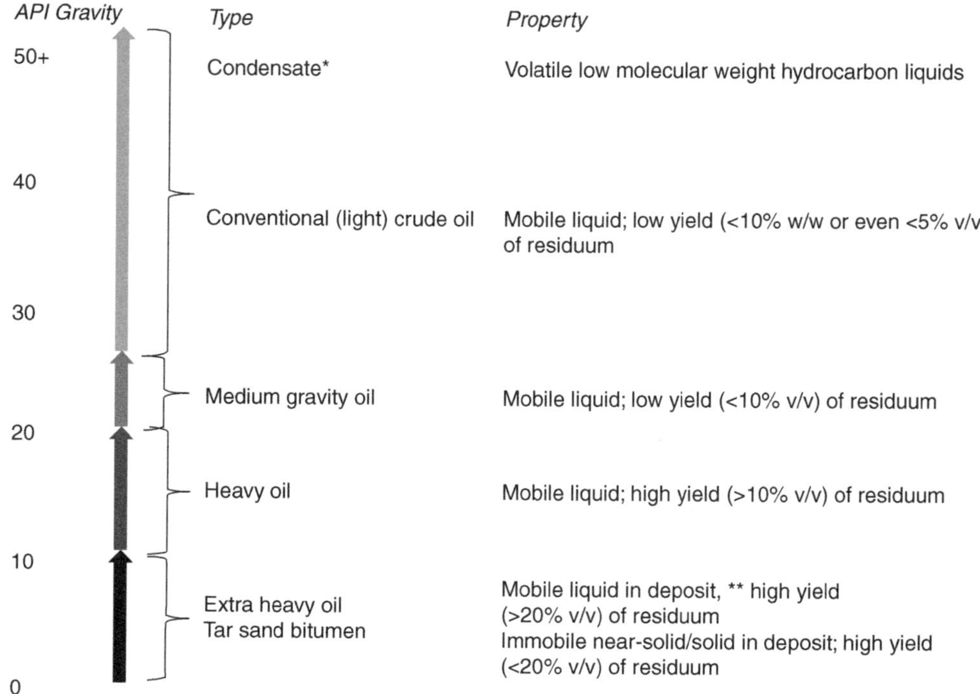

FIGURE 1.4 Arbitrary description of various feedstocks by API gravity. (*Included for comparison. **Typically, the temperature of the deposit is above the pour point of the material.)

is quite arbitrary although there have been attempts to rationalize the definition based upon API gravity (Figure 1.4) as well as on viscosity (Speight, 2014).

For many years, crude oil and heavy crude oil were very generally defined in terms of physical properties. For example, heavy crude oil was considered to be crude oil that had gravity somewhat less than 20° API with the heavy crude oils falling into the range 10°–15° API. For example, Cold Lake heavy crude oil has an API gravity equal to 12° and extra heavy crude oils, such as tar sand bitumen, usually have an API gravity in the range 5°–10° (Athabasca bitumen=8° API). Residua would vary depending upon the temperature at which distillation was terminated but usually vacuum residua are in the range 2°–8° API (Speight, 2000, 2014, 2017).

Heavy crude oil has a much higher viscosity (and lower API gravity) than conventional crude oil, and primary recovery of these crude oil types usually requires thermal stimulation of the reservoir. The generic term heavy crude oil is often applied to a crude oil that has less than 20° API and usually, but not always, a sulfur content higher than 2% w/w (Speight, 2000). Furthermore, in contrast to conventional crude oils, heavy crude oils are darker in color and may even be black. On the other hand, extra heavy crude oil is the material that occurs in the solid or near-solid state and is generally incapable of free flow under ambient conditions (tar sand bitumen).

Thus, the generic term heavy crude oil is often applied inconsistently to crude oil that has an API gravity of less than 20°. Other definitions classify heavy crude oil as heavy crude oil having an API gravity less than 22° API, or less than 25° API and usually, but not always, a sulfur content higher than 2% w/w (Ancheyta and Speight, 2007). Furthermore, in contrast to conventional crude oils, heavy crude oils are darker in color and may even be black. The term heavy crude oil has also been arbitrarily used to describe both the heavy crude oils that require thermal stimulation of recovery from the reservoir and the bitumen in bituminous sand (tar sand) formations from which the heavy bituminous material is recovered by a mining operation.

1.3.7 EXTRA HEAVY CRUDE OIL

Extra heavy crude oil is a non-descript term (related to viscosity) of little scientific meaning that is usually applied to tar sand bitumen-like material. The general difference is that extra heavy crude oil, which may have properties similar to tar sand bitumen in the laboratory but, unlike immobile tar sand bitumen in the deposit, has some degree of mobility in the reservoir or deposit (Tables 1.3 and 1.4) (Delbianco and Montanari, 2009; Speight, 2014). An example is the Zaca-Sisquoc extra heavy crude oil (sometimes referred to as the Zaca-Sisquoc bitumen) that has an API gravity on the order of 4.0°–6.0°. The reservoir has average depth 3500 feet, average thickness 1700 feet, average temperature 51°C–71°C (125°F–160°F), and sulfur in the range 6.8%–8% w/w (Isaacs, 1992; Villarroel and Hernández, 2013). The deposit temperature is certainly equal to or above the pour point of the oil (Isaacs, 1992). This renders the oil capable of being pumped as a liquid from the deposit because of the high deposit temperature (which is higher than the pour point of the oil). The same rationale applied to the extra heavy crude oil is found in the Orinoco deposits.

Thus, extra heavy crude oil is a material that occurs in the solid or near-solid state and generally has mobility under reservoir conditions. While this type of oil resembles tar sand bitumen and does not flow easily, extra heavy crude oil is generally recognized as having mobility in the reservoir compared to tar sand bitumen, which is typically incapable of mobility (free flow) under reservoir conditions. It is likely that the mobility of extra heavy crude oil is due to a high reservoir temperature (that is higher than the pour point of the extra heavy crude oil) or due to other factors that are variable and subject to localized conditions in the reservoir.

TABLE 1.3

Comparison of Selected Properties of Athabasca Tar Sand Bitumen (Alberta, Canada) and Zuata Extra Heavy Oil (Orinoco, Venezuela)

		Athabasca Bitumen	Zuata Extra Heavy Oil
Whole oil	API gravity	8	8
	Sulfur, % w/w	4.8	4.2
Resid (650F+)	% v/v	85	86
	Sulfur, % w/w	5.4	4.6
	Ni+V, ppm	420	600
	CCR, % w/w[a]	14	15

[a] Conradson carbon residue.

TABLE 1.4

Simplified Use of Pour Point and Reservoir/Deposit Temperature to Differentiate between Heavy Oil, Extra Heavy Oil, and Tar Sand Bitumen

Oil	Location	Temperature	Effect on Oil
Heavy crude oil	Reservoir or deposit	Higher than oil pour point	Fluid and/or mobile
	Surface/ambient	Higher than oil pour point	Fluid and/or mobile
Extra heavy oil	Reservoir or deposit	Higher than oil pour point	Fluid and/or mobile
	Surface/ambient	Lower than oil pour point	Solid to near-solid Immobile
Tar sand bitumen	Reservoir or deposit	Lower than oil pour point	Solid to near-solid Immobile
	Surface/ambient	Lower than oil pour point	Solid to near-solid Immobile

1.3.8 TAR SAND BITUMEN

The term bitumen (also, on occasion, referred to as native asphalt and extra heavy crude oil) includes a wide variety of reddish brown to black materials of semi-solid, viscous to brittle character that can exist in nature with no mineral impurity or with mineral matter contents that exceed 50% w/w. Bitumen is frequently found in filling pores and crevices of sandstone, limestone, or argillaceous sediments, in which case the organic and associated mineral matrix is known as rock asphalt (Abraham, 1945; Hoiberg, 1964). The expression tar sand is commonly used in the crude oil industry to describe sandstone reservoirs that are impregnated with a high-boiling, high-density viscous black crude oil that cannot be retrieved through a well by conventional production techniques (FE-76-4, above). However, the term tar sand is actually a misnomer; more correctly, the name tar is usually applied to the high-boiling product remaining after the destructive distillation of coal or other organic matter (Speight, 2013a).

Tar sand deposits are widely distributed throughout the world (Speight, 1997). Various deposits have been described as belonging to two types: (1) materials that are found in stratigraphic traps and (2) deposits that are located in structural traps. Inevitably, there are gradations, and combinations of these two types of deposits and a broad pattern of deposit entrapment are believed to exist. In general terms, the entrapment characteristics for the very large tar sand deposits all involve a combination of stratigraphic and structural traps, and there are no very large ($>4 \times 10^9$ bbl) oil sand accumulations either in purely structural or in purely stratigraphic traps.

Thus, in spite of the high estimations of the reserves of bitumen, the two conditions of vital concern for the economic development of tar sand deposits are the concentration of the resource, or the percent bitumen saturation, and its accessibility, usually measured by the overburden thickness. Recovery methods are based on either mining combined with some further processing or operation on the oil sands in situ. The mining methods are applicable to shallow deposits, characterized by an overburden ratio (i.e. overburden depth to thickness of tar sand deposit). For example, indications are that for the Athabasca deposit, no more than 10% of the in-place bitumen can be recovered by mining within the current concepts of the economics and technology of open-pit mining; this 10% portion may be considered as the proven reserves of bitumen in the deposit.

Bitumen is a naturally occurring material that is found in deposits where the permeability is low and passage of fluids through the deposit can only be achieved by prior application of fracturing techniques. Tar sand bitumen is a high-boiling material with little, if any, material boiling below 350°C (660°F) and the boiling range approximates the boiling range of an atmospheric residuum.

Tar sands have been defined in the United States (FE-76-4, United States Congress, 1976) as:

> …the several rock types that contain an extremely viscous hydrocarbon which is not recoverable in its natural state by conventional oil well production methods including currently used enhanced recovery techniques. The hydrocarbon-bearing rocks are variously known as bitumen-rocks oil, impregnated rocks, oil sands, and rock asphalt.

The recovery of the bitumen depends to a large degree on the composition and construction of the sands. Generally, the bitumen found in tar sand deposits is an extremely viscous material that is immobile under reservoir conditions and cannot be recovered through a well by the application of secondary or enhanced recovery techniques. By inference, heavy crude oil, which can be recovered by application of secondary or enhanced recovery techniques, falls into the crude oil family.

The bitumen in tar sand formations requires a high degree of thermal stimulation for recovery to the extent that some thermal decomposition may have to be induced. The current recovery operations of bitumen in tar sand formations involve the use of a mining technique.

It is incorrect to refer to native bituminous materials as tar or pitch. Although the word tar is descriptive of the black, viscous bituminous material, it is best to avoid its use with respect to natural materials and to restrict the meaning to the volatile or near-volatile products produced in the destructive distillation of such organic substances as coal (Speight, 2013a). In the simplest sense, pitch is the distillation residue of various types of tar.

Thus, alternative names, such as bituminous sand or oil sand, are gradually finding usage, with the former name (bituminous sands) more technically correct. The term oil sand is also used in the same way as the term tar sand, and these terms are used interchangeably throughout this text.

However, in order to define conventional crude oil, heavy crude oil, extra heavy crude oil, and tar sand bitumen, the use of a single physical parameter such as viscosity is not sufficient. Physical properties such as API gravity, elemental analysis, and composition fall short of giving an adequate definition. It is the properties of the bulk deposit and, most of all, the necessary recovery methods that form the basis of the definition of these materials. Only then, it is possible to classify crude oil, heavy crude oil, extra heavy crude oil, and tar sand bitumen.

Tar sands, also variously called oil sands or bituminous sands, are loose-to-consolidated sandstone or a porous carbonate rock, impregnated with bitumen, a high-boiling, high-density asphaltic material with an extremely high viscosity that is immobile under reservoir conditions and vastly different from conventional crude oil (Speight, 1990, 1997, 2008). It is therefore worth noting here the occurrence and potential supply of these materials. On an international note, the bitumen in tar sand deposits represents a potentially large supply of energy. However, many of the reserves are available only with some difficulty and that optional refinery scenarios will be necessary for conversion of these materials to liquid products because of the substantial differences in character between conventional crude oil and tar sand bitumen (Table 1.5).

TABLE 1.5

Comparison of the Properties of Tar Sand Bitumen (Athabasca) with the Properties of Conventional Crude Oil

Property	Bitumen (Athabasca)[a]	Crude Oil
Specific gravity	1.01–1.03	0.85–0.90
API gravity	5.8–8.6	25–35
Viscosity, cp		
38°C/100°F	750,000	<200
100°C/212°F	11,300	
Pour point, °F	>50	ca. −20
Elemental analysis (wt. % w/w):		
Carbon	83.0	86.0
Hydrogen	10.6	13.5
Nitrogen	0.5	0.2
Oxygen	0.9	<0.5
Sulfur	4.9	<2.0
Ash	0.8	0.0
Nickel (ppm)	250	<10.0
Vanadium (ppm)	100	<10.0
Composition wt. % w/w)		
Asphaltenes (pentane)	17.0	<10.0
Resins	34.0	<20.0
Aromatics	34.0	>30.0
Saturates	15.0	>30.0
Carbon residue (% w/w)		
Conradson	14.0	<10.0

[a] Extra heavy oil (e.g., Zuata extra heavy oil) has a similar analysis to tar sand bitumen but has mobility in the deposit because of the relatively high temperature of the deposit.

The elemental analysis of tar sand bitumen (and, hence, extra heavy crude oil) has also been widely reported (Meyer and Steele, 1981; Speight, 2014), but the data suffer from the disadvantage that identification of the source is too general and is often not specific to any site within the tar sand area. In addition, the analysis is quoted for separated bitumen, which may have been obtained by any one of several procedures and may therefore not be representative of the total bitumen on the sand (Speight, 2014, 2015).

Like conventional crude oil, of the data that are available, the elemental composition of oil sand bitumen is generally constant and, like the data for crude oil, falls into a narrow range (Speight, 1990):

Carbon: 83.4±0.5% w/w
Hydrogen: 10.4±0.2% w/w
Nitrogen: 0.4±0.2% w/w
Oxygen: 1.0±0.2% w/w
Sulfur: 5.0±0.5% w/wMetals (Ni and V): >1000 ppm

The major exception to these narrow limits is the oxygen content of bitumen, which can vary from as little as 0.2% w/w to as high as 4.5% w/w. This is not surprising, since when oxygen is estimated by difference, the analysis is subject to the accumulation of all of the errors in the other elemental data. In addition, bitumen is susceptible to aerial oxygen and the oxygen content is very dependent upon the sample history. Furthermore, the ultimate composition of the Alberta bitumen does not appear to be influenced by the proportion of bitumen in the oil sand or by the particle size of the tar sand minerals.

However, because of the diversity of available information and the continuing attempts to delineate various world tar sand deposits, it is virtually impossible to present accurate numbers that reflect the extent of the reserves in terms of the barrel unit. Indeed, investigations into the extent of many of the world tar sand deposits are continuing at such a rate that the numbers vary from 1 year to the next. Accordingly, the data quoted here must be recognized as approximate with the potential of being quite different at the time of publication.

1.3.9 ALTERNATE TERMINOLOGY

The definition of refinery feedstocks is often confusing and variable and has been made even more confusing by the introduction of other terms that add little, if anything to crude oil definitions and terminology (Speight, 2014, 2017, 2020a). In fact, there are different classification schemes based on (1) economic or (2) geological criteria. For example, the economic definition of conventional oil is: "conventional oil is oil which can be produced with current technology under present economic conditions". An issue arising from this definition is that it is not precise and the definition changes whenever the economic or technological aspects of oil recovery change. In addition, there are other classifications such as those based on (1) API gravity, which is used in the API scale to define conventional oil that is crude oil having a viscosity above 17° API or above 20° API and (2) the EPA method, which subdivides crude oil into four categories which are (1) Class A, (2) Class B, (3) Class C, and (4) Class D based on the behavior of the oil after a spill. As an additional note, it must be assumed that all of the oils are toxic to flora and fauna.

The Class A oil (light, volatile crude oil) penetrates porous surfaces, such as dirt and sand, and may be persistent in such a matrix. This type of crude oil does not tend to adhere to surfaces and flushing with water generally removes the oil from the surface. Most refined products and many of the highest quality light (low-density) crude oils can be included in this class.

The Class B oil (non-sticky oil) has a waxy or oily feel and adheres more firmly to surfaces than a Class A oil, although the Class B crude oil can be removed from surfaces by vigorous flushing. As the temperature rises, the tendency of the Class B crude oil to penetrate porous substrates increases and can be persistent. Evaporation of volatile constituents may convert a Class B oil to Class C or D residue. Medium to heavy paraffin-based oils fall into the Class B oil category.

Class C oil (heavy, sticky oil) is typically viscous, sticky or tarry, and brown or black. Flushing with water will not readily remove this material from a surface but the oil does not readily penetrate porous surfaces. The density of a Class C oil may be near to (or slightly above) the density of water and the oil may sink in aqueous systems. Weathering or evaporation of the volatile constituents of a Class C crude oil may produce a solid (or tarry) Class D oil. This class includes medium to heavy crudes as well as the heavier fuel oils.

Class D oil (typically a nonfluid oil) does not penetrate porous substrates, and it is usually black or dark brown in color. When heated, a Class D oil may melt and coat surfaces making cleanup very difficult. Residual oils, heavy crude oils, some high paraffin oils, and some weathered oils fall into this class.

However, these classifications are dynamic for spilled oils. Weather conditions and water temperature greatly influence the behavior of oil and refined crude oil products in the environment. For example, as volatiles evaporate from a Class B oil, it may become a Class C oil. If a significant temperature drop occurs (e.g., at night), a Class C oil may solidify and resemble a Class D oil. Upon warming, the Class D oil may revert back to a Class C oil.

The EPA classification scheme (Table 1.6) notes that even though the crude oil industry often characterizes crude oils according to the geographical source of the oil, e.g., Alaska North Slope Crude, classification of crude oil types by geographical source is not a useful classification scheme for response personnel who attend to spills of the oil because the classification by geological source offers little information in general: (1) the toxicity of the oil constituents, (2) the physical state, and (3) the changes that occur with time and weathering all of which are primary considerations in response to a crude oil spill. However, the EPA classifications are dynamic for spilled oils since weather conditions and temperature greatly influence the behavior of oil and refined crude oil

TABLE 1.6

EPA Classification of Crude Oils

Crude Oil Type	Characteristics
Class A: Light, Volatile Oils	Highly fluid
	Often clear
	Spread rapidly on solid or water surfaces
	Have a strong odor
	High evaporation rate
	Usually flammable
Class B: Non-Sticky Oils	Have a waxy or oily feel
	Adhere more firmly to surfaces than Class A oils
	Can be removed from surfaces by vigorous flushing
	Tendency to penetrate porous substrates
	Can be persistent
	Evaporation of volatiles may lead to a Class C or D residue
Class C: Heavy, Sticky Oils	Characteristically sticky or tarry
	Brown or black
	Flushing with water will not readily remove from surfaces
	Does not readily penetrate porous surfaces
	Density near that of water
	Often sink
	Evaporation of volatiles may produce tarry Class D oil
Class D: Nonfluid Oils	Do not penetrate porous substrates
	Usually black or dark brown in color
	When heated, may melt and coat surfaces

Source: https://www.epa.gov/emergency-response/types-crude-oil.

products in the environment. For example, as volatile constituents evaporate from a Class B oil, it may become a Class C oil. If a significant decrease in the ambient temperature occurs (such as at night), a Class C oil may solidify and resemble a Class D oil. Upon warming, the Class D oil may revert back to Class C oil.

These definitions are cited above as examples of two of the several potential classification methods that are available but they do not change the definition of refinery feedstocks as presented elsewhere (Speight, 2014, 2017, 2020a) and which has been used throughout this book.

1.4　ALTERNATE FEEDSTOCKS

Although it is projected that crude oil and heavy crude oil (as well as the extra heavy crude oil and tar sand bitumen) will be the refinery feedstock of choice for the next 50 years, reducing the national dependence on imported crude oil is of critical importance for long-term security and continued economic growth (Speight, 2021a, b). Supplementing crude oil consumption with renewable biomass resources is a first step toward this goal. The realignment of the chemical industry from one of petrochemical refining to a biorefinery concept is, at a given time, feasible and it has become a national goal of many oil-importing countries. However, clearly defined goals are necessary for increasing the use of biomass-derived feedstocks in industrial chemical production and it is important to keep the goal in perspective (Speight, 2021a, b). In this context, the increased use of biofuels should be viewed as one of a range of possible measures for achieving self-sufficiency in energy, rather than a panacea (Crocker and Crofcheck, 2006; Langeveld et al., 2010), although there are arguments against the rush to the large-scale production of biofuels (Giampietro and Mayumi, 2009).

While the previous sections have focused on the typical fossil fuel feedstock (i.e. crude oil) that is used in a modern refinery, it would be a serious omission if the so-called atypical feedstocks were not given some recognition and description since these materials could be common-place feedstocks in future refineries (Speight, 2020b). Thus, with any plan to reconfigure a refinery for the acceptance of alternate feedstocks, it is necessary to understand the current methods for the production of fuels (commonly referred to as biofuels), oil that are in active development.

Thus, a major aspect of designing a refinery for a variety of feedstocks from different sources is the composition of not only the feedstocks but also the composition and properties of the original source material, which dictates the properties of the feedstocks produced therefrom and the manner by which how these feedstocks will be processed, especially in the early stages of refinery operations when the alternate feedstocks must also pass through the dewatering and desalting processes. For example, a heavy oil refinery would differ somewhat from a conventional refinery and a refinery for tar sand bitumen would be significantly different from both in terms of the preliminary processing that is required (Parkash, 2003; Gary et al., 2007; Hsu and Robinson, 2017; Speight, 2014, 2017, 2020a).

1.4.1　COAL LIQUIDS

Coal can be upgraded (converted) through thermal decomposition by a variety of processes to produce a marketable and transportable product. Typically, the product of the conversion (synthetic crude oil) contains very few constituents that passed unchanged from the source (coal, oil shale, and tar sand bitumen) into the product (synthetic crude oil). In fact, coal liquids (produced by the liquefaction of coal) have long been cited as an alternate feedstock for crude oil refineries. The production of coal liquids includes the use of (1) direct coal liquefaction technology by processes, such as pyrolysis, and (2) indirect coal liquefaction technology (Fischer–Tropsch synthesis, F–T synthesis).

Coal liquids, which are the products produced from the thermal decomposition of coal, also fall into the category of synthetic crude oil, whether or not the liquids have been hydrotreated (Speight, 2013a). The composition can vary from a mixture containing a majority of hydrocarbon species to a mixture containing a majority of heteroatom species. Predominant heteroatom species contain oxygen, usually in the form of phenolic oxygen (e.g., C_6H_5OH) or ether oxygen (R_1OR_2).

The reaction produces gas, naphtha, middle distillate oil, and high boiling oil:

$$nC_{coal} + (n+1)H_2 \rightarrow CnH_{(2n+2)}$$

A number of catalysts have been developed over the years, including catalysts containing tungsten, molybdenum, tin, or nickel.

Liquid products from coal are for the typical crude oil-based feedstocks, even the heavier feedstocks, particularly since the coal-based liquids contain substantial amounts of phenol derivatives. Also, the composition of coal liquids produced from coal depends very much on the character of the coal and on the process conditions and, particularly, on the degree of hydrogen addition to the coal. In fact, the current concepts for refining the products of coal liquefaction processes have relied, for the most part, on already existing crude oil refineries, although it must be recognized that the acidity (i.e. phenol content) of the coal liquids and the potential incompatibility of the coal-based liquids with conventional crude oil (including heavy crude oil) may pose new issues within the refinery system (Speight, 2013b, 2014). These issues include the presence of polar (potentially water-soluble) low molecular weight product that may interfere with the dewatering–desalting system as well as mineral matter that has been entrained in the organic volatile material during the thermal process.

However, there is the need for caution when using coal-derived liquids in a crude oil refinery scenario. The direct coal liquefaction technology features conversion to a variety of products some of which (especially the phenol-based products) may be fully soluble or partially soluble in water and interfere with the dewatering and desalting processes, while the indirect coal liquefaction technology features the potential for the production of high yields of wax derivatives which may also interfere with the dewatering and desalting processes.

1.4.2 SHALE OIL

Shale oil is produced from a class of bituminous rocks that has achieved some importance which is the so-called oil shale (Scouten, 1990; Lee, 1991, 1996; Speight, 2012). These are argillaceous, laminated sediments of generally high organic content that can be thermally decomposed to yield appreciable amounts of oil, commonly referred to as shale oil. Oil shale does not yield shale oil without the application of high temperatures and the ensuing thermal decomposition that is necessary to decompose the organic material (kerogen) in the shale. The kerogen produces a liquid product (shale oil) by thermal decomposition at high temperature (>500°C, >930°F). The raw oil shale can even be used directly as a fuel akin to a low-quality coal. Indeed, oil shale deposits have been exploited as such for several centuries and shale oil has been produced from oil shale since the 19th century.

Oil shale does not contain oil (Scouten, 1990; Lee, 1991, 1996; Speight, 2012). It is an argillaceous, laminated sediment of generally high organic content (kerogen) that can be thermally decomposed to yield appreciable amounts of a hydrocarbon-based oil that is commonly referred to as shale oil. Oil shale does not yield shale oil without the application of high temperatures and the ensuing thermal decomposition that is necessary to decompose the organic material (kerogen) in the shale.

Kerogen is the complex carbonaceous macromolecular (organic) material that occurs in sedimentary rocks and shale. It is not, by any definition, a member of the crude oil family of the bitumen family. Because kerogen produces oil on high-temperature (>500°C, >830°F) thermal decomposition, it is believed by some to be the precursor to crude oil but it may also be a byproduct of the crude oil maturation process (Speight, 2014). The production of a crude oil-type product on thermal decomposition of the source material is not always an indicator of crude oil source material. It is included here because kerogen is the source of shale oil (Scouten, 1990; Lee, 1991, 1996; Speight, 2012) which could, at some future time, be a feedstock for refinery operations (Speight, 2011a).

The processes for producing liquids (*shale oil*) from oil shale involve heating (retorting) the shale to convert the organic kerogen to a raw shale oil. There are two basic oil shale retorting approaches: (1) mining followed by retorting at the surface and (2) in situ retorting, i.e. heating the shale in place underground (Speight, 2020a). Retorting essentially involves destructive distillation (*pyrolysis*) of oil shale in the absence of oxygen. Pyrolysis (temperatures above 900°F) thermally breaks down (*cracks*) the kerogen to release the hydrocarbon derivatives and then cracks the hydrocarbon derivatives into lower weight hydrocarbon molecules. Conventional refining uses a similar thermal cracking process, termed *coking*, to break down high molecular weight constituents of heavy crude oil, extra heavy crude oil, tar sand bitumen, and crude oil residua.

From the point of view of sending shale oil to a crude oil-based refinery, shale oil not only contains a large variety of hydrocarbon compounds but also has high nitrogen content compared to a nitrogen content of 0.2–0.3 wt. % for a typical crude oil. In addition, shale oil also has a high olefin and diolefin content (Speight, 2020a). It is the presence of these olefins and diolefins in conjunction with high nitrogen content that gives shale oil the characteristic difficulty in refining and the tendency to form insoluble sediment. Crude shale oil also contains appreciable amounts of arsenic, iron, and nickel that interfere with refining.

Most oil shale retorting processes produce a crude shale oil which contains appreciable quantities of water. This water is present as a stable water in oil emulsion. It is possible to resolve the emulsions by passing the emulsion through a fixed bed of suitable packing material. Several different packing materials have been tested, and on the basis of these results, a fine grade of steel wool appears to be most effective for crude shale oil (Tripp et al., 2016).

However, as with crude oil, shale oil composition varies from basin to basin and some of the common characteristics (Table 1.7) can lead to significant disruptions across the refining supply chain – transportation from the production site to shipment from the refinery. In fact, the paraffinic

---

TABLE 1.7

Factors[a] that Influence the Properties of Shale Oil

High Paraffin Content

Many shale oils contain waxes melting above 93°C (200°F) and can consequently create wax deposits that can foul pipelines, storage tanks, and process units starting with the dewatering and desalting operations

Hydrogen Sulfide Content

Shale oils contain hydrogen sulfide (H_2S) which is a corrosive gas that presents significant health and safety issues during transportation and dewatering/desalting operations

Light Paraffinic Constituents

If a light (low-density) paraffinic shale oil is blended with heavy, asphaltene-containing crude oil, the resulting blend can be expected to exhibit asphaltene separation thereby creating sludge/deposits that stabilize emulsions in the desalter unit and foul process equipment

Low Sulfur Content

Refiners typically blend a mix of moderate-to-high sulfur crudes with high acid containing crude oils to help reduce the risk of naphthenic acid corrosion. Replacing these moderate/high sulfur crudes with lower sulfur shale crudes can lead to an increase in the risk of fouling in the process units as well as causing corrosion by the naphthenic acids

Tramp Amines

When shale oil is treated with scavengers to curb the presence of hydrogen sulfide, the resulting tramp amines can affect the efficiency of downstream refining processes. For example, the amines can partition into the oil phase at the desalter, and once through the desalter, the amines can react with hydrogen chloride (HCl) in the atmospheric distillation column and the overhead system to deposit as corrosive salts

Variations in Composition

While shale oil can vary from basin to basin, shale oil from the same oil shale formation can differ widely. Also, oil can contain filterable solids and sludge

[a] Listed alphabetically.

---

nature and low sulfur content of shale oil can lead to cold flow, water separation, and lubricity issues in distillate products. Some light-ends products may experience corrosion problems, and heavy fuel oil stability can be negatively impacted by blending shale oil residual materials with asphaltenic stocks. Therefore, the composition of shale oil can impact each step in the refining process starting with the dewatering and desalting processes. For example, the paraffinic nature and low sulfur content of shale oil can lead to cold flow, water separation, and lubricity issues in distillate products.

Also, users of shale oil while offering some advantages to refiners can also offer disadvantages and refiners are well advised to take note of the disadvantages of the use of shale oil on refinery equipment.

For example, in storage tanks, solids, precipitated waxes, and destabilized asphaltene constituents can build up in refinery storage tanks. The sludge that builds up can lead to downstream fouling of cold preheat exchangers and emulsions in the desalters. Also, in the cold preheat train, fouling of the preheat exchangers before the desalter can occur with the presence of precipitated paraffins, waxes, asphaltenes, and solids present in shale oil. In the desalter, shale oil introduces a host of challenges in the desalting process, including formation of emulsions resulting from precipitated waxes, asphaltenes, and solids. As the emulsion builds, salt removal and dehydration can be impacted, leading to downstream fouling and corrosion.

In the hot preheat train, there can be significant risk to the hot preheat exchangers and furnaces in the form of fouling. Blending of shale oil with asphaltene-containing crude oils can lead to asphaltene precipitation. A shale oil with a high solid content will also contribute to higher rates of fouling and any difficulty in the desalter will contribute to higher levels of solids, water, and salt carryover.

In the atmospheric distillation tower overhead system, tramp amines from the use of scavengers for hydrogen sulfide removal will react with hydrogen chloride in the crude distillation tower and overhead system to deposit corrosive amine-hydrochloride salts. In addition, the use of low sulfur shale oil (thereby replacing high-sulfur crude oil) can lead to an increased risk of naphthenic acid corrosion – the presence of sulfur compounds in the feedstock can help passivate metal surfaces and prevent naphthenic acid corrosion.

In summary, shale oil is different from conventional crude oils, and several refining technologies have been developed to deal with this. The primary problems identified in the past were related to the presence of arsenic, nitrogen-containing constituents, and the waxy nature of the crude. Nitrogen and wax problems have been resolved using hydroprocessing approaches, essentially classical hydrocracking and the production of high-quality lube stocks, which require that waxy materials to be removed or isomerized. However, the arsenic problem remains.

Also, the composition of shale oil depends very much on the character of the kerogen in the oil shale which, like coal, include the presence of polar (potentially water-soluble) low molecular weight product that may interfere with the dewatering–desalting system as well as mineral matter that has been entrained in the organic volatile material during the thermal process which may pose similar property if used within the refinery system (Speight, 2010a, 2013b, 2014).

However, there is the need for caution when using shale oil in a crude oil refinery scenario. The production of liquids from shale oil features conversion of the kerogen to a variety of products some of which (especially the high polarity products) may be fully soluble or partially soluble in water and interfere with the dewatering and desalting processes. There is also the potential for the production of high yields of wax derivatives which may also interfere with the dewatering and desalting processes.

1.4.3 BIO-OIL

In the context of this book, bio-oil is any liquid product that is produced from biomass which is the term used to describe any material of recent biological origin, including plant materials such as trees, grasses, agricultural crops, and even animal manure. Thus, biomass (also referred to as bio-feedstock) refers to living and recently dead biological material which can be used as fuel or for industrial production of chemicals (Lee, 1996; Speight, 2020b).

Biomass is a promising feedstock to produce liquid products as a substitute for crude oil since it is readily available in high quantities and it is renewable. In terms of the chemical composition, the forms of biomass include (1) cellulose and related compounds which can be used for the production of paper and/or bioethanol and (2) long-chain lipid derivatives which can be used in cosmetics or for other specialty chemicals. Thus, biomass is a mixture of complex organic compounds that contain, for the most part, carbon, hydrogen, and oxygen, with small amounts of nitrogen and sulfur, as well as with traces of other elements including metals.

The biomass used as industrial feedstock can be supplied by agriculture, forestry, and aquaculture, as well as resulting from various waste materials. The biomass can be classified as follows: (1) agricultural feedstocks, such as sugarcane, sugar-beet, and cassava, (2) starch feedstocks, such as wheat, maize, and potatoes, (3) oil feedstocks, such as rapeseed and soy, (4) dedicated energy crops, such as short rotation coppice which includes poplar, willow, and eucalyptus, (5) high-yield perennial grass, such as miscanthus, and switchgrass, (6) non-edible oil plants, such as jatropha, camellia, and sorghum, and (7) lignocellulosic waste material, which includes forestry wood, straw, corn stover, bagasse, paper-pulp, and algal crops from land farming.

Consequently, there is a renewed interest in the utilization of plant-based matter (biomass) as a raw material feedstock for the chemicals industry. Plants accumulate carbon from the atmosphere via photosynthesis and the widespread utilization of these materials as basic inputs into the generation of power, fuels, and chemicals is a viable route to reduce greenhouse gas emissions.

Plants offer a unique and diverse feedstock for chemicals. Plant biomass can be gasified to produce synthesis gas: a basic chemical feedstock and also a source of hydrogen for a future hydrogen economy (Speight, 2020b). In addition, the specific components of plants such as carbohydrates, vegetable oils, plant fiber, and complex organic molecules known as primary and secondary metabolites can be utilized to produce a range of valuable monomers, chemical intermediates, pharmaceuticals, and materials. More generally, biomass feedstocks are recognized by the specific plant content of the feedstock or the manner in which the feedstock is produced (Speight, 2020a).

Furthermore, understanding the biomass composition is very important as a basis for the integration of the crude oil refinery with a biorefinery. The intermediate products from biomass as feedstock to a crude oil refinery should have a composition close to the current refinery feedstock. Biomass is typically composed of 65%–85% w/w sugar-based polymers (principally cellulose and hemicellulose), with another 10%–25% w/w corresponding to lignin. Other biomass minor components include triglyceride derivatives, sterol derivatives, alkaloid derivatives, terpene derivatives, terpenoid derivatives, and waxes (often collectively referred to as lipids), as well as inorganic minerals. However, biomass, like crude oil, originates from a multitude of sources with high variability and detailed characterization is an important aspect before the selection processing technology. The most important biomass characterization comprises proximate and ultimate composition, heating value, and the process of biomass production, collection, storing, transporting, and processing. However, the bio-oil resulting from the pyrolysis of biomass has the potential to be co-processed with crude oil feedstocks (unless compatibility issues surface) in the existing crude oil refinery unit (with some modifications) to produce a variety of products. It seems to be the most probable intermediate product of biorefinery for linkage with the crude oil refinery.

The bio-oil is produced by a pyrolysis of biomass in the relative absence of oxygen or at least in the presence of significantly less oxygen than required for complete combustion. The yields of bio-oil from biomass (organic condensate) vary in the range 40%–65% w/w, depending on the initial feedstock: lower for high-lignin content lignocellulosic biomass, since the presence of lignin depresses the production of liquids, and higher for more cellulosic materials. Also produced are gases 10%–30%), char (10%–20%), and water (10%–25%).

The bio-oil is formed together with gases (mainly carbon dioxide, methane, and higher hydrocarbon derivatives) as well as a solid carbonaceous residue. The properties of the bio-oils, as well as the composition, depending on both the specific starting feedstock and the conversion. The distillable oil (sometimes referred to as pyrolysis oil or bio-oil), as well as other products such as specialty chemicals (Table 1.8)

TABLE 1.8
Different Grades of Biomass

Grade	Description
Primary biomass	Produced directly by photosynthesis and includes all terrestrial plants now used for food, feed, fiber, and fuel wood
Secondary biomass	Differs from primary biomass feedstocks in that the secondary feedstocks are a byproduct of processing of the primary feedstocks in which there has been a substantial physical or chemical breakdown of the primary biomass and production of byproducts; processors may be factories or animals. Specific examples of secondary biomass include (1) sawdust from sawmills, (2) black liquor, which is a byproduct of paper making, (3) cheese whey, which is a byproduct of cheese making processes, (4) manure from concentrated animal feeding operations, and (5) vegetable oils used for biodiesel that are derived directly from the processing of oilseeds for various uses
Tertiary biomass	Includes post-consumer residues and wastes, such as fats, greases, oils, construction and demolition wood debris, other waste wood from the urban environments, as well as packaging wastes, municipal solid wastes, and landfill gases

(Speight, 2011b, 2019b, 2020a), has the potential to produce a variety of liquid products from the bio-oil intermediate that are more environmentally benign than crude oil-based fuels (Speight, 2008). Bio-oil exhibits a wide range of physical, chemical, and agricultural/process engineering properties.

In terms of the bulk composition, the bio-oil may consist of two phases which are (1) a non-aqueous phase, which comprises 35%–50% w/w of the total bio-oil and consists of different oxygen-containing derivatives, such as aliphatic alcohol derivatives, carbonyl derivatives, acid derivatives, phenol derivatives, hydroxy-aldehyde derivatives, hydroxy-ketone derivatives, and aromatic hydrocarbon derivatives, which include benzene, toluene, indene, and naphthalene and (2) an aqueous phase, comprising 15–30 wt% of the total bio-oil, in which several low molecular weight oxygenated organic derivatives are dissolved and include acetic acid, methanol, and acetone. More specifically, the high water content of bio-oil derives from water in the feedstock and dehydration reactions during the pyrolysis process. The presence of such a large amount of water can have a major consequence (potentially serious consequences) on the dewatering and desalting processes when the bio-oil is introduced into a crude oil refinery.

Due to the high oxygen content, bio-oil is often characterized by a low heating value; though this depends on the initial composition of the starting material, a high lignin feedstock can lead to a bio-oil with higher heating values because of the lower oxygen content of the lignin (Bridgwater, 1999; Czernik and Bridgwater, 2004; Melero et al., 2012).

Also, bio-oil is not fully miscible with some of the typical hydrocarbon feedstocks in a refinery because of the high polarity of oxygenated compounds in the biomass, which is chemically unstable, and displays low volatility, high viscosity, and high corrosivity. Nevertheless, the liquid nature of the bio-oil intermediate makes it easier to handle in the refinery rather than in the original form of the solid biomass.

Biomass	Pyrolysis, 550°C (1030°F), no air	
	Gases	H_2, CO, CO_2
	Liquids	Bio-oil (C_nH_{2n+2})
	Char	Coke

The utilization of biomass through the adoption of the conventional crude oil refinery systems and infrastructure to produce substitutes for refinery feedstocks and other chemicals currently derived from conventional fuels (coal, oil, natural gas) is one of the most favored methods to combat fossil fuel depletion as the 21st century matures. In a biorefinery, a solid biomass feedstock is converted through either a thermochemical process (such as gasification, pyrolysis) or a biochemical process (such as hydrolysis, fermentation) into a mixture of organic compounds, such as (1) a distillable oil, often referred to as bio-oil, and (2) inorganic compounds, such as carbon monoxide and hydrogen, that can be upgraded through catalytic reactions to high-value fuels or chemicals.

The production of bio-oil to replace oil and natural gas is in active development, focusing on the use of cheap organic matter (usually cellulose, agricultural and sewage waste) in the efficient production of liquid and gas biofuels which yield high net energy gain (Speight, 2020a). One advantage of biofuel over most other fuel types is that it is biodegradable and so relatively harmless to the environment if spilled.

Biomass also contains a variety of organic constituents that contain carbon, hydrogen, oxygen, often nitrogen and also small quantities of other atoms, including alkali metals, alkaline earth metals, and heavy metals. The alkali metals consist of the chemical elements lithium (Li), sodium (Na), potassium (K), rubidium (Rb), cesium (Cs), and francium (Fr). Together with hydrogen, they make up Group I of the Periodic Table (Figure 1.5).

On the other hand, the alkaline earth metals are the six chemical elements in Group 2 of the Periodic Table and are beryllium (be), magnesium (Mg), calcium (Ca), strontium (Sr), barium (Ba), and radium (Ra). These elements have very similar properties – they are shiny, silvery-white and are somewhat reactive at standard temperature and pressure.

Finally, the heavy metals are less easy to define but are generally recognized as metals with relatively high density, atomic weight, or atomic number. The common transition metals such as copper (Cu), lead (Pb), and zinc (Zn) are often classified as heavy metals but the criteria used for the definition and whether metalloids (types of chemical elements which have properties in between, or that are a mixture of, those of metals and nonmetals) are included vary depending on the context. These metals are often found in functional molecules such as the porphyrin derivatives which include chlorophyll and contain magnesium.

As with coal-based liquids and shale oil, the chemical composition of bio-oil is also very complex, and in general, is composed of water, organic constituents, and a small amount of ash. The organic components consist mainly of alcohols, furans, phenols, aldehydes, and organic acids which

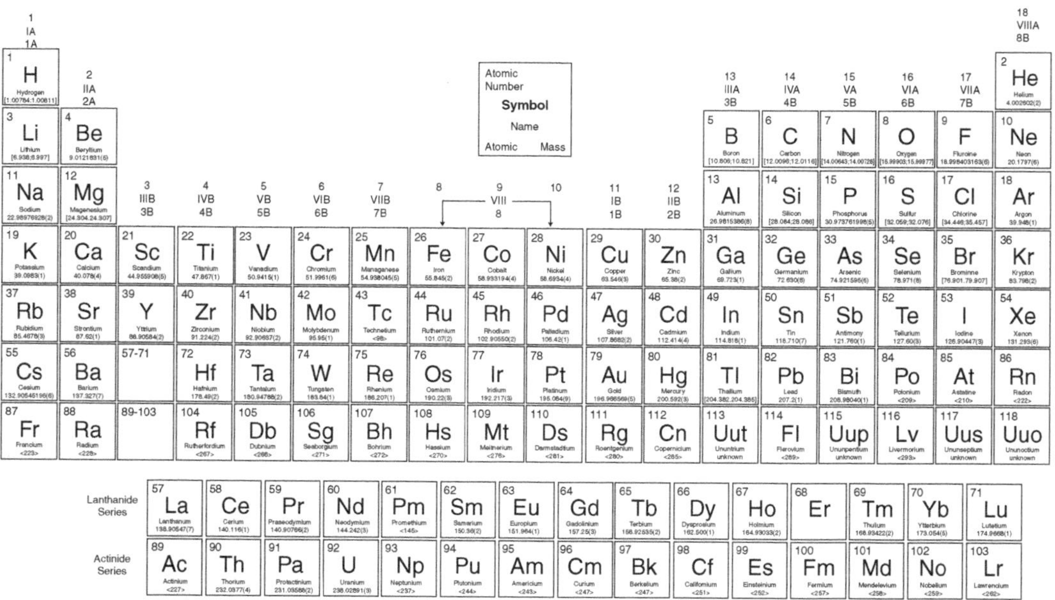

FIGURE 1.5 The periodic table of the elements.

(depending upon the molecular size) may have a marked solubility in water and which (with the also-produced wax derivatives) can interfere with the dewatering and desalting processes.

In summary, with the relatively low value of bio-oil as a fuel (because of the high oxygen content), it is more suited for the production of high-value chemicals and bio-products containing functional entities within the constituent molecules. Thus, a desirable option to that showed by pyrolysis conversion of biomass into bio-oil is the controlled chemical transformation of the biomass feedstock into simple, well-defined ones (often referred to as platform molecules or platform chemicals) that are suitable for the production of chemical products (Speight, 2011a, 2014, 2017, 2019b; Melero et al., 2012).

1.5 FEEDSTOCK EVALUATION

The evaluation of various refinery feedstocks is an important aspect of the pre-refining examination of the feedstock. Evaluation, in this context, is the determination of the physical and chemical characteristics of crude oil, heavy crude oil, extra heavy crude oil, and tar sand bitumen since the yields and properties of products or factions produced from these feedstocks vary considerably and are dependent on the concentration of various types of hydrocarbons as well as the amounts of the so-called heteroatom constituents (which are the molecular constituents containing nitrogen and/or oxygen and/or sulfur and/or metals).

In addition, the chemical composition of a feedstock is also an indicator of refining behavior (Parkash, 2003; Gary et al., 2007; Speight, 2014, 2015, 2017; Hsu and Robinson, 2017). Whether the composition is represented in terms of compound types or in terms of generic compound classes, it can enable the refiner to determine the nature of the reactions. Hence, chemical composition can play a large part in determining the nature of the products that arise from the refining operations. It can also play a role in determining the means by which a particular feedstock should be processed (Parkash, 2003; Gary et al., 2007; Speight, 2014, 2017; Hsu and Robinson, 2017).

The behavior of any of the aforementioned feedstocks (conventional crude oil, heavy crude oil, extra heavy crude oil, and tar sand bitumen) in dewatering, desalting, and distillation operations is dependent not only on the properties of the feedstock which are, in turn, dictated by constituents of the feedstock but also on the concentration of various types of constituents that are present (Speight, 2014, 2015, 2017).

Thus, it is the purpose of this chapter to present an outline of the composition of the aforementioned feedstocks with that of the behavior of the feedstock as in a refinery – especially in the present context in the dewatering and desalting operations – which may be assessed.

1.5.1 CRUDE OIL ASSAY

An efficient assay is derived from a series of test data that give an accurate description of crude oil quality and allow an indication of its behavior during refining. The first step is, of course, to assure adequate (correct) sampling by use of the prescribed protocols (ASTM D4057).

Thus, analyses are performed to determine whether each batch of crude oil received at the refinery is suitable for refining purposes. The tests are also applied to determine if there has been any contamination during wellhead recovery, storage, or transportation that may increase the processing difficulty (cost). The information required is generally crude oil dependent or specific to a particular refinery and is also a function of refinery operations and desired product slate. To obtain the necessary information, two different analytical schemes are commonly used and these are (1) an inspection assay and (2) a comprehensive assay (Speight, 2014, 2015).

Inspection assays usually involve determination of several key bulk properties of crude oil (e.g., API gravity, sulfur content, pour point, and distillation range) as a means of determining if *major* changes in characteristics have occurred since the last comprehensive assay was performed. For example, a more detailed inspection assay might consist of the following tests: API gravity (or density or relative density), sulfur content, pour point, viscosity, salt content, water and sediment content, and trace metals (or organic halides). The results from these tests with the archived data from a comprehensive assay provide an estimate of any changes that have occurred in the crude oil that may be critical to refinery operations. Inspection assays are routinely performed on all crude oils received at a refinery.

On the other hand, the comprehensive (or full) assay is more complex (as well as time-consuming and costly) and is usually only performed only when a new field comes on stream or when the inspection assay indicates that significant changes in the composition of the crude oil have occurred. Except for these circumstances, a comprehensive assay of a particular crude oil stream may not (unfortunately) be updated for several years. A full crude oil assay may involve at least determination of (1) carbon residue yield, (2) density (specific gravity), (3) sulfur content, (4) distillation profile (volatility), (5) metallic constituents, (6) viscosity, and (7) pour point as well as any tests designated necessary to understand properties and behavior of the crude oil under examination (Speight, 2014, 2015, 2017).

Using the data derived from the test assay, it is possible to assess crude oil quality that acquires a degree of predictability of performance during refining. However, knowledge of the basic concepts of refining will help the analyst understand the production and, to a large extent, the anticipated properties of the product, which in turn are related to storage, sampling, and handling the products.

1.5.2 Elemental Composition

The analysis of feedstocks for the percentages by weight of carbon, hydrogen, nitrogen, oxygen, and sulfur (elemental composition, ultimate composition) is perhaps the first method used to examine the general nature and perform an evaluation of a feedstock. The atomic ratios of various elements to carbon (i.e. H/C, N/C, O/C, and S/C) are frequently used for indications of the overall character of the feedstock. It is also of value to determine the amounts of trace elements, such as vanadium and nickel, in a feedstock not only because these elements can have serious deleterious effects on catalyst performance during refining by catalytic processes but also because these elements may have a degree of water solubility that will require careful examination of the waste water before disposal.

With few exceptions, the proportions of the elements (carbon, hydrogen, nitrogen, oxygen, sulfur, and metals) in crude oil (whatever and wherever the source) vary over fairly narrow limits:

Carbon, 83.0%–87.0% w/w
Hydrogen, 10.0%–14.0% w/w
Nitrogen, 0.1%–2.0% w/w
Oxygen, 0.05%–1.5% w/w
Sulfur, 0.05%–6.0% w/w
Metals (Ni and V), <1000 ppm

The narrow range of variation is quite surprising when the variation of the precursors is considered (Speight, 2014) and even more surprising when one considers the wide variation in physical properties from the low-density more mobile crude oils at one extreme to the heavier asphaltic crude oils at the other extreme (Charbonnier et al., 1969; Draper et al., 1977). In addition, when many localized or regional variations in maturation conditions are assessed, it is perhaps surprising that the ultimate compositions are so similar. Perhaps, this observation, more than any other observation, is indicative of the similarity in nature of the precursors from one site to another.

Also, it has become apparent, with the introduction of the heavier feedstocks into refinery operations, that these ratios are not the only requirement for predicting feedstock character before refining. The use of more complex feedstocks (in terms of chemical composition) has added a new dimension to refinery operations. Thus, although atomic ratios, as determined by elemental analyses, may be used on a comparative basis between feedstocks, there is now no guarantee that a particular feedstock will behave as predicted from these data. Product slates cannot be predicted accurately, if at all, from these ratios. Additional knowledge such as defining various chemical reactions of the constituents as well as the reactions of these constituents with each other also plays a role in determining the processability of a feedstock.

The elemental analysis of extra heavy crude oil and tar sand bitumen has also been widely reported (Speight, 1990, 2009), but the data suffer from the disadvantage that identification of the source is too general (i.e. Athabasca bitumen which covers several deposits) and is often not site specific. In addition, the analysis is quoted for separated bitumen, which may have been obtained by any one of several procedures and may therefore not be representative of the total bitumen on the sand. However, recent efforts have focused on a program to produce sound, reproducible data from samples for which the origin is carefully identified (Wallace et al., 1988).

1.5.3 CHEMICAL COMPOSITION

Crude oil and the other members of the crude oil family (presented above), heavy crude oil, extra heavy crude oil, and tar sand bitumen, contain an extreme range of organic functionality and molecular size. In fact, the variety is so great that it is unlikely that a complete compound-by-compound description for even a single crude oil would not be possible. As already noted, the composition of crude oil can vary with the location and age of the field in addition to any variations that occur with the depth of the individual well. Two adjacent wells are more likely to produce crude oil with very different characteristics (Speight, 2014, 2017, 2021a).

In very general terms (and as observed from elemental analyses), crude oil, heavy crude oil, extra heavy crude oil, bitumen, and residua are a complex composition of (1) hydrocarbons, (2) nitrogen compounds, (3) oxygen compounds, (4) sulfur compounds, and (5) metallic constituents. However, this general definition is not adequate to describe the composition of crude oil as it relates to the behavior of these feedstocks. Indeed, the consideration of hydrogen-to-carbon atomic ratio, sulfur content, and API gravity is no longer adequate to the task of determining behavior of the feedstock in a refinery.

Furthermore, the molecular composition of crude oil can be described in terms of three classes of compounds: saturates, aromatics, and compounds bearing heteroatoms (sulfur, oxygen, or nitrogen). Within each class, there are several families of related compounds: (1) saturated constituents include normal alkanes, branched alkanes, and cycloalkanes (paraffin derivatives, iso-paraffin derivatives, and naphthene derivatives in crude oil terms); (2) alkene constituents – i.e. olefins derivatives – are rare to the extent of being considered an oddity; (3) monoaromatic constituents range from benzene to multiple fused ring analogs, such as derivatives of naphthalene and phenanthrene; (4) thiol constituents – mercaptan constituents – contain sulfur as do thioether derivatives and thiophene derivatives; and (5) nitrogen-containing and oxygen-containing constituents are more likely to be found in polar forms (pyridines, pyrroles, phenols, carboxylic acids, amides, etc.) than in non-polar forms (such as ethers). The distribution and characteristics of these molecular species account for the rich variety of crude oils.

Feedstock behavior during refining – in the current context, this is the dewatering, desalting, and distillation operations – is better addressed through consideration of the molecular makeup of the feedstock (perhaps, by analogy, just as genetic makeup dictates human behavior).

Each of these feedstocks contains an extreme range of organic functionality and molecular size. In fact, the variety is so great that it is unlikely that a complete compound-by-compound description for even a single crude oil would not be possible. As already noted, the composition of crude oil can vary with the location and age of the field in addition to any variations that occur with the depth of the individual well. Two adjacent wells are more likely to produce crude oil with very different characteristics.

The hydrocarbon content of crude oil may be as high as 97% w/w (for example, in low-density paraffinic crude oil) or as low as 50% w/w or less as illustrated by the high-boiling, high-density asphaltic crude oils. Nevertheless, crude oils with as little as 50% w/w hydrocarbon components are still assumed to retain most of the essential characteristics of the hydrocarbon derivatives. It is, nevertheless, the non-hydrocarbon (sulfur, oxygen, nitrogen, and metal) constituents that play a large part in determining the processability of the crude oil (Rossini et al., 1953). But there is more to the composition of crude oil than the hydrocarbon content. The inclusion of organic compounds of sulfur, nitrogen, and oxygen serves only to present crude oils as even more complex mixtures, and the appearance of appreciable amounts of these non-hydrocarbon compounds causes some concern in the refining of crude oils. Even though the concentration of non-hydrocarbon constituents (i.e. those organic compounds containing one or more sulfur, oxygen, or nitrogen atoms) in certain fractions may be quite small, they tend to concentrate in the higher boiling fractions of crude oil. Indeed, their influence on the processability of the crude oil is important irrespective of their molecular size and the fraction in which they occur.

The presence of traces of non-hydrocarbon compounds can impart objectionable characteristics to finished products, leading to discoloration and/or lack of stability during storage. On the other hand, catalyst poisoning and corrosion are the most noticeable effects during refining sequences when these compounds are present. It is therefore not surprising that considerable attention must be given to the non-hydrocarbon constituents of crude oil as the trend in the refining industry, of late, has been to process more heavy crude oil as well as residua that contain substantial proportions of these non-hydrocarbon materials.

The occurrence of amphoteric species (i.e. compounds having a mixed acid/base nature) is not always addressed nor is the phenomenon of molecular size or the occurrence of specific functional types which can play a major role in the interactions between the constituents of a feedstock. All of these items are important in determining feedstock behavior during dewatering and desalting operations.

The presence of traces of non-hydrocarbon compounds in a feedstock can impart objectionable characteristics leading to inefficient dewatering and desalting outcomes, discoloration, and/or lack of stability during storage. It is therefore not surprising that considerable attention must be given to the non-hydrocarbon constituents of crude oil as the trend in the refining industry, of late, has been to process more heavy crude oil as well as extra heavy crude oil, tar sand bitumen, and the corresponding residua from these feedstocks as residua that contain substantial proportions of these non-hydrocarbon materials.

Furthermore, oxygen in the feedstocks may occur as a variety of potential water-soluble derivatives (such as ROH, ArOH, R^1OR^2, RCO_2H, $ArCO_2H$, RCO_2R, $ArCO_2R$, $R_2C{=}O$ as well as the cyclic furan derivatives, where R and R′ are alkyl groups and Ar is an aromatic group) (Speight, 2014).

Typically, carboxylic acids with fewer than eight carbon atoms per molecule are almost entirely aliphatic in nature. In addition to the carboxylic acids and phenolic compounds (ArOH, where Ar is an aromatic moiety), the presence of ketones ($>C{=}O$), esters [$>C({=}O)\text{-}OR$], ethers (R-O-R), and anhydrides [$>C({=}O)\text{-}O\text{-}(O{=})C<$] has been claimed for a variety of crude oils. However, the precise identification of these compounds is difficult because most of them occur in the higher molecular weight nonvolatile residua. They are claimed to be products of the air blowing of the residua, and their existence in virgin crude oil, heavy crude oil, or bitumen may yet need to be substantiated.

Metallic constituents are found in every crude oil and the concentrations have to be reduced to convert the oil to transportation fuel. Metals affect many upgrading processes and cause particular problems because they poison catalysts used for sulfur and nitrogen removal as well as other processes such as catalytic cracking. The trace metals Ni and V are generally orders of magnitude higher than other metals in crude oil, except when contaminated with co-produced brine salts (Na, Mg, Ca, Cl) or corrosion products gathered in transportation (Fe).

Two groups of elements appear in significant concentrations in the original crude oil associated with well-defined types of compounds. Zinc, titanium, calcium, and magnesium appear in the form of organometallic soaps with surface active properties adsorbed in the water/oil interfaces and act as emulsion stabilizers. However, vanadium, copper, nickel, and part of the iron found in crude oils seem to be in a different class and are present as oil-soluble compounds.

Thus, by careful selection of an appropriate technique, it is possible to obtain an overview of feedstock composition that can be used for behavioral predictions in terms of the bulk fractions (Figure 1.6).

Crude oil and the associated heavy feedstocks then appear more as a continuum than as four specific fractions. Such a concept has also been applied to the asphaltene fraction of crude oil in which asphaltenes are considered a complex state of matter based on molecular weight and polarity (Long, 1979, Speight, 1994, 2014, 2015, 2017).

In summary, the composition of crude oil is complex, and in addition to the complexity, there may also be phenol derivatives as well as amphoteric derivatives (organic compounds that exhibit acidic and basic character, such as hydroxyquinoline) (Figures 1.7 and 1.8).

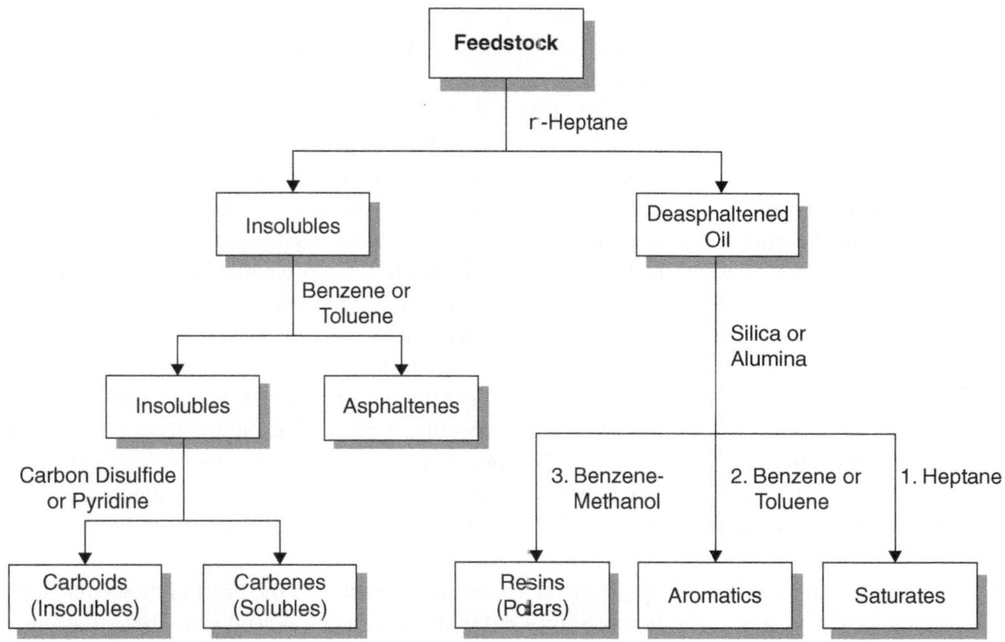

FIGURE 1.6 Separation scheme for various feedstocks. (*Since the procedure for the heavier feedstocks (extra heavy oil and tar sand bitumen as well as residua) is diffusion controlled, it is customary to mix these feedstocks with an equal volume of toluene before adding heptane – in such cases, the ratio of heptane to the actual feedstock is 1 part feedstock to 80 parts heptane to mitigate any effects from the added toluene (Speight, 2015). The carbene and carboid fractions are thermal products and do not occur naturally in crude oil.)

FIGURE 1.7 Phenol.

FIGURE 1.8 Hydroxyquinoline.

These types of derivatives also occur in heavy oil, in extra heavy oil, in tar sand bitumen, coal liquids, in shale oil, and in bio-oil and, in many cases, to a greater extent that in crude oil. The tendency for the solubility of such derivatives in water is often substantial (depending upon the chemical structure of these derivatives) and, hence, caution is advised when subjecting the alternate feedstocks to the dewatering and desalting processes.

1.5.4 Fractional Composition Using Solvents

The selection of any separation procedure depends primarily on the information desired in relation to the feedstock. In addition, the variety of fractions isolated by application of any of the following methods and the potential for the differences in composition of the fractions make it even more essential that the method is described accurately and that it reproducible not only in any one laboratory but also between various laboratories.

The use of solvents invokes the concept of the solubility (or insolubility) of a solute in the chosen solvent. The solubility of a solute is the maximum quantity of solute that can dissolve in a certain quantity of solvent or quantity of solution at a specified temperature. The main factors that have an effect on solubility are (1) the nature of the solute and solvent, (2) the temperature, and (3) the pressure. The rate of solution is a measure of how fast a substance dissolves and some of the factors determining the rate of solution are (1) the size of the particles, (2) whether or not the mixture is stirred, (3) the amount of solute already dissolved, and (4) the temperature.

In order for a solvent to dissolve a solute, the particles of the solvent must be able to separate the particles of the solute and occupy the intervening spaces. Polar solvent molecules can effectively separate the molecules of other polar substances. This happens when the positive end of a solvent molecule approaches the negative end of a solute molecule. A force of attraction then exists between the two molecules. The solute molecule is pulled into solution when the force overcomes the attractive force between the solute molecule and its neighboring solute molecule. Ethyl alcohol and water are examples of polar substances that readily dissolve in each other. Polar solvents can generally dissolve solutes that are ionic.

Solvent methods have also been applied to crude oil fractionation on the basis of molecular weight. The major molecular weight separation process used in the laboratory as well as in the refinery is solvent precipitation. Solvent precipitation occurs in a refinery in a deasphalting unit (Parkash, 2003; Gary at al., 2007; Speight, 2014, 2017; Hsu and Robinson, 2017) and is essentially an extension of the procedure for separation by molecular weight, although some separation by polarity might also be operative. The deasphalting process is usually applied to the higher molecular weight fractions of crude oil such as atmospheric and vacuum residua for the production of asphalt or demetallized deasphalted oil.

1.5.4.1 Asphaltene Separation

The systematic separation of crude oil by treatment with solvents has been practiced for several decades. If chosen carefully, solvents effect a separation between the constituents of conventional crude oil, heavy crude oil, extra heavy crude oil, tar sand bitumen, and crude oil residua according to differences in molecular weight and aromatic character. The nature and the quantity of the components separated depend on the conditions of the experiment, namely, the degree of dilution temperature and the nature of the solvent.

On the basis of the solubility in a variety of solvents, it has become possible to distinguish among various constituents of crude oil, heavy crude oil, and tar sand bitumen. High paraffin crude oil may contain only small portions of asphaltenes. Crude oil generally does not contain carboids and carbenes that are, for the purposes of this text, considered to be the products of thermal processes. Hence, residua from cracking distillation of from cracking processes may contain 2% w/w or more of the carbene and carboid fractions.

Thus, the separation of crude oil into two fractions, asphaltenes and maltenes, is conveniently achieved by means of low molecular weight paraffin hydrocarbon derivatives, which were recognized to have selective solvency for hydrocarbon derivatives, and simple relatively low molecular weight hydrocarbon derivatives. The more complex, higher molecular weight compounds are precipitated particularly well by the addition of 40 volumes of n-pentane or n-heptane in the methods generally preferred at present (Speight, 1994, 2014, 2019b) although hexane is used on occasion, which is not in keeping with the recommended standards methods using pentane or heptane (Speight, 2014, 2015). It is no doubt a separation of the chemical components with the most complex structures from the mixture, and this fraction, which should correctly be called n-pentane asphaltenes or n-heptane asphaltenes, is qualitatively and quantitatively reproducible (Figure 1.6).

Variation in the solvent type also causes significant changes in asphaltene yield. The solvent power of the solvents (i.e. the ability of the solvent to dissolve asphaltenes) increases in the order

$$2\text{-Methyl paraffin (iso-paraffin)} < \text{n-paraffin} < \text{terminal olefin}$$

Cycloparaffin derivatives (naphthene derivatives) have a remarkable effect on asphaltene yield and give results totally unrelated to those from any other non-aromatic solvent (Speight, 2014, 2015). For example, when cyclopentane, cyclohexane, or their methyl derivatives are employed as precipitating media, only approximately 1% w/w of the material remains insoluble.

In any method used to isolate asphaltenes as a separate fraction, standardization of the technique is essential. For many years, the method of asphaltene separation was not standardized, and even now, it remains subject to the preferences of the standard organizations of different countries. The use of both n-pentane and n-heptane has been widely advocated, and although n-heptane is becoming the deasphalting liquid of choice, this is by no means a fixed rule. And it must be recognized that large volumes of solvent may be required to effect a reproducible separation, similar to amounts required for consistent asphaltene separation. It is also preferable that the solvents be of sufficiently low boiling point that complete removal of the solvent from the fraction can be effected and, most importantly, the solvent must not react with the feedstock. Hence, there has been a preference for hydrocarbon liquids. Although standard methods are available, they are not unanimous in the particular hydrocarbon liquid or in ratio of hydrocarbon liquid to feedstock. (Speight, 1994, 2014, 2015)

However, it must be recognized that these methods were developed for use with feedstocks other than extra heavy crude oil and tar sand bitumen. Therefore, adjustments in the methods may be necessary to ensure efficient separation.

Insofar as industrial solvents are very rarely one compound, it was also of interest to note that the physical characteristics of two different solvent types, in this case benzene and n-pentane, are additive on a mole-fraction basis and also explain the variation of solubility with temperature (Speight, 2014, 2015). The data also show the effects of blending a solvent with the bitumen itself and allowing the resulting solvent-heavy crude oil blend to control the degree of bitumen solubility. Varying proportions of the hydrocarbon alter the physical characteristics of the oil to such an extent that the amount of precipitate (asphaltenes) can be varied accordingly within a certain range.

At constant temperature, the quantity of precipitate first increases with the increasing ratio of solvent to feedstock and then reaches a maximum. In fact, there are indications that when the proportion of solvent in the mix is <35% w/w, little or no asphaltenes are precipitated (Speight, 2014, 2015). When pentane and the lower molecular weight hydrocarbon solvents are used in large excess, the quantity of precipitate and the composition of the precipitate change with the increasing temperature. One particular example is the separation of asphaltenes from using n-pentane. At ambient temperatures (21°C, 70°F), the yield of the asphaltene fractions is 17% w/w; but at 35°C (95°F), the yield of the asphaltene fraction is on the order of 22.5% w/w using the same feedstock–pentane ratio. This latter precipitate is in fact asphaltenes plus resins; similar effects have been noted with other hydrocarbon solvents at temperatures up to 70°C (160°F). These results are self-explanatory when it is realized that the heat of vaporization ΔH^v and the surface tension γ, from which the solubility parameters are derived, both decrease with the increasing temperature.

Contact time between the hydrocarbon and the feedstock (especially feedstocks such as heavy crude oil, extra heavy crude oil, and tar sand bitumen) also plays an important role in asphaltene separation (Speight, 1994, 2014, 2019b). The yields of the asphaltene fraction reach a maximum after approximately 8 hours, which may be ascribed to the time required for the asphaltene particles to agglomerate into particles of a filterable size as well as the diffusion-controlled nature of the process since the heavier feedstocks also need time for the hydrocarbon to penetrate their mass.

1.5.4.2　Fractionation

After removal of the asphaltene fraction, further fractionation of crude oil is also possible by variation of the hydrocarbon solvent. For example, liquefied gases, such as propane and butane, precipitate as much as 50% w/w of the residuum or bitumen. The precipitate is a black, tacky, semi-solid material, in contrast to the pentane-precipitated asphaltenes, which are usually brown, amorphous solids. Treatment of the propane precipitate with pentane then yields the insoluble brown, amorphous asphaltenes and soluble, near-black, semi-solid resins, which are, as near as can be determined, equivalent to the resins isolated by adsorption techniques (Speight, 2014).

There are also claims that solvent treatment at low temperatures (−4°C to −20°C, −4°F to 25°F) assists in the fractionation of the maltenes. The hydrocarbon solvents pentane and hexane have been claimed adequate for this purpose but may not be successful with maltenes from bitumen or from residua. The author has had considerable success using acetone at −4°C (25°F) for the fractionation of maltenes from material other than cracked residua. Other miscellaneous fractionation procedures involving the use of solvents are available (Speight, 2014, 2015).

The disadvantage of an all-solvent separation technique is that, in some instances, low temperatures (such as 0°C to −10°C and 32°F to 14°F) are advocated as a means of effecting oil fractionation with solvents (Speight, 2014, 2015). Such requirements may cause inconvenience in a typical laboratory operation by requiring a permanently cool temperature during the separation. Second, it must be recognized that large volumes of solvent may be required to effect a reproducible separation in the same manner as the amounts required for consistent asphaltene separation (ASTM D2007, ASTM D4124, ASTM D893). Finally, it is also essential that the solvent be of sufficiently low boiling point so that complete removal of the solvent from the product fraction can be effected. Although not specifically included in the three main disadvantages of the all-solvent approach, it should also be recognized that the solvent must not react with the feedstock constituents. In addition, caution is still required to ensure that there is no interaction between the solvent and the solute.

1.5.4.3　Adsorption Methods

The most common industrial adsorbents are activated clay, carbon, silica gel, and alumina, because they present enormous surface areas per unit weight and result in fractions given operational (common or trivial) names which are subject to the separation procedure: (1) the saturates fraction, (2) the aromatics fraction, (3) the resin fraction, and (4) the asphaltene fraction (Figure 1.6). However, each fraction is a complex mixture of a variety of chemical constituents and the common (trivial) names are not meant to indicate specific chemical types (Speight, 2014, 2015).

By definition, the saturate fraction consists (or should consist) of paraffin derivatives and cycloparaffin derivatives (naphthene derivatives). The single-ring naphthene derivatives, or cycloparaffin derivatives, present in crude oil are primarily alkyl-substituted cyclopentane and cyclohexane. The aromatics fraction consists of those compounds containing an aromatic ring and those varying from monoaromatic derivatives (containing one benzene ring in a molecule) to diaromatic derivatives (substituted naphthalene) to triaromatic (three-ring) derivatives (i.e. substituted phenanthrene derivatives). The resins fraction contains low molecular weight functional (polar) species.

1.5.4.4　General Methods

Separation of crude oil, heavy crude oil, extra heavy crude oil, tar sand bitumen, and residua by adsorption chromatography essentially commences with the preparation of a porous bed of finely divided solid, the adsorbent (Hoiberg, 1964). The adsorbent is usually contained in an open tube (column chromatography); the sample is introduced at one end of the adsorbent bed and induced to flow through the bed by means of a suitable solvent.

It is essential that, before application of the adsorption technique to any refinery feedstock, the asphaltene fractions should first be completely removed, for example, by any of the methods outlined in the previous section. The prior removal of the asphaltenes is essential insofar as they are usually difficult to remove from the earth or clay and may actually be irreversibly adsorbed on the adsorbent. The proportions of each fraction are subject to the ratio of fuller's earth to n-pentane or n-hexane soluble materials.

Other methods of fractionation by the use of adsorbents include separation of the maltene fraction by elution with n-heptane from silica gel into two fractions named aromatic derivatives and non-aromatic derivatives and is, in fact, a separation into two broad groups called resins and oils in other methods. The silica gel method may also be modified to produce three fractions: (1) non-aromatic derivatives eluted with n-heptane, (2) aromatic derivatives eluted with benzene, and (3) compounds that contain oxygen as well as sulfur and nitrogen eluted with pyridine. Prior separation of the asphaltene fraction renders the procedure especially suitable and convenient for use with extra heavy crude oil and tar sand bitumen. Other modifications include successive elution with n-pentane, benzene, carbon tetrachloride, and ethanol.

Alumina has also been used as an adsorbent and involves (1) precipitation of asphaltene fraction with normal pentane or with normal heptane, (2) elution of oils from alumina with pentane, and (3) elution of resins from alumina with a methanol–benzene mixture (Speight, 1994, 2014, 2015, 2019b). In fact, the choice of the adsorbent appears to be arbitrary, as does the choice of various solvents or solvent blend. The use of ill-defined adsorbents, such as earths or clays, is a disadvantage in that certain components of the crude oil may undergo changes (for example, polymerization) caused by the catalytic nature of the adsorbent and can no longer be extracted quantitatively. Furthermore, extraction of the adsorbed components may require the use of solvents of comparatively high solvent power, such as chloroform or pyridine, which may be difficult to remove from the product fractions.

1.5.4.5 USBM–API and SARA Methods

There are two procedures that have received considerable attention over the years and these are (1) the United States Bureau of Mines–American Petroleum Institute, also known as the USBM–API method and (2) the saturates–aromatic derivatives–resins–asphaltenes method, also known as the SARA method (Speight, 2014, 2015). This latter method is often also called the saturates–aromatic derivatives–polars–asphaltenes method, also known as the SAPA method. These two methods are used for representing the standard methods of crude oil fractionation. Other methods are also noted, especially when the method has added further meaningful knowledge to compositional studies.

The SARA method is essentially an extension of the API method that allows more rapid separations by placing the two ion-exchange resins and the ferric chloride ($FeCl_3$)-clay-anion-exchange resin packing into a single column (Speight, 2014, 2015). The adsorption chromatography of the non-polar part of the same is still performed in a separation operation. Since the asphaltene content of crude oil (and synthetic fuel) feedstocks is often an important aspect of processability, an important feature of the SARA method is that the asphaltenes are separated as a group. Perhaps, more important is that the method is reproducible and applicable to a large variety of feedstocks.

Another issue related to the use of any adsorption-based fractionation scheme is the nature of the adsorbent. In the early reports of crude oil fractionation (Speight, 2104, 2015), clay often appeared as an adsorbent to effect the separation of the feedstock into various constituent fractions. However, clay (fuller's earth, Attapulgus clay, etc.) is often difficult to define with any degree of precision from one batch to another. Variations in the nature and properties of the clay can, and will, cause differences not only in the yields of composite fractions but also in the distribution of the compound types in those fractions. In addition, irreversible adsorption of the more polar constituent to the clay can be a serious problem when further investigations of the constituent fractions are planned.

One option for resolving this problem has been the use of more standard adsorbents, such as alumina and silica. These materials are easier to define and are often accompanied by guarantees of composition and type by various manufacturers. They also tend to irreversibly adsorb less of the feedstock than clay. Once the nature of the adsorbent is guaranteed, reproducibility becomes a reality. Without reproducibility, the analytic method does not have credibility.

1.5.4.6 ASTM Methods

There are three ASTM methods that provide for the separation of a feedstock into four or five constituent fractions, and it is interesting to note that as the methods have evolved, there has been a change from the use of pentane to heptane to separate the asphaltene fraction (ASTM D2007, ASTM D4124) (Speight, 2014, 2015). This is, in fact, in keeping with the production of a more consistent fraction that represents these higher molecular weight, more complex constituents of crude oil (Speight, 1994, 2014, 2015, 2019a).

In summary, the terminology used for the identification of various methods might differ. However, in general terms, group-type analysis of crude oil is often identified by the acronyms for the names: PONA (paraffin derivatives, olefin derivatives, naphthene derivatives, and aromatic derivatives), PIONA (paraffin derivatives, iso-paraffin derivatives, olefin derivatives, naphthene derivatives, and aromatic derivatives), PNA (paraffin derivatives, naphthene derivatives, and aromatic derivatives), PINA (paraffin derivatives, iso-paraffin derivatives, naphthene derivatives, and aromatic derivatives), or SARA (saturates, aromatic derivatives, resins, and asphaltenes). However, it must be recognized that the fractions produced by the use of different adsorbents will differ in content and will also be different from fractions produced by solvent separation techniques.

1.5.5 FRACTIONATION BY DISTILLATION

Distillation is a means of separating chemical compounds (usually liquids) through differences in their respective vapor pressures. In the mixture, the components evaporate, such that the vapor has a composition determined by the chemical properties of the mixture. Distillation of a given component is possible if the vapor has a higher proportion of the given component than the mixture. This is caused by the given component having a higher vapor pressure and a lower boiling point than the other components.

Fractionation of refinery feedstocks by distillation is an excellent means by which the volatile constituents can be isolated and studied. However, the nonvolatile residuum, which may actually constitute from 1% to 60% w/w of the crude oil, cannot be fractionated by distillation without the possibility of thermal decomposition, and as a result, alternative methods of fractionation have been developed. The distillation process separates the low boiling (light or lower molecular weight) and higher boiling (heavier or higher molecular weight) constituents by virtue of their volatility and involves the participation of a vapor phase and a liquid phase. These are, however, physical processes that involve the use of two liquid phases, usually a solvent phase and an oil phase.

By the nature of the process, it is theoretically impossible to completely separate and purify the individual components of crude oil when the possible number of isomers is considered for the individual carbon numbers that occur within the paraffin family. When other types of compounds are included, such as the aromatic derivatives and heteroatom derivatives, even though the maturation process might limit the possible number of isomeric permutations (Tissot and Welte, 1978), the potential number of compounds in crude oil is still (in a sense) astronomical.

However, crude oil can be separated into a variety of fractions on the basis of the boiling points of the crude oil constituents. Such fractions are primarily identified by their respective boiling ranges and, to a lesser extent, by chemical composition. However, it is often obvious that as the boiling ranges increase, the nature of the constituents remains closely similar and it is the number of the substituents that caused the increase in boiling point.

1.5.5.1 Gases and Naphtha

Methane is the main hydrocarbon component of crude oil gases with lesser amounts of ethane, propane, butane, isobutane, and C_4^+ hydrocarbon derivatives. Other gases, such as hydrogen, carbon dioxide, hydrogen sulfide, and carbonyl sulfide, are also present.

Saturated constituents with lesser amounts of monoaromatic and diaromatic derivatives dominate the naphtha fraction. While naphtha covers the boiling range of gasoline, the raw crude oil naphtha constituents typically have a low octane number. However, most raw naphtha is processed further and combined with other process naphtha and additives to formulate commercial gasoline.

Within the saturated constituents in crude oil gases and naphtha, every possible paraffin from methane (CH_4) hydrocarbon to n-decane (n-$C_{10}H_{22}$, normal decane) is present. Depending upon the source, one of these low-boiling paraffin derivatives may be the most abundant compound in a crude oil. The iso-paraffin derivatives begin at C4 with isobutane as the only isomer of n-butane. The number of isomers grows rapidly with carbon number and there may be increased difficulty in dealing with multiple isomers during analysis.

In addition to aliphatic molecules, the saturated constituents consist of cycloalkane derivatives (naphthene derivatives) with predominantly five-carbon or six-carbon rings. Methyl derivatives of cyclopentane and cyclohexane which are commonly found at higher levels than the parent unsubstituted structures may be present (Tissot and Welte, 1978). Fused ring dicycloalkane derivatives such as cis-decahydronaphthalene (cis-decalin) and trans-decahydronaphthalene (trans-decalin) and hexahydroindan are also common but bicyclic naphthene derivatives separated by a single bond, such as cyclohexyl cyclohexane, are not.

The aromatic constituents in crude oil naphtha begin with benzene, but its C_1 to C_3 alkylated derivatives are also present (Tissot and Welte, 1978). Each of the alkyl benzene homologs through twenty isomeric C4 alkyl benzenes has been isolated from crude oil along with various C_5-derivatives. Benzene derivatives having fused cycloparaffin rings (naphthene-aromatic derivatives) such as indane and tetralin have been isolated along with their methyl derivatives. Naphthalene is included in this fraction, while the 1-methyl and 2-methyl naphthalene derivatives and higher homologs of fused two-ring aromatic derivatives appear in the mid-distillate fraction.

Sulfur-containing compounds are the only heteroatom compounds to be found in this fraction. Usually, the total amount of the sulfur in this fraction accounts for less than 1% w/w of the total sulfur in the crude oil. In naphtha from high-sulfur (sour) crude oil, 50%–70% w/w of the sulfur may be in the form of mercaptans (thiols). Over 40 individual thiols have been identified, including all the isomeric C_1 to C_6 compounds plus some C_7 and C_8 isomers plus thiophenol. In naphtha from low sulfur (sweet) crude oil, the sulfur is distributed between sulfides (thioethers) and thiophenes. In these cases, the sulfides may be in the form of both linear (alkyl sulfides) and five-ring or six-ring cyclic (thiacyclane) structures. The sulfur structure distribution tends to follow the distribution hydrocarbon derivatives, such as naphthenic oils, that may have a high content of cycloalkane derivatives as well as a high content of saturated cycloalkane ring systems that also contain a sulfur atom in the ring. Typical alkyl thiophene derivatives in naphtha have multiple short side chains or exist as naphthene–thiophene derivatives. Methyl and ethyl disulfides have been confirmed to be present in some crude oils in analyses that minimized their possible formation by oxidative coupling of thiols (Aksenova and Kayanov, 1980).

1.5.5.2 Middle Distillates

Saturated species are the major component in the mid-distillate fraction of crude oil but aromatic derivatives, which now include simple compounds with up to three aromatic rings, and heterocyclic compounds are present and represent a larger portion of the total. Kerosene, jet fuel, and diesel fuel are all derived from raw middle distillate which can also be obtained from cracked and hydroprocessed refinery streams.

Within the saturated constituents, the concentration of n-paraffin derivatives decreases regularly from C_{11} to C_{20}. Two isoprenoid species (pristane=2,6,10,14-tetramethylpentadecane and phytane=2,6,10,14-tetramethyl hexadecane) are generally present in crude oils in sufficient concentration to be seen as irregular peaks alongside the n-C_{17} and n-C_{18} peaks in a gas chromatogram. These isoprene derivatives, believed to arise as fragments of ancient precursors, have relevance as simple biomarkers to the genesis of crude oil. The distribution of pristane and phytane relative to their neighboring n-C_{17} and n-C_{18} peaks has been used to aid the identification of crude oils and to detect the onset of biodegradation. The ratio of pristane to phytane has also been used for the assessment of the oxidation and reduction environment in which ancient organisms were converted into crude oil.

Mono-cycloparaffin and di-cycloparaffin derivatives with five or six carbons per ring constitute the bulk of the naphthene derivatives in the middle distillate boiling range, which is decreasing in concentration as the carbon number increases (Tissot and Welte, 1978) and the alkylated naphthene derivatives may have a single long side chain as well as one or more methyl or ethyl groups.

The most abundant aromatic derivatives in the mid-distillate boiling fractions are dimethyl and trimethyl naphthalene derivatives. Other one and two-ring aromatic derivatives are undoubtedly present in small quantities as either naphthene or alkyl homologs in the C_{11} to C_{20} range. In addition to these homologs of alkylbenzenes, tetralin, and naphthalenes, the mid-distillate contains fluorene derivatives and phenanthrene derivative. The phenanthrene structure appears to be favored over that of anthracene structure (Tissot and Welte, 1978) and this appears to continue through the higher boiling fractions of crude oil (Speight, 2014, 2015).

The five-membered heterocyclic constituents in the mid-distillate range are primarily the thiacyclane derivatives, benzothiophene derivatives, and dibenzothiophene derivatives with lesser amounts of dialkyl-, diaryl-, and aryl-alkyl sulfides (Aksenova and Kayanov, 1980). Alkylthiophenes are also present. As with the naphtha fractions, these sulfur species account for a minimal fraction of the total sulfur in the crude.

Although only trace amounts (usually ppm levels) of nitrogen are found in the middle distillate fractions, both neutral and basic nitrogen compounds have been isolated and identified in fractions boiling below 345°C (650°F) (Hirsch et al., 1974). Pyrrole derivatives and indole derivatives account for approximately two-thirds of the nitrogen, while the remainder is found in the basic alkylated pyridine and alkylated quinoline compounds.

The saturate constituents contribute less to the VGO than the aromatic constituents but more than the polars that are now present at percentage rather than trace levels. VGO is occasionally used as a heating oil but most commonly it is processed by catalytic cracking to produce naphtha or extraction to yield lubricating oil.

Within the VGO, saturates, distribution of paraffin derivatives, iso-paraffin derivatives, and naphthene derivatives is highly dependent upon the crude oil source. Generally, the naphthene constituents account for approximately 60% w/w of the saturate constituents but the overall range of variation is from <20% w/w to >80% w/w. In most samples, the n-paraffin derivatives from C_{20} to C_{44} are still present in sufficient quantity to be detected as distinct peaks in gas chromatographic analysis. Some (but not all) crude oils show a preference for odd-numbered alkane derivatives. Both the distribution and the selectivity toward odd-numbered hydrocarbon derivatives are considered to reflect differences in the petrogenesis of the crude oil.

The bulk of the saturated constituents in VGO consist of iso-paraffin derivatives and especially naphthene species although isoprenoid compounds, such as squalene (C_{30}) and lycopene (C_{40}), have been detected. Analytical techniques show that the naphthene derivatives contain from one to more than six fused rings accompanied by alkyl substitution. For monoaromatic and diaromatic derivatives, the alkyl substitution typically involves methyl and ethyl substituents. Hopanes and steranes have also been identified and are also used as internal markers for estimating biodegradation of crude oils during bioremediation processes.

The aromatic derivatives in VGO may contain one to six fused aromatic rings that may bear additional naphthene rings and alkyl substituents in keeping with their boiling range. Monoaromatic derivatives and diaromatic derivatives account for approximately 50% w/w of the aromatic derivatives in crude oil VGO samples. Analytical data show the presence of up to four fused naphthenic rings on some aromatic compounds. This is consistent with the suggestion that these species originate from the aromatization of steroids. Although present at lower concentration, alkyl benzenes and naphthalenes show one long side chain and multiple short side chains.

The fused ring aromatic compounds (having three or more rings) in crude oil include phenanthrene, chrysene, and picene as well as fluoranthene, pyrene, benzo(a)pyrene, and benzo(ghi)perylene.

The most abundant reported individual phenanthrene compounds appear to be the 3-derivatives. In addition, phenanthrene derivatives outnumber anthracene derivatives by as much as 100:1. In addition, chrysene derivatives are favored over pyrene derivative.

Heterocyclic constituents are significant contributors to the VGO fraction. In terms of sulfur compounds, thiophene and thiacyclane sulfur predominate over sulfide sulfur and there are constituents that contain more than one sulfur atom. The benzothiophenes and dibenzothiophenes are the prevalent thiophene forms of sulfur.

In the VGO range, the nitrogen-containing compounds include higher molecular weight pyridines, quinolines, benzoquinoline derivatives, amides, indoles, carbazole, and molecules with two nitrogen atoms (diaza compounds) with three and four aromatic rings are especially prevalent (Green et al., 1989). Typically, approximately one-third of the compounds are basic, i.e. pyridine and its benzologs, while the remainder is present as neutral species (amides and carbazoles). Although benzoquinoline and dibenzoquinoline derivatives found in crude oil are rich in sterically hindered structures, hindered and unhindered structures have been found to be present at equivalent concentrations in source rocks. This has been rationalized as geo-chromatography in which the less polar (hindered) structures moved more readily to the reservoir and are less likely to be adsorbed onto the rock.

Oxygen levels in the VGO parallel the nitrogen content. Thus, the most commonly identified oxygen compounds are the carboxylic acids and phenols, collectively called naphthenic acids.

1.5.5.3 Vacuum Residua

This fraction, the vacuum bottoms ($565°C^+$, $1050°F^+$), is the most complex of crude oil. Vacuum residua contain the majority of the heteroatoms originally in the crude oil and molecular weight of the constituents range, as near as can be determined subject to method dependence, up to several thousand. The fraction is so complex that the characterization of individual species is virtually impossible, no matter what claims have been made or will be made. Separation of vacuum residua by group type becomes difficult and confused because of the multi-substitution of aromatic and naphthenic species as well as by the presence of multiple functionalities in single molecules.

Classically, n-pentane or n-heptane precipitation is used as the initial step for the characterization of vacuum residuum (Speight, 2014, 2015; ASTM, 2021). The insoluble fraction, the pentane-asphaltene fraction or the heptane-asphaltene fraction, may be as much as 50% w/w of a vacuum residuum. The pentane-soluble or heptane-soluble portion (maltenes) of the residuum is then fractionated chromatographically into solubility or adsorption classes for characterization. However, in spite of claims to the contrary, the method is not a separation by chemical type. Kit is a separation by solubility and adsorption.

The separation of the asphaltene constituents does, however, provide a simple way to remove the highest molecular weight and most polar constituents but the asphaltene fraction is so complex that compositional detail based on average parameters is of questionable value. The use of ion-exchange chromatography has offered indications of the chemical types within the high molecular weight fractions (Green et al., 1989).

For the vacuum residua (i.e. the $565°C^+$, $1050°F^+$ fractions) of crude oil, the levels of nitrogen and oxygen may begin to approach the concentration of sulfur. These elements consistently concentrate in the most polar fractions to the extent that every molecule contains more than one heteroatom.

At this point, structural identification is somewhat fruitless and characterization techniques are used to confirm the presence of the functionalities found in lower boiling fractions such as acids, phenols, non-basic (carbazole-type) nitrogen, and basic (quinoline-type) nitrogen.

Molecular models have been proposed based on the observed functionalities, apparent molecular weight, and elemental analysis of the fraction but whether or not these models offer insights into the nature and behavior of the asphaltene constituents remains open to speculation and question. In fact, the continued insistence that the behavior of the asphaltene fraction can be explained by a so-called average structure (whatever that means) is a major omission and misrepresentation of the investigative work that proves otherwise (Speight, 1994, 2014, 2019a).

The nickel and vanadium that are concentrated into the vacuum residuum appear to occur in two forms: (1) porphyrin derivatives and (2) non-porphyrin derivatives (Reynolds, 1998). Because the metalloporphyrins can provide insights into crude oil maturation processes, they have been studied extensively and families of related structures have been identified. On the other hand, the non-porphyrin metals remain not clearly identified although metal derivatives in these compounds can exist in a tetrapyrrole (porphyrin-type) environment.

It is more likely that, in a specific molecule in the residuum, the heteroatom-containing compounds are arranged in different functional groups, which contribute to very complex molecular constituents. In addition, when the potential number of different molecular combinations is considered, the chances of determining every structure in the high molecular weight and extremely molecularly complex fractions are minimal to zero. Because of this seemingly insurmountable task, it may be better to determine ways of utilizing the residuum rather than attempting to determine (at best questionable) molecular structures.

1.6 POTENTIAL FOR CORROSION AND FOULING OF EQUIPMENT

Corrosion is a major problem that involves the deterioration of a material as a result of the interaction(s) with the surroundings, and it can occur at any point or at any time during crude oil recovery, transportation, and refining. The first step in attempting to mitigate corrosion involves removal from the crude oil of the constituents that are the cause of the corrosion and this is where the dewatering and desalting processes are valuable assets for the wellhead and the refinery. Furthermore, corrosion processes not only affect the chemical properties of metal or metal alloys but also cause changes in the physical properties and mechanical behavior of the metals.

Although the subjects of corrosion of refinery equipment and fouling of refinery equipment are dealt with in detail elsewhere in this text (Chapter 6), the definition of these phenomena is necessary here in order to assist in the definition of the role of dewatering and desalting as the text progresses.

Corrosion is a phenomenon (initiated by the use of untreated feedstocks) that involves the deterioration of equipment. Furthermore, corrosion processes not only influence the chemical properties of a metal or metals alloy but also generate changes in the physical properties and the mechanical behavior. Of particular relevance in the current context is the potential for the occurrence of acidic corrosion which is typically (but not always) due to the presence of naphthenic acids which is a generic name used for all of the organic acids present in crude oils, especially high-acid crude oils (Chapters 1 and 6). These acids can be represented by simple chemical formulas – such as $R(CH_2)_n COOH$, where R is a cyclopentane ring or cyclohexane ring and n is typically greater than 12 – but the actual structures of the acids may be much more complex. Any such acids in the refinery feedstock must be removed before the feedstock enters the refinery proper (i.e. the distillation units and other downstream units).

Fouling as it pertains to the use of crude oil (and crude oil products) relates to the formation of deposits in processing units and pipes (as well as encrustation, deposition, scaling, scale formation, slagging, and sludge formation) which has an adverse effect on operations. Thus, fouling is the accumulation of unwanted material within a refinery unit or on the solid surfaces of the unit to the detriment of function. For example, when it does occur during refinery operations, the major effects include (1) loss of heat transfer as indicated by charge outlet temperature decrease and pressure drop increase, (2) blocked process pipes, (3) under-deposit corrosion and pollution, and (4) localized hot spots in reactors, all of which culminate in production losses and increased maintenance costs.

Moreover, crude oil refining has grown increasingly complex in the last three to four decades with the necessity of acceptance of low-quality crude oils (as well as the heavy crude oils, extra heavy crude oils, and tar sand bitumen) as refinery feedstocks which have presented new challenges to the refining industry. Improving processes and increasing the efficiency of energy use are key to meeting the challenges and maintaining the viability of the crude oil refining industry. Furthermore, in this context, feedstocks that have not been subject to the dewatering and desalting processes render the feedstock as a means of causing corrosion to a variety of refinery units (Chapter 6).

For example (heavy crude oils, extra heavy crude oils, and tar sand bitumen aside), opportunity crude oils are often dirty and need cleaning before refining by removal of undesirable constituents such as high sulfur compounds, high nitrogen compounds, and aromatic compounds (especially the polynuclear aromatic compounds). A controlled visbreaking treatment would offer the potential for mitigation of corrosion by removing these undesirable constituents as insoluble coke precursors or coke. If these constituents were not removed, there is a high potential for problems further downstream in the refinery when they appear as sediment or coke.

On the other hand, high-acid crude oils cause corrosion in the atmospheric and vacuum distillation units. In addition, overhead corrosion is caused by the mineral salts, magnesium, calcium, and sodium chloride which are hydrolyzed to produce volatile hydrochloric acid causing a highly corrosive condition in the overhead exchangers. Therefore, these salts present a significant contamination in opportunity crude oils. Other contaminants in opportunity crude oils which are shown to accelerate the hydrolysis reactions are inorganic clays and organic acids.

In addition to taking preventative measure for the refinery to process these high-margin crude oils without serious deleterious effects on the equipment, refiners will need to develop programs for detailed and immediate feedstock evaluation so that they can understand the qualities of a crude oil very quickly and it can be valued appropriately. There is also the need to assess the potential impact of contaminants, like metals or acidity, in crudes so that the feedstock can be correctly valued and management of the crude processing can be planned.

Also, a major issue that occurs in distillation units and decreases efficiency is corrosion of the metal components found throughout the process line of the hydrocarbon refining process (Speight, 2014, 2017). Corrosion causes the failure of parts in addition to dictating the shutdown schedule of the unit, which can cause shutdown of the refinery. Attempts to block such corrosive influences will be a major issue of future refineries.

Changes in the characteristics of conventional crude oil can be specified through the application of a suite of standard test methods (ASTM, 2022) and will trigger changes in refinery configurations and corresponding investments. The future crude slate is expected to consist of larger fractions of both heavier, sourer crudes and extra-light inputs, such as natural gas liquids (NGLs). There will also be a shift toward extra heavy crude oil and tar sand bitumen, such as the Venezuelan extra heavy crude oil and the bitumen from the Canadian tar sands. These changes will require investment in upgrading, either at field level to process difficult-to-transport heavy crude oil, extra heavy crude oil, tar sand bitumen, coal liquids, shale oil into synthetic crude oil shale either at a field site or at a remote refinery (Speight, 2014, 2017, 2020b).

As a first step in the refining process, to reduce corrosion, plugging, and fouling of equipment and to prevent poisoning the catalysts in processing units, these contaminants must be removed by desalting (dehydration). However, the desalting operation does not always remove all of the corrosive elements and hydrogen chloride may be a product of the thermal treatment that occurs as part of the distillation process. Inadequate desalting can cause fouling of heater tubes and heat exchangers throughout the refinery. Fouling restricts product flow and heat transfer and leads to failures due to increased pressures and temperatures. Corrosion, which occurs due to the presence of hydrogen sulfide, hydrogen chloride, naphthenic (organic) acids, and other contaminants in the crude oil, also causes equipment failure. Neutralized salts (ammonium chlorides and sulfides), when moistened by condensed water, can cause corrosion.

Corrosion occurs in various forms in the distillation section of the refinery and is manifested by events such as pitting corrosion from water droplets, embrittlement from chemical attack if the dewatering and desalting unit has not operated efficiently, and stress corrosion cracking from sulfide attack.

High-temperature corrosion of distillation units will continue to be a major concern to the refining industry. The presence of naphthenic acid and sulfur compounds considerably increases corrosion in the high temperature parts of the distillation units and equipment failures have become a critical safety and reliability issue. The difference in process conditions, materials of construction, and blend processed in each refinery and especially the frequent variation in crude diet increases the problem of correlating corrosion of a unit to a certain type of crude oil. In addition, interdependent parameters influence the high-temperature crude corrosion process.

Naphthenic acid corrosion is differentiated from sulfidic corrosion by the nature of the corrosion (pitting and impingement) and by its severe attack at high velocities in crude distillation units. Crude feedstock heaters, furnaces, transfer lines, feed and reflux sections of columns, atmospheric and vacuum columns, heat exchangers, and condensers are among the type of equipment subject to this type of corrosion.

From a materials standpoint, carbon steel can be used for refinery components. Carbon steel is resistant to the most common forms of corrosion, particularly from hydrocarbon impurities at temperatures below 205°C (400°F), but other corrosive chemicals and high-temperature environments prevent its use everywhere. Common replacement materials are low alloy steel containing chromium and molybdenum, with stainless steel containing more chromium dealing with more corrosive environments. More expensive materials commonly used are nickel, titanium, and copper alloys. These are primarily saved for the most problematic areas where extremely high temperatures or very corrosive chemicals are present.

Attempts to mitigate corrosion will continue to use a complex system of monitoring, preventative repairs, and careful use of materials. *Monitoring methods* include both off-line checks taken during maintenance and online monitoring. Off-line checks measure corrosion after it has occurred, telling the engineer when equipment must be replaced based on the historical information he has collected.

Blending of refinery feedstocks (with the inherent danger of phase separation and incompatibility) will continue to be used to mitigate the effects of corrosion as well as blending, inhibition, materials upgrading, and process control. Blending will be used to reduce the naphthenic acid content of the feed, thereby reducing corrosion to an acceptable level. However, while blending of heavy and light crude oils can change shear stress parameters and might also help reduce corrosion, there is also the potential for incompatibility of heavy and light crude oils.

In summary, the potential for corrosion and fouling may be estimated using a series of standard test methods which include (1) water content (ASTM D4006, ASTM D4007, ASTM D4377, ASTM D4928), (2) salt content (ASTM D3230), and (3) bottom sediment/water (BS&W) (ASTM D96, ASTM D473, ASTM D1796, ASTM D4007) indicating the concentrations of aqueous contaminants, present in the crude either originally or picked up by the crude during handling and storage. Water and salt content of crude oils produced at the field can be very high, forming sometimes its major part. The salty water is usually separated at the field, usually by settling and draining, and surface active agents electrical emulsion breakers (desalters) are sometimes employed. If the salts and water are not removed, fouling and corrosion of the distillation unit will occur.

REFERENCES

Abraham, H. 1945. *Asphalts and Allied Substances.* Van Nostrand Scientific Publishers, New York.

Aksenova, V.S., and Kanayanov, V.F. 1980. Regularities in composition and structures of native sulfur compounds from petroleum. *Proceedings of 9th International Symposium on Organic Sulfur Chemistry,* Riga, USSR. June 9–14.

Ancheyta, J., and Speight, J.G. 2007. Heavy oils and residua. In: *Hydroprocessing of Heavy Oils and Residua,* J. Ancheyta, and J.G. Speight (Editors). CRC-Taylor & Francis Group, Boca Raton, FL. 2007. Chapter 1.

ASTM. 2022. Annual Book of Standards. ASTM International. West Conshohocken, PA.

ASTM D96. 2021. Standard Test Method for Water and Sediment in Crude Oil by Centrifuge Method (Field Procedure). Annual Book of Standards. ASTM International. West Conshohocken, PA.

ASTM D473. 2021. Standard Test Method for Sediment in Crude Oils and Fuel Oils by the Extraction Method. Annual Book of Standards. ASTM International. West Conshohocken, PA.

ASTM D664. 2021. Standard Test Method for Acid Number of Petroleum Products by Potentiometric Titration. Annual Book of Standards. ASTM International. West Conshohocken, PA.

ASTM D893. 2021. Standard Test Method for Insolubles in Used Lubricating Oils. Annual Book of Standards. ASTM International. West Conshohocken, PA.

ASTM D974. 2021. Standard Test Method for Acid and Base Number by Color-Indicator Titration. Annual Book of Standards. ASTM International. West Conshohocken, PA.

ASTM D1796. 2021. Standard Test Method for Water and Sediment in Fuel Oils by the Centrifuge Method (Laboratory Procedure). Annual Book of Standards. ASTM International. West Conshohocken, PA.

ASTM D2007. 2021. Standard Test Method for Characteristic Groups in Rubber Extender and Processing Oils and Other Petroleum-Derived Oils by the Clay-Gel Absorption Chromatographic Method. Annual Book of Standards. ASTM International. West Conshohocken, PA.

ASTM D3230. 2021. Standard Test Method for Salts in Crude Oil (Electrometric Method). Annual Book of Standards. ASTM International. West Conshohocken, PA.

ASTM D3279. 2021. Standard Test Method for n-Heptane Insolubles. Annual Book of Standards. ASTM International. West Conshohocken, PA.

ASTM D4006. 2121. Standard Test Method for Water in Crude Oil by Distillation. Annual Book of Standards. ASTM International. West Conshohocken, PA.

ASTM D4007. 2021. Standard Test Method for Water and Sediment in Crude Oil by the Centrifuge Method (Laboratory Procedure). Annual Book of Standards. ASTM International. West Conshohocken, PA.

ASTM D4124. 2021. Standard Test Method for Separation of Asphalt into Four Fractions. Annual Book of Standards. ASTM International. West Conshohocken, PA.

ASTM D4377. 2021. Standard Test Method for Water in Crude Oils by Potentiometric Karl Fischer Titration. Annual Book of Standards. ASTM International. West Conshohocken, PA.

ASTM D4928. 2021. Standard Test Method for Water in Crude Oils by Coulometric Karl Fischer Titration. Annual Book of Standards. ASTM International. West Conshohocken, PA.

Baugh, T.D., Grande, K.V., Mediaas, H., Vindstad, J.E., and Wolf, N.O. 2005. The discovery of high molecular weight naphthenic acids (ARN Acid) responsible for calcium naphthenate deposits. *SPE International Symposium on Oilfield Scale*, Aberdeen, United Kingdom, May 11–12.

Bridgwater, A.V. 1999. Principles and practice of biomass fast pyrolysis processes for liquids. *Journal of Analytical and Applied Pyrolysis*, 51(1–2): 3–22.

Charbonnier, R.P., Draper, R.G., Harper, W.H., and Yates, A. 1969. Analyses and Characteristics of Oil Samples from Alberta. Information Circular IC 232. Department of Energy Mines and Resources, Mines Branch, Ottawa, Canada.

Crocker, M., and Crofcheck, C. 2006. Reducing national dependence on imported oil. Energeia Vol. 17, No. 6. Center for Applied Energy Research, University of Kentucky, Lexington, Kentucky.

Czernik, S., and Bridgwater, A.V. 2004. Overview of applications of biomass fast pyrolysis oil. *Energy & Fuels*, 18(2): 590–598.

Delbianco, A., and Montanari, R. 2009. *Encyclopedia of Hydrocarbons, Volume III/New Developments: Energy, Transport, Sustainability.* Eni S.p.A., Rome.

Draper, R.G., Kowalchuk, E., and Noel, G. 1977. Analyses and characteristics of crude oil samples performed between 1969 and 1976. *Report ERP/ERL 77-59 (TR). Energy, Mines, and Resources, Ottawa, Canada.* I.K. Freidlina, and A.E. Skorova (Editors). 1980. Organic Sulfur Chemistry.), Pergamon Press, New York.

Gary, J.G., Handwerk, G.E., and Kaiser, M.J. 2007. *Petroleum Refining: Technology and Economics*, 5th Edition. CRC Press, Taylor & Francis Group, Boca Raton, FL.

Giampietro, M., and Mayumi, K. 2009. *The Biofuel Delusion: The Fallacy of Large-Scale Agro-Biofuel Production.* Earthscan, Washington, DC.

Green, J.A., Green, J.B., Grigsby, R.D., Pearson, C.D., Reynolds, J.W., Sbay, I.Y., Sturm Jr., O.P., Thomson, J.S., Vogh, J.W., Vrana, R.P., Yu, S.K.Y., Diem, B.H., Grizzle, P.L., Hirsch, D.E., Hornung, K.W., Tang, S.Y., Carbongnani, L., Hazos, M., and Sanchez, V. 1989. Analysis of Heavy Oils: Method Development and Application to Cerro Negro Heavy Petroleum, NIPER-452 (DE90000200). Volumes I and II. Research Institute, National Institute for Petroleum and Energy Research (NIPER), Bartlesville, Oklahoma.

Hirsch, D.E., Cooley, J.E., Coleman, H.J., and Thompson, C.J. 1974. Qualitative Characterization of Aromatic Concentrates of Crude Oils from GPC Analysis. Report 7974. Bureau of Mines, U.S. Department of the Interior, Washington, DC.

Hoiberg, A.J. 1964. *Bituminous Materials: Asphalts, Tars, and Pitches.* John Wiley and Sons, New York.

Hsu, C.S., and Robinson, P.R. 2017. *Practical Advances in Petroleum Processing, Volumes 1 and 2.* Springer, New York.

Isaacs, C.M. 1992. Preliminary Petroleum Geology Background and Well Data for Oil Samples in the Cooperative Monterey Organic Geochemistry Study, Santa Maria and Santa Barbara-Ventura Basins, California. Open-File Report No. USGS 92–539-F. United States Geological Survey, Reston, Virginia.

Kane, R.D., and Cayard, M.S. 2002. A Comprehensive Study on Naphthenic Acid Corrosion. Paper no. 02555. Corrosion 2002. NACE International, Houston, TX.

Langeveld, H., Sanders, J., and Meeusen, M. (Editors). 2010. *The Biobased Economy.* Earthscan, Washington, DC.

Lee, S. 1991. *Oil Shale Technology.* CRC Press, Taylor & Francis Group, Boca Raton, FL.

Lee, S. 1996. *Alternative Fuels.* CRC Press, Taylor & Francis Group, Boca Raton, FL.

Long, R.B. 1979. The concept of asphaltenes. *Preprints Division of Petroleum Chemistry American Chemical Society,* 24(4): 891.

Melero, J.A., Iglesias, J., and Garcia, A. 2012. Biomass as renewable feedstock in standard refinery units: Feasibility, opportunities and challenges. *Energy & Environmental Science,* 5: 7393–7420.

Meyer, R.F., and Steele, C.F. (Editors). 1981. *The Future of Heavy Crude and Tar Sands,* McGraw-Hill, New York.

Parkash, S. 2003. *Refining Processes Handbook.* Gulf Professional Publishing, Elsevier, Amsterdam.

Reynolds, J.G. 1998. Metals and heteroatoms in heavy crude oils. In: *Petroleum Chemistry and Refining,* J.G. Speight (Editor). Taylor & Francis Publishers, Washington, DC. Chapter 3. Page 63–102.

Rossini, F.D., Mair, B.J., and Streif, A.J. 1953. Hydrocarbons from Petroleum. Reinhold, New York.

Sayles, S., and Routt, D.M. 2011. Unconventional Crude Oil Selection and Compatibility. Digital Refining, March. https://cdn.digitalrefining.com/data/articles/file/488759177.pdf.

Scouten, C.S. 1990. Oil shale. In: *Fuel Science and Technology Handbook.* Marcel Dekker Inc., New York. Chapters 25–31. Page 795–1053.

Shalaby, H.M. 2005. Refining of Kuwait's heavy crude oil: Materials challenges. *Proceedings of Workshop on Corrosion and Protection of Metals.* Arab School for Science and Technology. December 3–7, Kuwait.

Speight, J.G. 1990. Tar sands. In: *Fuel Science and Technology Handbook.* Marcel Dekker Inc., New York. Chapter 12.

Speight, J.G. 1994. Chemical and physical studies of petroleum asphaltenes. In: *Asphaltenes and Asphalts. I. Developments in Petroleum Science, 40.* T.F. Yen and G.V. Chilingarian (Editors). Elsevier, Amsterdam, Chapter 2.

Speight, J.G. 1997. *Kirk-Othmer Encyclopedia of Chemical Technology,* 4th Edition. John Wiley & Sons Inc., Hoboken, NJ, Volume 23, p. 717.

Speight. J.G. 2000. *The Desulfurization of Heavy Oils and Residua,* 2nd Edition. Marcel Dekker, New York.

Speight, J.G. 2008. *Handbook of Synthetic Fuels: Properties, Processes, and Performance.* McGraw-Hill, New York.

Speight, J.G. 2009. *Enhanced Recovery Methods for Heavy Oil and Tar Sands.* Gulf Publishing Company, Houston, TX.

Speight, J.G. 2011a. *An Introduction to Petroleum Technology, Economics, and Politics.* Scrivener Publishing, Beverly, MA.

Speight, J.G. (Editor). 2011b. *The Biofuels Handbook.* The Royal Society of Chemistry, London.

Speight, J.G. 2012. *Shale Oil Production Processes.* Gulf Professional Publishing, Elsevier, Oxford.

Speight, J. G. 2013a. *The Chemistry and Technology of Coal,* 3rd Edition. CRC Press, Taylor & Francis Group, Boca Raton, FL.

Speight, J.G. 2013b. *Heavy and Extra Heavy Oil Upgrading Technologies.* Gulf Professional Publishing, Elsevier, Oxford.

Speight. J.G. 2014. *The Chemistry and Technology of Petroleum,* 5th Edition. CRC Press, Taylor & Francis Group, Boca Raton, FL.

Speight, J.G. 2015. *Handbook of Petroleum Product Analysis,* 2nd Edition. John Wiley & Sons Inc., Hoboken, NJ.

Speight. J.G. 2017. *Handbook of Petroleum Refining.* CRC Press, Taylor & Francis Group, Boca Raton, FL.

Speight, J.G. 2019a. The asphaltene fraction – demystified. *Petroleum and Chemical Industry International,* 2(4). https://opastonline.com/wp-content/uploads/ 2019/10/the-asphaltene-fraction-demystified-pcii-19.pdf.

Speight, J.G. 2019b. *Handbook of Petrochemical Processes.* CRC Press, Taylor & Francis Group, Boca Raton, FL.

Speight, J.G. 2020a. *Synthetic Fuels Handbook: Properties, Processes, and Performance*, 2nd Edition. McGraw-Hill, New York.

Speight, J.G. 2020b. *Refinery of the Future*, 2nd Edition. Gulf Professional Publishing, Elsevier, Cambridge, MA.

Speight, J.G. 2020c. *Global Climate Change Demystified.* Scrivener Publishing, Beverly, MA.

Speight, J.G. 2021a. *Refinery Feedstocks.* CRC Press, Taylor & Francois Group, Boca Raton, FL.

Speight, J.G. 2021b. *Chemistry and Technology of Alternate Fuels.* World Scientific Publishing Co., PTE Ltd., Singapore and Hackensack, NJ.

Tissot, B.P., and Welte, D.H. 1978. *Petroleum Formation and Occurrence.* Springer-Verlag, New York.

Tripp, J.G., Putman, M.W., Van Fossen, H., and Long, R. 2016. Dewatering of Crude Shale Oil by Means of Solid Packing Materials. https://edx.netl.doe.gov/dataset/dewatering-of-crude-shale-oil-by-means-of-solid-packing-materials.

United States Congress. 1976. Public Law FE-76-4. United States Library of Congress, Washington, DC.

Villarroel, T., and Hernández, R. 2013. Technological developments for enhancing extra heavy oil productivity in fields of the faja petrolifera del orinoco (FPO), Venezuela. *Proceedings. AAPG Annual Convention and Exhibition*, Pittsburgh, PA, May 19–22. American Association of Petroleum Geologists, Tulsa, Oklahoma.

Wachtmeister, H., Linnea Lund, L., Aleklett, K., and Höök, M.M. 2017. Production decline curves of tight oil wells in eagle ford shale. *Natural Resources Research*, 26(3): 365–377.

Wallace, D., Starr, J., Thomas, K.P., and Dorrence, S.M. 1988. Characterization of Oil Sands Resources. Alberta Oil Sands Technology and Research Authority, Edmonton, Alberta, Canada.

2 Dewatering and Desalting

2.1 INTRODUCTION

Crude oil, heavy crude oil, extra heavy crude oil, and tar sand bitumen are recovered from a reservoir (crude oil and heavy crude oil) or a deposit (extra heavy crude oil and tar sand bitumen) mixed with a variety of substances such as gases, water, mineral salts, and dirt (minerals) (Atta, 2013; Speight, 2014a, 2017). The initial step in handling the crude oil is the wet-crude handling facilities since the crude oil will likely contain up to 15% v/v water, which may exist in an emulsified form.

In the natural state, crude oil, heavy oil, extra heavy oil, and tar sand bitumen are not homogeneous materials and the physical characteristics differ depending on where the material was produced. In the natural, unrefined state, crude oil, heavy oil, extra heavy oil, and tar sand bitumen range in density and consistency from very thin, lightweight, and volatile fluidity to an extremely thick, semi-solid (Speight, 2104a, 2017, 2021). In addition, any of these feedstocks from different geographical locations will naturally have unique properties that affect not only the refining conversion processes but also the pretreatment processes.

Thus, crude oil, heavy crude oil, extra heavy crude oil, and tar sand bitumen are not (within the individual categories) uniform materials and the chemical and physical (fractional) composition of each of these refinery feedstocks can vary not only with the location and age of the reservoir or deposit but also with the depth of the individual well within the reservoir or deposit. On a molecular basis, the three feedstocks are complex mixtures containing (depending upon the feedstock) hydrocarbons with varying amounts of hydrocarbonaceous constituents that contain sulfur, oxygen, and nitrogen as well as constituents containing metallic constituents (such as porphyrin derivatives), particularly those containing vanadium nickel, iron, and copper. The hydrocarbon content may be as high as 97% w/w, for example, in a light crude oil or less than 50% w/w in heavy crude oil and bitumen (Speight, 2104a, 2021).

Typically, the minerals (more commonly referred to as salt) are dissolved in the water, while the remainder may exist as very fine crystals. This amount of salt creates acids at high temperature, which are corrosive to the transportation and refining equipment. To scale down the corrosion effect on those types of equipment, the inorganic salt substance in crude oil is reduced through a desalting system.

In fact, it is important to reduce the amount of salt in the feedstock before processing the crude in the atmospheric distillation unit and in any subsequent downstream processing units of a refinery because (1) salt causes corrosion in the equipment, (2) salt fouls inside the equipment which not only negatively impacts the heat transfer rates in the exchangers and furnace tubes but also affects the hydraulics of the system by increasing the pressure drops and hence requiring more pumping power to the system, (3) salt also plugs the fractionator trays and causes reduced mass transfer thereby causing reduced separation efficiency, and (4) the salt in the crude usually has a source of metallic compounds, which could cause poisoning of catalyst in hydrotreating and other refinery units.

Thus, corrosion is related to mineral constituents; however, other constituents such as nonmetallic substances, which include ceramics and polymers, can also be damaged when they are attacked by chemicals. In this regard, there is an increase in the chemical corrosion due to acid attack and the desalting system is used to cut the influence of acid corrosion within the distinctive forms of equipment used in crude oil refining.

The presence of chloride salts of the alkali metals can result in the formation of hydrogen chloride (HCl) which will corrode the equipment such as the top of the distillation column. In a characteristic atmospheric distillation unit column, hydrolysis of these alkaline earth chlorides can occur

DOI: 10.1201/9781003184959-2

with the production of hydrogen chloride (which become hydrochloric acid) in the presence of stripping steam in the distillation column or the presence of wash water in an upstream desalter unit or the presence of water in the crude oil. On the other hand, the magnesium chloride ($MgCl_2$) and calcium chloride ($CaCl_2$) salts are subject to hydrolysis in the feedstock heater, and in the bottom section of the distillation column, the sodium chloride salt is stable and capable of commencing various corrosion reactions. On the other hand, the hydrochloric acid vapor is unstable and moves through crude column along with the volatile constituents (of the feedstock). If untreated, the hydrogen chloride vapor will condense in the crude column overhead system and cause corrosion in that area of the distillation column.

Thus, refining actually commences with the production of fluids from the well or reservoir and is followed by pretreatment operations that are applied to the crude oil either at the refinery or prior to transportation. Pipeline operators, for instance, are insistent upon the quality of the fluids put into the pipelines; therefore, any crude oil to be shipped by pipeline or, for that matter, by any other form of transportation must meet rigid specifications with regard to water and salt content. In some instances, sulfur content, nitrogen content, and viscosity may also be specified. Field separation, which occurs at a field site near the recovery operation, is the first attempt to remove the gases, water, and dirt that accompany crude oil coming from the ground.

Production → Gas-oil separation and treatment → Transportation → Refinery

The separator at the field site may be no more than a large vessel that provides a non-turbulent zone that allows for gravity separation of the production fluids into three phases: (1) gases, (2) crude oil, and (3) water which not only contain soluble inorganic materials but also contain entrained dirt.

Thus, at the production facility, the bulk of the water with dissolved salts is removed from the crude oil by gravity separation (i.e. settling) or by centrifuging so that the crude oil meets the specification for pipelining. But even then, the crude oil can arrive at the refinery containing traces of water (sometimes more than traces depending upon the pipeline specification for transporting the crude oil) and salt through the salt water produced with the crude feedstock (Parkash, 2003; Gary et al., 2007; Speight, 2014a, 2017, 2021; Hsu and Robinson, 2017). Thus, dewatering and desalting are required at the refinery so that the feedstock meets the specifications necessary for the refinery equipment.

In the modern refinery, the usual practice is to blend crude oils of similar characteristics, although fluctuations in the properties of the individual crude oils may cause significant variations in the properties of the blend over a period of time. Blending several crude oils prior to refining can eliminate the frequent need to change the processing conditions that may be required to process each of the crude oils individually. However, blending of different components prior to the dewatering–desalting operations can cause fouling if the components of crude oils in the blend are not tested for compatibility prior to the blending operation. The same situation is true for the ensuing downstream units in which the incompatibility of the blend components can also cause major issues leading to fouling (deposition of an insoluble phase) and unit shutdown (Chapter 3).

However, simplification of the refining procedure through blending of two or more crude oils is not always the end result. Incompatibility of different crude oils, which can occur if, for example, a paraffinic crude oil is blended with heavy asphaltic oil, can cause sediment formation in the unrefined feedstock or in the products, thereby complicating the refinery process. Thus, pretreatment of the crude oil is necessary and it involves in the removal of harmful impurities which occupies an important place among the main processes associated with the production, collection, and transportation of oil to refineries or export.

Adverse effects of the impurities in crude oil can result in reduced equipment reliability. To prevent corrosion, plugging, and fouling of equipment, electrical desalting plants are often installed in crude oil production units in order to remove water-soluble salts from a stream of crude oil. The refiners often wash the crude oil with fresh water, add chemical (demulsified), and use electrical desalting vessel to remove the added water and most of the inorganic contaminants from the crude oil.

Prior to the refinery operations, a certain amount of pretreatment typically occurs at the recovery site (i.e. at the wellhead), and the purpose of this preliminary pretreatment is to meet the specification for the crude oil as stated by the pipeline company. There is still a need for further removal of impurities by dewatering and desalting the crude oil at the refinery. In the general sense, crude oil obtained from the underground reservoir contains a variety of contaminants such as water, water-soluble materials (such as naphthenic acids and soluble salts), and rock fragments. Production fluids, hydrogen sulfide (H_2S) scavengers, and corrosion inhibitors may also be present.

These contaminants can be divided into two groups which are (1) hydrophilic impurities, also called lipophobic impurities, which are insoluble in the crude oil and include water, water-soluble inorganic salts as well as solid salts, mechanical impurities such as sand and clay, and (2) hydrophobic impurities, also called lipophilic impurities. The hydrophobic impurities are soluble in the crude oil and include organometallic compounds (such as metalloporphyrin derivatives) and organic acids, among which are the most undesirable organochlorine compounds. Also, the hydrophilic impurities are in the water phase which are dispersed in the oil in the form of drops of water and, therefore, the degree of pretreatment has a significant effect on the efficiency and reliability of transportation by pipeline, tanker, rail, and even road transportation.

Thus, once the crude oil arrives at the refinery, any contaminants that were not removed at the wellhead must be removed before the crude oil enters the refinery properly. The quality of crude oil (and hence the type and amount of contaminants) can vary from shipment to shipment. If the refinery is not prepared to deal with the contaminants in the crude oil, serious consequences can ensue as a result of fouling and corrosion of process equipment. By definition, corrosion is the deterioration of a material, usually a metal, because of a reaction with its environment. Deteriorated equipment can cause process leaks which can result in fire and chemical hazards, posing a catastrophic threat to personnel. The consequences of fouling in the refinery are similar to those of corrosion with costs including equipment cleaning and increased energy costs.

The dewatering and desalting units are at the front-end of the oil refinery because of the need to remove contaminants from the refinery feedstocks. The processes are necessary to ensure the good quality of the crude oil by minimizing or eliminating harmful substances such as water, salts, and even mechanical impurities. The other front-end process, distillation, is a more physical process and is characterized by mass-thermal transfer of materials, which leads to the obtaining of fractions. The distillation is carried out consecutively in two ways: atmospheric distillation and vacuum distillation. Also, the distillation section makes it possible to obtain more low-boiling products and there is a reduction of the residuum, the refinement of which requires more expensive processes such as hydrocracking or catalytic cracking.

Thus, the first processes (desalting and dewatering) in a refinery (Figure 2.1) are focused on the cleanup of the feedstock, particularly the removal of the troublesome brine constituents (Speight, 2104a, 2017).

This is followed by distillation to remove the volatile constituents with the concurrent production of a residuum that can be used as a cracking (coking) feedstock or as a precursor to asphalt.

However, at any stage of the crude oil train (i.e. from the production well to the refinery), corrosion and fouling can interfere with the production equipment and there may be the need for periodic shutdown of process units for mechanical cleaning and maintenance due to corrosion and/or fouling resulting in throughput losses. The crude oils feedstocks to the refinery and the blends of these feedstocks play an important role in refinery profitability, but the risks are high because there are several factors that determine the formation of deposits in the pipelines and pipes of heat exchangers. The major issue dealing with corrosion and fouling relates to the presence of contaminants, such as inorganic salts, sediments, asphaltene constituents, waxes, stable emulsions, or inorganic solids, and the presence of metals (Murphy et al., 1992; Ben Mahmoud and Aboujadeed, 2017).

Furthermore, although corrosion and fouling are complex processes, which may involve several parallel mechanisms, it is assumed that the corrosion and subsequent fouling takes place through the destabilization of some components of the refinery feedstock. Crude preheat train and vacuum

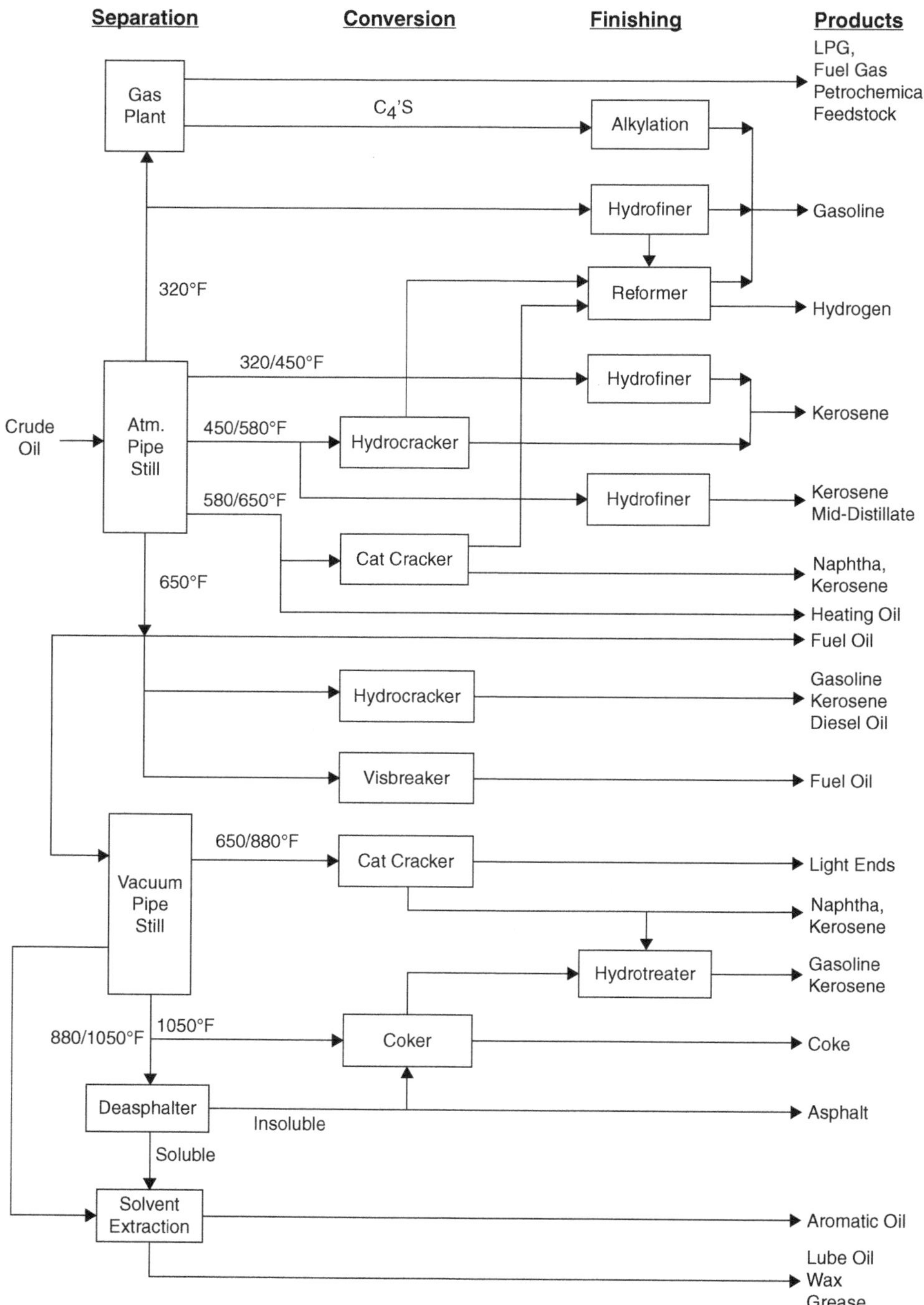

FIGURE 2.1   General schematic of a refinery. (Speight, J.G. 2017. *Handbook of Petroleum Refining.* CRC Press, Taylor & Francis Group, Boca Raton, FL. Figure 7.1, p. 252.)

bottom heat exchangers can corrode and/or plug, and as a result, require chemical or mechanical cleaning, otherwise, throughput has to reduced, leading to a loss of production.

As an introduction, to the dewatering and desalting processes – the importance of these processes cannot be overestimated – in the modern refinery, the removal of water and contaminants generally comprises two steps which are (1) dehydration and (2) desalting. Both processes are essential to avoid corrosion and fouling of refineries equipment such as heat exchangers and pumps. Electrostatic treatment is typically preceded by 2- or 3-phase bulk separation in the upstream process, and electrostatic coalescers are applied as the final separation step to break up emulsion and reduce the remaining water fraction. Desalting is a process to remove salt contents from crude oil. Desalting often occurs in two stages, such as dehydration followed by desalting as salt is soluble in water. Salts are dissolved by the addition of wash water between dehydrator and desalter, which ensures a high dilution rate. The dilution is necessary since dehydration alone is usually not sufficient to reach typical salinity values in the export crude. Hence, both the processes are essential to yield premium quality petroleum products.

Thus, dewatering and desalting operations are necessary to remove salt from the brines that are present with the crude oil after recovery (Abdel Aal et al., 2015). The remnant brine could be defined as that part of the salty water that comes with the oil and cannot be reduced any further by the dehydration methods and is referred to as basic sediments and water (BS&W). Typically, this remnant water exists in the crude oil as a dispersion of very fine droplets highly emulsified in the bulk of oil. The main mineral salts found are the chlorides of sodium (NaCl), calcium ($CaCl_2$), and magnesium ($MgCl_2$).

These salts can decompose in the heater to form hydrochloric acid and cause corrosion of the fractionator overhead equipment. In order to remove the salt, water is injected into the partially preheated crude and the stream is thoroughly mixed so that the water extracts the salt from the oil. The mixture of oil and water is separated in a desalter, which is a large vessel in which the water settles out of the oil, a process that may be accelerated by the addition of chemicals or by electrical devices. The salt-laden water is automatically drained from the bottom of the desalter.

Failure to remove the brine to acceptable levels can result in unacceptable levels of hydrogen chloride produced during downstream refining. The hydrogen chloride will cause corrosion to equipment even to the point of weakening the equipment that can lead to explosions and fires.

It is the purpose of this chapter to present the process options that can be used to mitigate all or part of the corrosion-fouling problem and are in the form of the dewatering and desalting processes that prepare the feedstock for introduction to the distillation section of the refinery and then to a variety of conversion processes which are often, incorrectly, referred to as refining proper.

2.2   DEWATERING

Even though distillation (Chapter 4) is often considered to be the first step in crude oil refining, it should be recognized that crude oil that is contaminated by salt water either from the well or during transportation to the refinery must be treated to remove the emulsion. If salt water is not removed, the materials of construction of the heater tubes and column intervals are exposed to chloride ion attack and the corrosive action of hydrogen chloride, which may be formed at the temperature of the column feed.

The process of dewatering of crude oil commences at the production process since production wells rarely produce either anhydrous or low water oil. However, although preliminary treatment of crude oil is carried out at the wellhead so that the oil can meet the specification set by the pipeline company for transportation of the oil, small amounts of inorganic salts (such as magnesium chloride, $MgCl_2$ and calcium chloride, $CaCl_2$) may remain in the water that is still associated with the crude oil that is sent to the refinery. Also, as crude oil production from the reservoir increases, the water content of the oil produced increases at different rates.

The dewatering processes depend on the type of crude oil and the amount associated with the crude oil; any of the following typical processes for dewatering (and desalting) crude oil are combined using methods such as thermal methods, chemical methods, electrical methods, and mechanical methods as well as combinations of these technologies. Also, a dual function is the main objective of the dehydration step: (1) to ensure that the remaining free water is totally removed from the bulk of oil and (2) to apply the necessary methods to break the oil emulsion. In general, free water removed by gravity in the separator is limited to large water droplets. The crude oil leaving the separator would normally contain free water droplets that are smaller in size in addition to the water emulsified in oil (Abdel-Aal et al., 2018). However, the produced crude oil is not "clean" and still contains sediment and produced water, mineral salts, and other impurities which require the further steps of dewatering and desalting.

The chemical treatment method for the use of a demulsifier must be introduced as early as possible (even at the bottom of the well) to give an increased contact time and the interaction of the demulsifier with the emulsion for maximum emulsion destruction. The injection of the demulsifying agent before the pump ensures proper contact with the crude oil and minimizes the formation of an emulsion. Preference is often given to colloidal surface active substances, among which are anionic, cationic, and nonionic types. The number of steps required for the process is determined by the characteristics of the initial oil emulsion and the salts contained therein for the desalination of oil emulsion independently of the stage.

Thus, the first step in crude oil processing, even before the crude oil enters the refinery occurs at the wellhead (Abdel-Aal et al., 2015), is the separation of the fluids into natural gas, crude oil, and water using a gas–oil separator. The separators can be horizontal, vertical, or spherical and are generally classified into two types based on the number of phases to separate: (1) two-phase separators, which are used to separate gas from oil in oil fields or gas from water for gas fields and (2) three-phase separators, which are used to separate the gas from the liquid phase and water from oil. The liquid (oil, emulsion) leaves at the bottom through a level control or an exit valve. The gas leaves the vessel at the top, passing through a mist extractor to remove the small liquid droplets in the gas.

At the well or the refinery, the main types of equipment used for dewatering crude oil are separators which separate gas from the crude oil in two or three steps under slight pressure or dilution. The separators of the first stage simultaneously play the role of buffer reservoirs and are usually located on the deposit. The separators of the second and third stages are mostly on the territory of central collection and distribution points (the sites of pretreatment and pumping of oil). The vertical separators are more productive in comparison with horizontal separators and are suitable for situations where the production capacity is high as well as when the emulsion contains many solid particles. The horizontal separators are suitable for processing small volumes of material, as well as liquids with a high content of dissolved gas. To achieve maximum efficiency when using horizontal natural gas and crude oil separators, the crude oil is mixed, the temperature is increased, and the pressure is reduced.

Separators can also be categorized according to their operating pressure: (1) low-pressure units can tolerate pressures on the order of 10–180 psi, (2) medium-pressure separators operate from 230 to 700 psi, and (3) high-pressure units can tolerate pressures of 975–1500 psi. Even after this type of separation, there is the need to clean the crude oil before proceeding with the separation of crude oil into its various constituents. This is often referred to as desalting and dewatering in which the goal is to remove water and the constituents of the brine that accompany the crude oil from the reservoir to the wellhead during recovery operations.

The emulsions formed in the crude oil industry are typically water-in-oil emulsions (often referred to in the industry as regular emulsions) in which the oil is the continuous or external phase and the dispersed water droplets form the dispersed or internal phase. The water droplets increase in size with time after the demulsifier is added. The role of the demulsifier is to change the interfacial properties and to destabilize the surfactant-stabilized emulsion film in the demulsification process. In the beginning, there are small and uniform drops while larger drops begin to form after which two or more large drops continue to form a larger drop (Atta, 2013). The droplet size grows fast and the droplet number reduces after demulsifier is added, and in many cases, the coalescence of water droplets destroyed emulsions.

By the way of clarification, there are three terms related to stability commonly encountered in crude oil emulsion which are (1) flocculation, (2) coalescence, and (3) breaking.

Flocculation refers to the mutual attachment of individual emulsion drops to form flocs or loose assemblies and may, in many cases, be a reversible process, overcome by the input of much less energy than what was required in the original emulsification process. It should be noted that the word flocculation is also well established when dealing with consolidation of solid suspensions such that the flocculated solid phase gradually goes to the bottom. Coalescence refers to the joining of two or more drops to form a single drop of greater volume, but with smaller interfacial area. Although coalescence will result in significant microscopic changes in the condition of the dispersed phase, it may not immediately result in a macroscopically apparent alteration of the system. On the other hand, the breaking of an emulsion refers to a process in which gross separation of the two phases occurs. In such an event, the identity of individual drops is lost, along with the physical and chemical properties of the emulsion, and it represents a loss of stability of the emulsion.

2.2.1 Process Description

To dehydrate most crude oils (especially those crude oils with a lower API gravity, <30°–35° API) to water levels where they are acceptable to refineries, heating is needed. This has several benefits which are (1) the oil density is reduced relative to the water density, providing better gravity separation, (2) the viscosity is reduced which reduces the flow resistance of water droplets falling through the oil layer, providing better gravity separation, and (3) heat improves coalescing and assists in breaking emulsions that would otherwise prevent separation to the required degree. The heating process is generally carried out using a submerged fire tube either in a separate or combined vessel.

After the heating option, the process is initiated by mixing the raw crude oil with an appropriate amount of washing water (also known as dilution water) as a volume percent of the crude oil processed on the order of 3%–10% (w/w) depending on the API gravity of the crude oil. The heavier (more viscous) the feedstock, the more the water that is required. Typically, demulsifiers are added to the feedstock in this process step.

The water associated with crude oil contains salts and other water-soluble materials. At the production facility, the bulk of the water with dissolved salts is removed from the crude oil by gravity separation (i.e. settling) or by centrifuging. The emulsion might also be heated to 35°C to −80°C (95°F–176°F) to facilitate water separation.

Typically, the removal of this unwanted water is relatively straightforward and involves the use of a wash tank or a heater. Removal of this same water along with salt concentration reduction presents a completely different set of problems. These can be overcome with relative ease if the operator is willing to spend many thousands of dollars each year for water, fuel, power, and chemical additives. Conversely, if some time is spent in the initial design stages to determine the best methods for water removal to achieve lower BS&W remnants, mixing, and injection, then a system can be designed that will operate at a greatly reduced annual cost.

In order to promote the effective mixing between the organic and aqueous phases and to ensure the proper dilution of the salts and minerals in the aqueous phase, a mixing valve is used. This is a common globe valve which causes a pressure drop and, as a result, a shear stress over the droplets that promotes an intimate water and oil mixture. The main aspect that needs to be considered is the pressure drop which varies according to the flow through the valve (automatic differential pressure controllers could be used).

In addition to the mixing valve, upstream premixing device could be used, such as spray nozzles at the point of water injection or static mixers, between the water injection point and the mixing valve. A high-pressure differential in the mixing valve promotes smaller droplets, which is positive because it improves the contact among the phases; however, very small droplets could yield a more stable emulsion, which could cause problems in the separating vessel. Therefore, it is important to balance both effects in the selection of the operation pressure drop.

This mixture then passes to the desalter, a horizontal cylindrical tank that provides a sufficiently long residence time to separate the water and oil mixture in two phases. Some of the water droplets may have diameters that are so small that they could not be separated by gravity; so, an electrostatic field between two electrodes installed into the desalter is used to promote coalescence.

In the case of electrostatic dehydration, an electrical field is used to excite droplets of brine within the bulk oil phase so that they collide with other droplets and coalesce into larger globules that separate under gravity. Typically, an electrostatic coalescer employs two horizontal grids connected to a high-voltage alternating current (AC) supply, and the emulsion flows upward through the grids. The water coalesces and flows downward, while the clean oil leaves via the top of the vessel. Distributors are used to ensure even "plug" flow up the vessel and grids.

The destabilization of water-in-crude oil emulsions is an important aspect of the dewatering–desalting process. For example, during production of the crude oil, it is essential to remove water from crude oil in order to reduce corrosion of pipelines and hence optimize the pipeline use. In the refinery, potentially corrosive salts are removed from the crude by deliberate emulsification of fresh (wash) water to produce 5%–10% v/v water-in-crude oil emulsions, which are subsequently broken using electrical dehydrators with added demulsifying chemicals. Various chemicals have been used as demulsifiers, such as nonionic surfactants based on ethylene and propylene oxides (Mohammed et al., 1994). In the case of water-in-crude emulsions, a balanced optimum formulation is attained by adding demulsifiers which are hydrophilic to the lipophilic natural surfactants contained in the crude oil.

2.2.2 Equipment

Dewatering crude oil is the process of removing water from crude oil before it is charged to the atmospheric distillation tower. Typically, this is achieved by storing the crude oil in a tank, allowing the water and oil to naturally separate, and then pumping out the layer of water from underneath the oil. On the other hand, heavy crude oil and extra heavy crude oil do not flow easily and it is always a challenge to remove water from heavy crude oil (<20° API) due to its near-water density (10° API) and high viscosity which dictates that the separation will be a slow process.

The preliminary treatment of crude oil involves treatment in the wet-crude handling facilities in which a dual function is the main objective is to ensure that the remaining free water that is associated with the crude oil is removed from the bulk of oil and to apply whatever methods are necessary to break the oil emulsion. In general, free water is removed by gravity in a separator but this process is limited to large water droplets. However, many crude oil streams contain free water droplets that are smaller in size than the typical droplets and require further treatment.

All unrefined crude oil stored in tanks has a percentage of water entrained within it, and while stored in tanks, separation naturally occurs with water collecting at the bottom of the tank beneath the oil. The two fluids are very distinct except for a "black water" or "rag" interface layer which is an emulsion of mixed oil and water. To dewater the tank, water is drawn off of the bottom of the tank and is then sent off to water treatment.

Dewatering is the process of removing water from crude oil before it is charged to the atmospheric distillation unit. Typically, this is done by letting the crude rest in a tank, allowing the water and oil to naturally separate, and then pumping out the layer of water from underneath the oil. It also has the purpose of destabilizing water-in-crude oil emulsions, and during production, it is essential to remove water from crude oil in order to reduce the potential for corrosion of pipelines. In the refinery, potentially corrosive salts must be removed from the crude oil by emulsification of fresh (wash) water to produce 5%–10% w/w water-in-crude oil emulsions, which are subsequently broken using electrical dehydrators with added demulsifying chemicals. However, caution is advised when deciding on the use of an electric desalting operation since desalting heavy crude oil in an electrostatic desalter designed for normal crude oils creates interface level problem which can result in more oil in desalter effluent.

The electrostatic dehydrator dewatering unit is divided into AC, direct current (DC), and AC-DC dehydrator. AC electrostatic dehydrator is projected to dominate the market, owing to its simplicity and reliability.

The electric dehydrator has the highest efficiency with the best treatment ability relying on the function of electric field for dewatering. There are many types for electric dehydrator, such as

(1) the pipe type dehydrator, (2) the storage tank type dehydrator, (3) the vertical cylindrical type dehydrator, and (4) the spherical type dehydrator. With the development of the crude oil industry, the horizontal cylindrical type of electric dehydrator has been widely adopted. The effective combination of the horizontal electric dehydrator, the oil–gas separator, and the settlement dehydrator is to improve the dewatering process.

Typically, in the days of conventional crude oils when a single crude oil was the refinery feedstocks, removal of any unwanted water has been fairly straightforward in which wash tanks and/ or a heater were employed. By removal of this along with any soluble or suspended inorganic compounds (i.e. soluble or suspended salts), concentration reduction presents a completely different set of problems. However, with the prominence of heavier crude oil as refinery feedstocks and blending of several such feedstocks with one or more conventional crude oils, the dewatering equipment has had to evolve to accommodate such changes.

In the process of dewatering crude oil by means of settlement and separation, the separation of the crude oil and water is achieved by passing the wet oil through equipment, which takes advantage of the principle of light (lower density) crude oil and the heavier (higher density) water. Alternatively, in the process that involves dewatering of crude oil through chemical demulsification, the oil and water in the emulsified state are separated by the use of a chemical agent (a demulsifier). This process (chemical demulsification) is the most widely adopted one during the dewatering of crude oil. Dewatering through electric demulsification is used to separate crude oil and water through the function of electric ion.

2.3 DESALTING

The desalter is a process unit in a crude oil refinery that removes salt (inorganic materials) from the crude oil (Nasehi et al., 2018). In this process, the salt is dissolved in the water in the crude oil and not in the crude oil itself. The salt content after the desalter is usually measured in pounds of salt per thousand barrels (PTB) of crude oil. Moreover, the removal of salt from crude oil for refinery feedstocks is necessary, especially if the salt content exceeds the limit of 20 pounds of salt PTB. The most economical place for the desalting process is usually in the refinery. However, when pipeline requirements are imposed, field plants are employed in order to desalt the oil prior to transportation of the crude oil. However, the desalting principles applied are the same whether desalting takes place at the field (recovery) site or at the refinery.

Crude oil introduced into the refinery contains many undesirable impurities, such as sand, inorganic salts, drilling mud, polymer, and any corrosion byproducts. The purpose of the desalting process is to remove these undesirable impurities, especially salts and water, from the crude oil prior to the introduction of the feedstock to the distillation section.

The salt content in the crude oil varies depending on the source of the crude oil. When a mixture from many crude oil sources is processed in a refinery, the salt content can vary greatly. The inorganic salts can be decomposed in the crude oil preheat exchangers and heaters. As a result, hydrogen chloride gas is formed which condenses to liquid hydrochloric acid at overhead system of distillation column that may cause serious corrosion of equipment. To avoid corrosion due to the presence of salts in the crude oil, corrosion control can be used. But the byproduct from the corrosion control of oil field equipment consists of particulate iron sulfide and oxide. Precipitation of these materials can, for example, cause plugging of heat exchanger trains, tower trays, and heater tubes. In addition, these materials can cause corrosion to any surface they are precipitated on. The sand or silt can cause significant damage due to abrasion or erosion to pumps and pipelines. The calcium naphthenate compounds in the crude unit residue stream if not removed can result in the production of lower-grade coke and deactivation of catalysts such as the catalyst in a fluid catalytic cracking unit. Thus, the benefits of the desalting process include (1) increased throughput of the refinery feedstock; (2) less plugging, scaling, and coking of heat exchanger and furnace tubes; (3) less corrosion in heat exchangers and fractionators; (4) less erosion by solids in control valves, exchanger, furnace, and pumps.

The desalting process is a water-washing operation performed at the production field and at the refinery site for additional crude oil cleanup (Parkash, 2003; Gary et al., 2007; Speight, 2014a, 2017; Hsu and Robinson, 2017). If the crude oil from the separators contains water and dirt, water washing can remove much of the water-soluble minerals and entrained solids. If these crude oil contaminants are not removed, they can cause operating problems during refinery processing, such as equipment plugging and corrosion as well as catalyst deactivation.

Desalters remove inorganic contaminants from crude oil to reduce fouling and corrosion in other processing units and desalting can be accomplished through two methods which are (1) chemical desalting and (2) electrical desalting. In the chemical desalting process, water and demulsifiers are added to the crude oil which is held in the storage tank for settling to occur.

The desalter is divided into single stage, double stage, and triple stage with the double stage desalter anticipated to become more prominent because of the acceptance of more heavy feedstocks (heavy crude oil, extra heavy crude oil, tar sand bitumen) and unconventional feedstocks (such as bio-oil) in refineries. In the electrical desalting process, high voltage is applied to concentrate the suspended water particles at the bottom of the tank. In both methods, chemicals such as ammonia (NH_3) are often used to reduce the impact of corrosion. The majority of desalting chemicals are either demulsifiers or used to adjust the pH of the wastewater.

The salts and minerals often present in the oil are mainly the chlorides of magnesium, calcium, and sodium (which originate in the brine that accompanies the crude oil out of the well) with sodium chloride being the abundant type. These salts cause corrosion of equipment. All chemical compounds based on chlorine, except for sodium chloride (NaCl), hydrolyze at high temperature to hydrogen chloride (HCl) which dissolves in the emulsion water producing hydrochloric acid, an extremely corrosive acid:

$$CaCl_2 + 2H_2O \rightarrow Ca(OH)_2 + 2HCl$$

$$MgCl_2 + 2H_2O \rightarrow Mg(OH)_2 + 2HCl$$

Any remaining salts are neutralized by the injection of sodium hydroxide which reacts with the calcium and magnesium chloride to produce sodium chloride, which does not readily hydrolyze to the corrosive hydrogen chloride:

$$CaCl_2 + 2NaOH \rightarrow Ca(OH)_2 + 2NaCl$$

The practice of desalting crude oil is an old process and can occur at the wellhead or (depending on the level of inorganic contaminants in the crude oil) at the refinery (Parkash, 2003; Gary et al., 2007; Speight, 2014a, 2017, 2021; Hsu and Robinson, 2017). The specifications for inorganic content and water are rigid because of their negative effect in downstream processes due to corrosion and catalyst deactivation. In the case of water-in-crude-oil emulsions, a balanced optimum formulation is attained by adding to the lipophilic natural surfactants contained in the crude oil, demulsifiers which are hydrophilic.

To the shipper and the refiner (especially the refiner), the non-crude oil constituents of greatest concerns of the impurities in crude oil are (1) inorganic salts and (2) sand and silt.

Crude oil containing inorganic salts when heated in preheat exchangers and heaters will lead to the decomposition of these salts. As a result, hydrogen chloride gas is formed which condenses to liquid hydrochloric acid at overhead system of distillation columns causing damages and corrosion of equipment. Also, sand and/or silt can cause significant damage to pumps and pipes that results in abrasion or erosion.

Mitigation of corrosion is necessary in order to preserve the refinery equipment. However, the byproduct from the corrosion control of refinery oil equipment consists of particulate iron sulfide and iron oxide which can lead to phase separation (precipitation) of these waste products on heat exchanger trains, tower trays in the distillation unit, and also in heater tubes thereby causing serious fouling effects in heat transfer.

More generally, the source and the type of wastes resulting from these processes are (1) free water plus other impurities, (2) desalter wash water which contains salt, sludge, rust, clay, and varying amounts of emulsified oil, also referred to as oil-under carry, (3) gas out, (4) dilution water for the desalter, and (5) treated and desalted oil to refining (Abdel Aal et al., 2015; Johnson and Affam, 2019).

By definition, the purpose of the desalter which can be either a mechanical unit or an electrical unit is to remove crude oil contaminants that will have harmful effects in downstream equipment and process units. The more the desalting is optimized, the greater the amount of contaminants that will be removed. Feeding the desalter a consistent stream of crude oil will assist in the efficiency of the process as well as maintain the temperature high. The process involves the following steps: (1) water is added to crude oil, (2) the mixture passes through a mixing device, (3) the mixture enters the desalter and separates the oil and water, (4) the crude oil exits the top of the desalter, and (5) the water exits the bottom of the desalter. The amount of water, injection point of the water, and the quality of the water are important. The wash water volume should be between 6% and 10% v/v of the crude oil fed to the unit. If the feedstock is a heavier (higher viscosity) crude oil, a high volume of wash water may be required. A higher water rate provides more opportunity for water to contact salts and solids in crude oil (McDaniels and Olowu, 2016).

The wash water should be injected as far upstream of the desalter as possible. If the crude charge pump has the capacity, it is a best practice to inject at least part of the wash water into the pump suction to maximize contact between the water and contaminants. This should be done with caution, however, because a tighter emulsion can form between the water and oil that may be difficult to break in the desalter. A robust chemical emulsion breaker must be used (McDaniels and Olowu, 2016).

The wash water should be oxygen-free, low in hardness, salts, solids, sulfide derivatives, and contain less than 30 ppm of ammonia. Preferred wash water sources are stripped sour water and crude and vacuum unit condensate. Not only does water help with solids and salt removal but sufficient water also aids in the separation of water and oil in the desalter. For the water and oil to separate in the desalter, two things must happen: flocculation and coalescence. Flocculation occurs when fluid droplets become attracted to each other. Coalescence means to come together and form one group or mass. When water droplets combine to form larger water droplets, they coalesce and fall vertically in the desalter.

Three general approaches have been taken to the desalting of crude oil (Figure 2.2). Numerous variations of each type have been devised, but the selection of a particular process depends on the type of salt dispersion and the properties of the crude oil.

The three stages of desalting are (1) adding dilution water to crude oil feedstock, (2) mixing dilution water with the feedstock using a mixer, and (3) dehydration of the feedstock in a settling tank to separate sediment and water (S&W) from the feedstock. Briefly, the crude oil is to be desalted.

In this process, the crude oil feedstock that is to be desalted is heated to a specified temperature and mixed with fresh water, which dilutes the salt. The mixture is then pumped into a settling tank where the salt water separates from the oil and is drawn off. An electrostatic field is applied by electrodes in the settling tank, inducing polarization of the water droplets floating in the larger volume of oil. This results in the water droplets associating and settling to the bottom of the tank

More specifically, the salt or brine suspensions may be removed from crude oil by heating (90°C–150°C, 200°F–300°F) under pressure (50–250 psi) that is sufficient to prevent vapor loss and then allowing the material to settle in a large vessel. Alternatively, coalescence is aided by passage through a tower packed with materials such as sand, gravel, and any related materials.

The common removal technique is to dilute the original brine with fresher water so that the salt content of water that remains after separation treatment is acceptable on the order of 10 pounds PTB

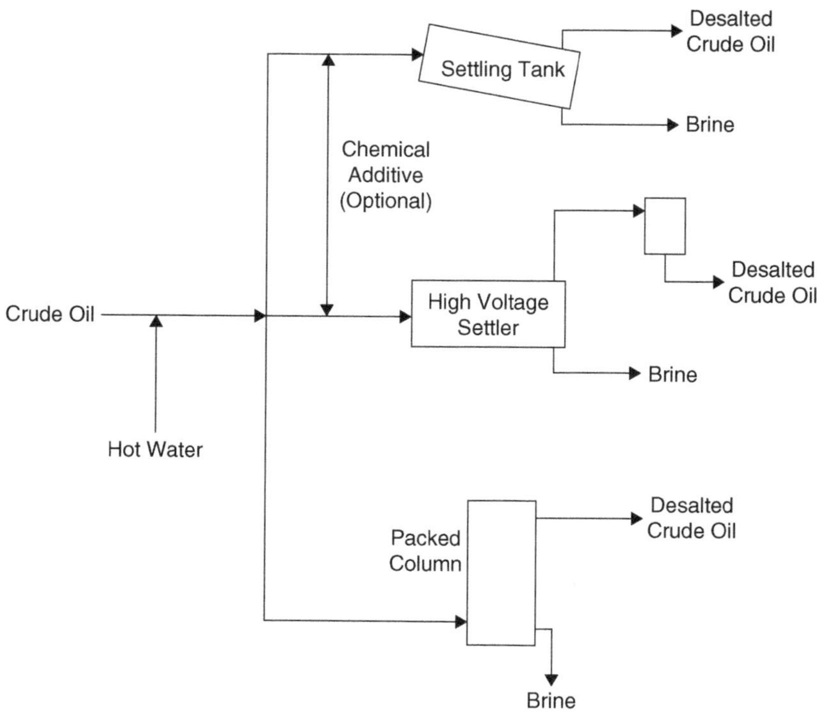

FIGURE 2.2 General methods for desalting crude oil. (Speight, J.G. 2017. *Handbook of Petroleum Refining.* CRC Press, Taylor & Francis Group, Boca Raton, FL. Figure 7.2, p. 254.)

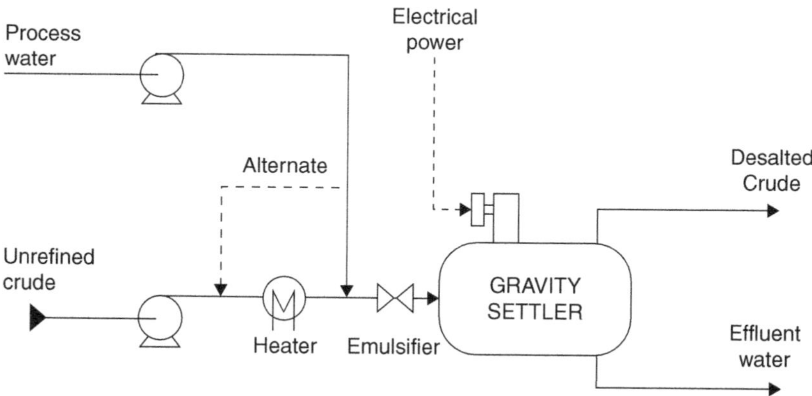

FIGURE 2.3 An electrostatic desalting unit. (OSHA Technical Manual, Section IV, Chapter 2: Petroleum Refining Processes. http://www.osha.gov/dts/osta/otm/otm_iv/otm_iv_2.html.)

of crude oil or less. In areas where fresh water supplies are limited, the economics of this process can be critical. However, crude oil desalting techniques in the field have improved with the introduction of the electrostatic coalescing process (Figure 2.3). Even when adequate supplies of fresh water are available for desalting operations, preparation of the water for dilution purposes may still be expensive.

Requirements for dilution water ratios based on water salinity calculations can be calculated as a material balance, and by combining the arithmetic mean of material balance and water injection and dispersion for contact efficiency, very low dilution water use rates can be achieved. This can

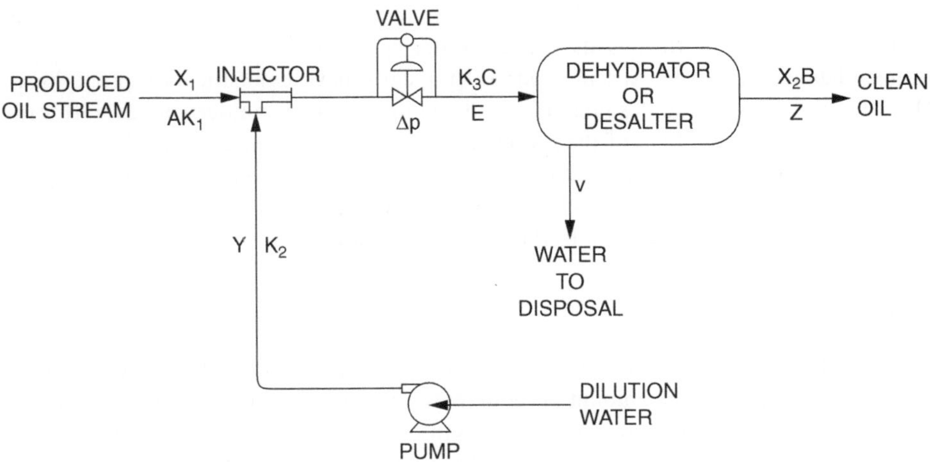

FIGURE 2.4 A single-stage desalting system. (Speight, J.G. 2017. *Handbook of Petroleum Refining*. CRC Press, Taylor & Francis Group, Boca Raton, FL. Figure 7.4, p. 256.)

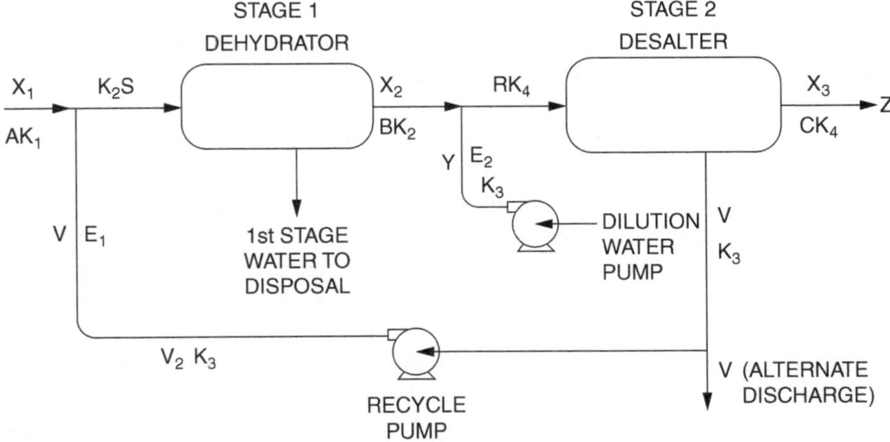

FIGURE 2.5 A two-stage desalting system. (Speight, J.G. 2017. *Handbook of Petroleum Refining*. CRC Press, Taylor & Francis Group, Boca Raton, FL. Figure 7.5, p. 258.)

be highly significant in an area where production rates of 100,000 bbl/day are common and fresh water supply is limited.

Desalter units will produce a dehydrated stream containing like amounts of BS&W from each stage. Therefore, BS&W can be considered as *pass through volume* and the dilution water added is the amount of water to be recycled. The recycle pump, however, generally is oversized to compensate for difficult emulsion conditions and upsets in the system. Dilution water calculations for a two-stage system using recycle are slightly more complicated than for a single-stage process (Figure 2.4) or a two-stage process without recycle (Figure 2.5) (Pereira et al., 2015).

In addition, the asphaltene constituents hinder separation of the oil phase and the water phase during dewatering and desalting because they (the asphaltene constituents) gather at the oil/water interface and undergo self-association, hence forming a rigid film at the oil/water interface (Liu et al., 2003). Furthermore, solids such as high molecular paraffin and clay are adsorbed and the mechanical strength of the interfacial film becomes more rigid than ever (Kim and Wasan, 1996). When the crude oil blend is incompatible or near-incompatible and the asphaltenes are unstable, the process of self-association at the oil/water interface may be accelerated resulting in the formation of a rigid film that deteriorates the desalting and dewatering efficiency.

The naphthenic acid derivatives in crude oil typically have emulsification properties, and as the pH of the water inside the desalter rises, the sodium naphthenate derivatives will form stable emulsions. To mitigate the effect of the presence of sodium naphthenate derivative, it is critical to maintain an effluent desalter water that is acidic (pH < 7). Furthermore, when there is a high concentration of naphthenic acid derivatives in the water, the potential for forming a rag layer (a stable water-in-oil emulsion) is high and can lead to an increase in fouling.

2.3.1 Process Description

Salt in crude oil is in most cases found dissolved in the remnant water within the crude oil which is attributed to two factors: (1) the amount of remnant water that is left in oil after normal dehydration and (2) the salinity or the initial concentration of salt in the source of this water. Desalting of crude oil will eliminate or minimize problems resulting from the presence of mineral salts in crude oil. In other words, the main function of the desalter is to remove salts that are soluble in water from the crude oil. However, contaminants such as clay, silt, rust, and miscellaneous debris also need to be removed.

The soluble salts contain some metals that can poison catalysts used in the process of refining which are dissolved in the water phase. These salts often deposit chloride derivatives on the heat transfer equipment of the distillation units and cause fouling effects. In addition, some chlorides will decompose under high temperature forming corrosive hydrochloric acid:

$$MgCl_2 + 2H_2O \rightarrow Mg(OH)_2 + 2HCl$$

Thus, an important aspect of the crude oil pretreating program is the reduction of the salt content of the crude oil. Depending on the downstream process, a variable limit of the salt content of the crude oil can require additional treatment beyond dehydration in a desalter unit. A two-stage system (a dehydration stage and a desalting stage) is recommended with emphasis placed on electrostatic desalting of the crude oil.

In the process, clean dilution or wash water is injected into the crude oil feed to the desalter through a mixing device to dilute the brine to a level where the target salt content can be achieved by the downstream dehydration unit. In difficult applications, this wash water can be recovered and recycled in a two-stage dehydration and desalting process.

More specifically, wash water, also called dilution water, is mixed with the crude oil coming from the dehydration stage. The wash water, which could be either fresh water or water with lower salinity than the remnant water, mixes with the remnant water, thus diluting its salt concentration. The mixing results in the formation of a water–oil emulsion and the crude oil is then (with the emulsion) dehydrated after which the separated water is disposed of through a water treatment and disposal system.

The mixing step in the desalting of crude oil is normally accomplished by pumping the crude oil (which is the continuous phase) and wash water (which is the dispersed phase) separately through a mixing device which is typically a throttling valve (which is a type of valve that can be used to start, stop, and regulate the flow of fluid). The degree of mixing depends on the interfacial area produced between the two phases. Higher interfacial area will cause a high degree of mixing. A useful device for such a purpose is the application of multiple-orifice-plate mixers but the process configuration (in order to maximize the efficiency of the desalting operation) does depend on the intimate mixing of remnant water with dilution water.

In the two-stage desalting system, dilution water is added in the second stage and all, or part, of the disposed water in the second stage is recycled and used as the dilution water for the first desalting stage. Two-stage desalting systems are normally used to minimize the wash water requirements.

The main objective of a desalting plant is to break the films surrounding the small water droplets, coalescing droplets to form larger drops, and allowing water drops to settle out during or after coalescing. Thus, the operating process variables in the desalter are (1) feedstock flow rate, (2) temperature, (3) pressure, (4) mixing valve pressure drop, (5) wash water rate, and (6) the desalting voltage. The temperature of the feedstock charged to the desalter is a very important aspect of the process since a lower temperature reduces desalting efficiency because of increased viscosity of oil, while a higher temperature reduces desalting efficiency due to the major electrical conductivity of the crude. The pressure in the vessel must be continued at a high value to avoid loss of crude oil pressure, which results in risky condition, undependable operation, and a loss of desalting efficiency. In addition, the voltage within the desalter is on the order of 16,000–30,000 volts AC which provides the coalescing force and the electric current has no participation on this effect. Crude oils that are highly conductive require the application of higher power to achieve the necessary coalescence.

The two most typical methods of crude-oil desalting are (1) chemical separation and (2) electrostatic separation, and both methods use hot water as the extraction agent. In the chemical desalting process, water and chemical surfactants (demulsifiers) are added to the crude oil, heated so that salts and other impurities dissolve into the water or attach to the water, and then held in a tank where they settle out. Electrical desalting is the application of high-voltage electrostatic charges to concentrate suspended water globules in the bottom of the settling tank. Surfactants are added only when the crude has a large amount of suspended solids. Both methods of desalting are continuous.

A third and less-common process involves filtering heated crude using diatomaceous earth. In this process, the heated crude is filtered using diatomaceous earth. In this method, the crude oil is heated to between 65°C and 175°C (150°F and 350°F) to reduce its viscosity and surface tension for easier mixing and separation of its water content (Nasehi et al., 2018). The temperature is limited by the vapor pressure of the crude oil feedstock. In addition, improvement of the contact between the crude oil feedstocks and the dewatering–desalting equipment may prove to be beneficial for the viscous crude oils (Forero et al., 2001).

The three methods of desalting may involve the addition of other chemicals to improve the separation efficiency and ammonia is often used to reduce corrosion while caustic soda or acid may be added to adjust the pH of the water wash. Wastewater and contaminants are discharged from the bottom of the setting tank to the wastewater treatment facility. The desalted crude is continuously drawn from the top of the settling tanks and sent to the crude distillation (fractionating) tower.

Also, the salt or brine suspensions may be removed from crude oil by heating (90°C–150°C, 200°F–300°F) under pressure (50–250 psi) that is sufficient to prevent vapor loss and then allowing the material to settle in a large vessel. Alternatively, coalescence is aided by passage through a tower packed with sand, gravel, etc. Thus, in the process, the feedstock passes through the cold preheat train and is then pumped to the desalters by crude booster pumps. The recycled water from the desalters is injected in the crude oil containing sediments and produced salty water which enters in the static mixer, which is a crude, water disperser, in order to maximize the interfacial surface area for the optimal contact between both liquids. The wash water is injected as near as possible emulsifying device to avoid the first separation with the feedstock. The still mixers are installed upstream the emulsifying devices to improve the contact between the salt in the crude oil and the wash water injected in the line. The oil/water mixture is homogeneously emulsified in the emulsifying device which is another important aspect to ensure contact between the salt production water contained in the oil and the wash water. Then, the emulsion enters the desalters where it separates into two phases by electrostatic coalescence.

Salts in crude oil are mainly in the form of magnesium, calcium, and sodium chlorides, with sodium chloride being the most abundant one. These salts can be found in two forms: dissolved in emulsified water droplets in the crude oil, as a water-in-oil emulsion, or crystallized and suspended solids. The negative effect of these salts in downstream processes can be summarized as follows: (1) salt deposit formation as scales where water-to-steam phase change takes place and (2) corrosion by hydrochloric acid formation. Hydrochloric acid is formed by magnesium and calcium chlorides' decomposition at high temperatures (on the order of 350°C, 660°F). Thus

$$CaCl_2 + 2H_2O \rightarrow Ca(OH)_2 + 2HCl$$

$$MgCl_2 + 2H_2O \rightarrow Mg(OH)_2 + 2HCl$$

In addition, other metals in inorganic compounds present in reservoir dirt and sand produce catalyst poisoning in downstream processes such as hydrotreaters and cat crackers as they are chemically adsorbed on the catalyst surface. The objective of desalting process is to remove chloride salts and other minerals from the crude oil by water washing. Depending on the desired salt content in the desalted crude oil, a one- or two-step process could be applied. By desalting, a considerable percentage of suspended solids (sand, clay, or soil particles, or even particles produced from corrosion of pipelines and other upstream equipment) are removed. Simplified desalting process flow diagrams are given for different configurations: (1) one-step and (2) two-step. The process starts by mixing the raw crude oil with an appropriate amount of washing water also known as dilution water. The washing water as a volume percent of the crude oil processed could oscillate between 3% and 10%, depending on the API gravity of the crude oil – the heavier the crude oil, the more the water required. Demulsifiers are added to the crude oil in this process step.

The common approach to desalting crude oil involves the use of two-stage desalting system (Figure 2.5) in which dilution water is injected between stages after the stream water content has been reduced to a very low level by the first stage. Further reduction is achieved by adding the second-stage recycle pump. The *second-stage water* is much lower in sodium chloride (NaCl) than the produced stream inlet water due to the addition of dilution water. By recycling this water to the first stage, both salt reduction and dehydration are achieved in this stage. The water volume to be recycled is assumed to be the same as dilution water injection volume.

A very low BS&W content at the first-stage desalter exit requires a high percentage of dilution water to properly contact dispersed, produced water droplets and achieve desired salt concentration reduction. This percentage dilution water varies with the strength of the water/oil emulsion and oil viscosity. Empirical data show that the range is from 4.0% to as high as 10%. Obviously, this indicates that the mixing efficiency of 80% is not valid when low water contents are present. Additional field data show that the usage rates of low dilution water can be maintained and still meet the required mixing efficiencies. When 99.9% of the produced water has been removed, the remaining 0.1% consists of thousands of very small droplets more or less evenly distributed throughout the oil. To contact them would require either a large amount of dilution water dispersed in the oil or a somewhat smaller amount with better droplet dispersion. Whatever the required amount is, it can be attained without exceeding the dilution water rates shown in the earlier example.

Water contained in each desalter unit is an excellent source of volume ratio makeup through the use of a recycle option (Figure 2.6).

Emulsions may also be broken by the addition of treating agents, such as soaps, fatty acids, sulfonates, and long-chain alcohols. When a chemical is used for emulsion breaking during desalting, it may be added at one or more of three points in the system. First, it may be added to the crude oil before it is mixed with fresh water. Second, it may be added to the fresh water before mixing with the crude oil. Third, it may be added to the mixture of crude oil and water. A high-potential field across the settling vessel also aids coalescence and breaks emulsions in which case dissolved salts and impurities are removed with water.

If the oil entering the desalter is not hot enough, it may be too viscous to permit proper mixing and complete separation of water and oil, and some of the water may be carried into the fractionator. If, on the other hand, the oil is too hot, some vaporization may occur, and the resulting turbulence can result in improper separation of oil and water. The desalter temperature is therefore quite critical, and normally, a bypass is provided around at least one of the exchangers so that the

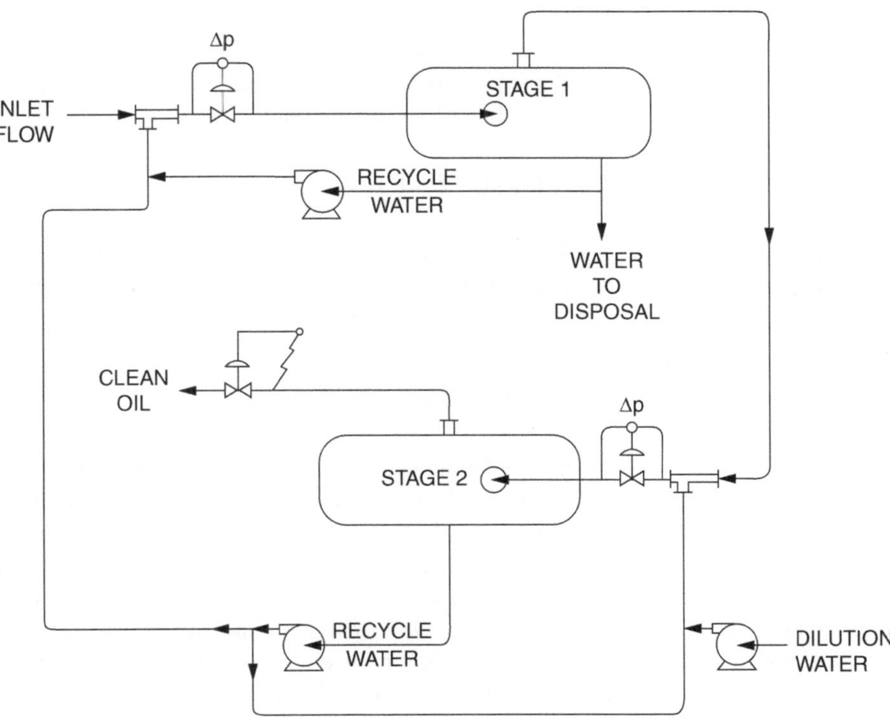

FIGURE 2.6 A two-stage desalting system with recycle. (Speight, J.G. 2017. *Handbook of Petroleum Refining*. CRC Press, Taylor & Francis Group, Boca Raton, FL. Figure 7.6, p. 258.)

temperature can be controlled. The optimum temperature depends upon the desalter pressure and the quantity of light material in the crude, but is normally approximately 120°C (250°F), 100°C (212°F), which is being lower for low pressures and light crude oils. The average water injection rate is 5% of the charge.

Regular laboratory analyses are necessary to monitor the desalter performance, and the desalted crude should normally not contain more than one kilogram of salt per 1000 barrels of feed.

Good desalter control is indicated by the chloride content of the overhead receiver water and should be on the order of 10–30 ppm chlorides. If the desalter operation appears to be satisfactory but the chloride content in the overhead receiver water is greater than 30 ppm, then caustic should be injected at the rate of 1–3 lbs per 1000 barrels of charge to reduce the chloride content to the range of 10–30 ppm; however, this addition of caustic is feedstock and process dependent. Salting out will occur below 10 and severe corrosion will occur above 30 ppm. Another controlling factor on the overhead receiver water is pH. This should be controlled between pH 5.5 and 6.5. Ammonia injection into the tower top section can be used as a control for this. In addition to electrical methods, desalting may also be achieved by using the concept of a packed column (Figure 2.2) that facilitates the separation of the crude oil and brine through the agency of an adsorbent.

In addition to the mixing valve, upstream premixing device could be used, such as spray nozzles at the point of water injection or static mixers, between the water injection point and the mixing valve.

High delta pressure in the mixing valve promotes smaller droplets, which is positive because it improves the contact among the phases; however, very small droplets could yield a more stable emulsion, which could cause problems in the separating vessel. Therefore, it is very important to balance both effects in the selection of the operation pressure drop. Then, this mixture goes to the desalter, and a horizontal cylindrical tank provides long enough residence time to separate the water and oil mixture in two phases. Some water droplets diameters are so small that they could not be separated by gravity; so, an electrostatic field between two electrodes installed into the desalter is used to promote coalescence. Due to the dipolar nature of the water molecule, this electric field promotes an attraction with the other water molecules in the neighbor droplets promoting coalescence.

Either AC or DC fields may be used and potentials from 12,000 to 35,000 V. The attraction force (F) between the water droplets is given by the following equation:

$$F = \left(K_s \cdot \varepsilon^2 \cdot d^6 \right) / s4$$

In this equation, ε is the voltage gradient, d is the droplet diameter, s is the distance between drop centers, and Ks is a constant for the system. Finally, after coalescence, water droplets settle according to the following equation:

$$Settling\,rate = \left[\left(k\,\rho_{water} - \rho_{oil} \right) d^2 \right] / \mu_{oil}$$

In this equation, k is a constant, d is the droplet diameter, ρ is the density, and μ is the viscosity. In a one-step configuration, 90% w/w of salt removal can be achieved. For higher salt removal percentages, a two-step configuration should be required. As far as the two-step process configuration is concerned, two demulsifier injection points are used, both before the mixing valve in the first and second step. In addition, fresh water is fed to the second stage, and effluent water from this stage is recycled to the first one. With this configuration, a 99% w/w salt removal could be achieved.

Finally, flashing the crude oil feed can frequently reduce corrosion in the principal distillation column. In the flashing operation, desalted crude is heat exchanged against other heat sources that are available to recover maximum heat before crude is charged to the heater, which ultimately supplies all the heat required for operation of the atmospheric distillation unit. Having the heater transfer temperature offset the flow of fuel to the burners allows control of the heat input. The heater transfer temperature is merely a convenient control, and the actual temperature, which has no great significance, will vary from 320°C (610°F) to temperatures as high as 430°C (805°F), depending on the type of crude oil and the pressure at the bottom of the fractionating tower.

As described above, dewatering followed by desalting of crude oil is a significant process to be carried out before refining the crude oil to produce various products. Thus, the dewatering–desalting is a key process for the preparation of crude oil as the feedstock for crude distillation units by the removal of undesirable constituents including all types of mineral salts. The salt content of crude oil in pounds of salt PTB is reduced by adjustment of the process parameters (Table 2.1) in order to meet the desired specification.

The performance of the desalter is generally considered good (or, at least, satisfactory) when the salt content of the desalted feedstock is less than one pound of salt PTB (lb/1000 bbls) crude. When large percentages of heavy feedstocks are processed, the desalted crude salt content can be chronically high (3–5 lb/1000 bbls) or the desalter can have periodic upsets, leading to extremely high salt content for short intervals. In either case, high salt content of the desalted feedstock can cause a significant increase in corrosion (Collins and Barletta, 2012).

In addition, the principal cause of crude unit overhead corrosion is hydrogen chloride vapor evolved from certain inorganic chloride salts when raw crude is heated prior to distillation. These non-water extractable chlorides that are organic chlorides or inorganic chlorides are not dissolved in emulsified water, removed by desalters, or indicated in normal, extractable chloride measurement methods. The exact identity of the non-desaltable chlorides varies with the specific feedstock, but potential sources include oil-wetted inorganic salt crystals, due to thermal stresses experienced in production. The presence of these materials can result in higher crude unit atmospheric column overhead system chloride loadings (in the form of hydrogen chloride), increased overhead neutralizer demand, and high potential for corrosion and the accompanying fouling.

Upstream of the desalter, the temperature is low enough that salt corrosion is not a major issue. The crude unit desalter uses electrostatic precipitation to extract most of the water and reduce the

TABLE 2.1

Typical Parameters[a] to be Considered for the Dewatering–Desalting Process

Parameter	Comment
Demulsifiers	The addition of demulsifiers aids in complete electrostatic coalescence and settling of emulsified oil drops, hence an oil desalting
	Demulsifiers are especially important when heavy crude oils are processed
Pressure	A high pressure drop across the mixing valve will result in a good mixing between wash water and the emulsion leading to a decrease in the salt content in the oil phase
	If the pressure drop (which is feedstock dependent) is excessive, the formed emulsion may be hard to break
	The optimum pressure drop is estimated to be about 22 psi bar for light crudes and 7 psi for heavy crudes
Temperature	The temperature affects the settling of water droplets by the effect on oil physical properties, in particular oil viscosity
	Heavy (viscous) crude oil requires higher desalting temperatures in order to keep products on specification and to minimize waste products
Wash water ratio	Heavy (viscous) crude oils require a high wash water ratio to increase electrical coalescence or residual products containing higher salt ratio will be formed
Water level	Water–crude interface level in the separators should be kept constant in electrical desalters
	Any changes will upset the electrical field and influence electrical coalescence of water drops
	Reduced performance may lead to waste products of crude oil and be carried away with separated water

[a] Listed alphabetically rather than by any effect.

chloride content. However, the desalter does not remove all the chlorides, resulting in some chloride contamination in the atmospheric heater. In fact, the native chlorides themselves are not the cause of corrosion. Instead, at 175°C–205°C (350°F–400°F), calcium chloride ($CaCl_2$) and magnesium chloride ($MgCl_2$) thermally decompose (disassociate). The chlorine rapidly reacts with hydrocarbons, producing hydrogen chloride. At normal crude atmospheric heater operating conditions, nearly all of the magnesium chloride and much of the calcium chloride will hydrolyze. While sodium chloride (NaCl) may start to hydrolyze at temperatures as low as 230°C (450°F), temperatures above 425°C (800°F) are often required for significant cracking. Decomposition of sodium chloride is not a substantial chloride source for most crude atmospheric tower overhead systems.

Hydrogen chloride in the absence of water does not significantly corrode carbon steel (CS). The overhead system of the crude tower condenses water. The water absorbs the hydrogen chloride, creating hydrochloric acid. At this point, many outcomes are possible, depending on the location of where the first drop of water forms. The first drop of water will occur at the point where film temperatures reach the dew point. Bulk temperatures are often reviewed, but this will miss many problems, as surface temperatures are what count. The initial drop of water may have very high hydrogen chloride concentrations, making it extremely corrosive. The water also absorbs ammonia (NH_3). Hydrogen chloride combines with ammonia or with ammonium hydroxide and forms ammonium chloride.

$$HCl + NH_3 \rightarrow NH_4Cl$$

$$HCl + NH_4OH \rightarrow NH_4Cl + H_2O$$

In situations where the water return to the vapor state, solid deposits of ammonium chloride form, thus creating the potential for under-deposit corrosion.

By lowering salt content, crude desalting becomes an important tool for controlling overhead corrosion. Compared to other refinery operations, crude desalting is a simple and straightforward process: water is added to the raw crude charge and inorganic salts are extracted into the water. The extract brine is separated and removed, while the now desalted crude charge is sent to preheat exchangers and the fired crude heater.

However, a portion of the salt leaving the desalter hydrolyzes to hydrogen chloride in the atmospheric and vacuum heaters – the amount of hydrolysis depends on the heater temperature, the type of salt, and the presence of other compounds such as naphthenic acids (as is the case with high-acid crudes) (Speight, 2014b). It has become the custom to inject caustic downstream of the desalter to convert chlorides that hydrolyze to more stable sodium chloride (NaCl) thereby reducing the level of hydrogen chloride in the crude overhead system. However, if the desalted crude salt content is high (3–6 lb/1000 bbls), even additional caustic is not believed to improve reliability (Collins and Barletta, 2012).

Furthermore, the sodium chloride that remains in the desalted feedstock and which does not hydrolyze in the atmospheric heater will, more than likely, be converted to hydrogen chloride in the vacuum heater. Hence, corrosion rates and fouling in the vacuum column overhead system can be high, relative to conventional feedstocks. Corrosion in the vacuum column overhead system and salt laydown in the top of the vacuum column has become much more common with heavy feedstock processing and are also common when processing opportunity crudes (Collins and Barletta, 2012).

Thus, poor desalting leads to a high corrosion rate in the atmospheric crude and vacuum column overhead systems. Corrosion has also been observed in related equipment such as top pump-around systems and product rundown systems. Rapid laydown of amine chloride salts ($RNH_3{}^+Cl^-$) in the top of the atmospheric crude column is relatively common and, increasingly, the internals in the top of vacuum columns foul with chloride salts. In the vacuum column, light vacuum gas oil pump-around systems and vacuum pre-flash column top pump-around systems have shown a high metal loss. Moreover, the chlorides can make their way to downstream hydrotreating equipment, where higher corrosion rates will also occur. Other consequences of poor desalting include severe exchanger fouling from poor filterable solids removal in conjunction with poor exchanger design.

In the refinery, desalter operations may suffer from issues related to the properties of the oil that originates in tight shale formations. Solids loading can be highly variable, leading to large shifts in solids removal performance. Sludge from the tank farm may cause severe upsets, including growth of stable emulsion bands and intermittent increases of oil in the brine water. In addition, agglomerated asphaltene constituents can enter from storage tanks or can flocculate in the desalter rag layer, leading to oil slugs in the effluent brine.

In summary, the desalter operation is dependent on the operating temperature via the viscosity and density. A higher temperature promotes settling via the reduced viscosity but it also increases the electrical conductivity of the mixture. A high conductivity will result in increased voltage gradients between the electrodes, increasing the electricity cost, and in the worst case, cause short-circuiting and desalter breakdown. A low temperature decreases the settling velocity and can reduce the unit throughput. Both the above effects are undesirable, so upper and lower temperature bounds are often specified for desalter operation.

2.3.2 Equipment

Generally, the desalting process is completed in the following steps which are (1) dilution water injection and dispersion, (2) emulsification of diluted water in oil, (3) distribution of the emulsion in the electrostatic field, (4) electrostatic coalescence, and (5) water droplet settling (Nasehi et al., 2018).

In this process, the crude oil passes through the cold preheat train and is then pumped to the desalters by crude charge pumps. The recycled water from the desalters is injected into the crude oil-containing sediments and produced salty water. This fluid enters in the static mixer which is a crude/water disperser, maximizing the interfacial surface area for optimal contact between both liquids.

The wash water should be injected as near as possible emulsifying device to avoid a first separation with crude oil. Wash water can come from various sources including relatively high salt sea water and stripping water. The static mixers are installed upstream of the emulsifying devices to improve the contact between the salt in the crude oil and the wash water injected in the line. The oil/water mixture is homogenously emulsified in the emulsifying device which is used to emulsify the dilution water injected upstream in the oil. The emulsification is important for contact between the salty production water contained in the oil and the wash water after which the emulsion enters the desalter where it separates into two phases by electrostatic coalescence.

The electrostatic coalescence is induced by the polarization effect resulting from an external electric source. Polarization of water droplets pulls them out from oil–water emulsion phase. Salt being dissolved in these water droplets is also separated along the way. The produced water is discharged to the water treatment system (effluent water) which can also be used as wash water for mud washing process during operation.

A desalting unit can be designed with single stage or two stages. In the refineries, the two-stage desalting system is normally applied, which consists of two electrostatic coalescers.

More specifically, the desalting process at the refinery can be accomplished through a one-stage process or a two-stage process using one or two electrostatics desalters. To increase the efficiency of the operation, demulsifiers are added to the crude oil. Two-stage installations are used to reduce the flow of fresh water when washing crude oil. The desalter of this design achieves 90% salt removal. However, 99% salt removal is possible with two-stage desalters. A second stage is also essential since desalter maintenance requires a lengthy amount of time to remove the dirt and sediment which settle at the bottom. Therefore, the crude unit can be operated with a one-stage desalter while the other is cleaned.

However, the common approach to desalting crude oil involves use of the two-stage desalting system (Figure 2.6) in which dilution water is injected between stages after the stream water content has been reduced to a very low level by the first stage. Also, the emphasis is placed on electrostatic desalting of crude oil.

In the process, wash water, also called dilution water (which could be either fresh water or water with lower salinity than the remnant water), is mixed with the crude oil from the dehydration stage. The wash water mixes with the remnant water, thus diluting its salt concentration but results in the formation of a water–oil emulsion. The oil (and emulsion) is then dehydrated using the dehydration technique. The separated water is disposed of through the field-produced water treatment and disposal system. In the two-stage desalting system (which is normally used to minimize the wash water requirements), dilution water is added in the second stage and all, or part, of the disposed water in the second stage is recycled and used as the dilution water for the first desalting stage. Further reduction is achieved by adding the second-stage recycle pump. The second-stage water is much lower in sodium chloride (NaCl) than the produced stream inlet water due to the addition of dilution water. By recycling this water to the first stage, both salt reduction and dehydration are achieved. The water volume to be recycled is assumed to be the same as dilution water injection volume.

A very low BS&W content at the first-stage desalter exit requires a high percentage of dilution water to properly contact dispersed, produced water droplets and achieve desired salt concentration reduction. This percentage dilution water varies with the strength of the water/oil emulsion and oil viscosity. Empirical data show that the range is from 4.0% to as high as 10%. Obviously, this indicates that the mixing efficiency of 80% is not valid when low water contents are present. Additional field data show that the use rates for low dilution water can be maintained while still achieving the required mixing efficiencies. When 99.9% of the produced water has been removed, the remaining 0.1% consists of thousands of very small droplets more or less evenly distributed throughout the oil. To contact them would require either a large amount of dilution water dispersed in the oil or a somewhat smaller amount with better droplet dispersion. Whatever the required amount, it can be attained without exceeding the dilution water rates shown in the earlier example.

Water contained in each desalter unit is an excellent source of volume ratio makeup through the use of a recycle option (Figure 2.6). The amount of water recycled to the first stage must be the same as the dilution water injection rate to maintain the water level in the second-stage unit.

The volume of water recycled to the second-stage inlet can be any amount since it immediately rejoins the controlled water volume in the lower portion of the desalter unit (*internal recycle*). An additional pump may be used for recycling first-stage water to first-stage inlet. The first-stage internal recycle is not necessarily required for each installation and is dependent upon the amount of produced water present in the inlet stream. In terms of water requirements, *internal recycle* may be ignored since it does not add salt or water volume to the stream process.

Emulsions may also be broken by the addition of treating agents, such as soaps, fatty acids, sulfonates, and long-chain alcohols. In fact, a common method of emulsion treatment is the addition of demulsifiers which are chemicals that are designed to neutralize the stabilizing effect of emulsifying agents. Demulsifiers are surface active compounds that, when added to the emulsion, migrate to the oil/water interface, rupture or weaken the rigid film, and enhance water droplet coalescence. Optimum emulsion breaking with a demulsifier requires a properly selected chemical for the given emulsion; adequate quantity of this chemical; adequate mixing of the chemical in the emulsion; and sufficient retention time in separators to settle water droplets. It may also require the addition of heat, electric grids, and coalescers to facilitate or completely resolve the emulsion.

When a chemical is used for emulsion breaking during desalting, it may be added at one or more of three points in the system. First, it may be added to the crude oil before it is mixed with fresh water. Second, it may be added to the freshwater before mixing with the crude oil. Third, it may be added to the mixture of crude oil and water. A high-potential field across the settling vessel also aids coalescence and breaks emulsions in which case dissolved salts and impurities are removed with the water.

Selection of the appropriate demulsifier is crucial to emulsion breaking which may be selected from chemical groups such as (1) solvents, (2) surface active ingredients, and (3) flocculants.

Solvents, such as benzene, toluene, xylene, short-chain alcohols, and heavy (high-boiling, high-density) aromatic naphtha, are generally carriers for the active ingredients of the demulsifier. Some solvents change the solubility conditions of the natural emulsifiers (e.g., asphaltenes) that are accumulated at the oil/brine interface. These solvents dissolve the indigenous surface active agents back into the bulk phase, affecting the properties of the interfacial film that can facilitate coalescence and water separation.

Surface active ingredients are chemicals that have surface active properties characterized by hydrophilic–lipophilic balance (HLB) values and the scale for the HLB varies from 0 to 20. A low HLB value refers to a hydrophilic or water-soluble surfactant. In general, natural emulsifiers that stabilize a water-in-oil emulsion exhibit an HLB value in the range of 3–8. Thus, demulsifiers with a high value of the HLB value will destabilize these emulsions. The demulsifiers act by total or partial displacement of the indigenous stabilizing interfacial film components (polar materials) around the water droplets. This displacement also brings about a change in properties such as interfacial viscosity or elasticity of the protecting film, thus enhancing destabilization. In some cases, demulsifiers act as a wetting agent and change the wettability of the stabilizing particles, leading to a breakup of the emulsion film.

Flocculants are chemicals that flocculate the water droplets and facilitate coalescence. A process for selecting the appropriate demulsifier chemicals includes the following steps: (1) characterization of the crude oil and contaminants includes the API gravity of the crude oil, type and composition of oil and brine, inorganic solids, amount and type of salts, contaminant type and amounts; (2) evaluation of operational data includes production rates, treating-vessel capabilities (residence time, electrostatic grids, temperature limitations, etc.), operating pressures and temperatures, chemical dosage equipment and injection points, sampling locations, maintenance frequency, and wash water rates; and (3) evaluation of emulsion-breaking performance: past experience and operating data including oil, water, and solids content during different tests; composition and quality of interface fluids; operating costs; and amounts of water generated and its disposal.

Electrostatic grid systems are also used for emulsion treatment. When a nonconductive liquid (oil) that contains a dispersed conductive liquid (water) is subjected to an electrostatic field, one of three physical phenomena causes the conductive particles or droplets to combine in which (1) the water droplets become polarized and tend to align themselves with the lines of electric force. In so doing, the positive and negative poles of the droplets are brought adjacent to each other, (2) an induced electric charge attracts the water droplets to an electrode such as in a direct current field, the droplets tend to collect on the electrodes or bounce between the electrodes, forming larger and larger droplets until eventually they settle by gravity, and (3) the electric field distorts and thus weakens the film of emulsifier surrounding the water droplets.

If the oil entering the desalter is not hot enough, it may be too viscous to permit proper mixing and complete separation of water and oil, and some of the water may be carried into the fractionator. If, on the other hand, the oil is too hot, some vaporization may occur, and the resulting turbulence can result in improper separation of oil and water. The desalter temperature is therefore quite critical, and normally, a bypass is provided around at least one of the exchangers so that the temperature can be controlled. The optimum temperature depends upon the desalter pressure and the quantity of light material in the crude, but is normally approximately 120°C (250°F), 100°C (212°F), which is being lower for low pressures and light crude oils. The average water injection rate is 5% of the charge – regular laboratory analyses will monitor the desalter performance, and the desalted crude should normally not contain more than 1 kg of salt per 1000 barrels of feed.

Desalter control is indicated by the chloride content of the overhead receiver water and should be on the order of 10–30 ppm chlorides. If the desalter operation appears to be satisfactory but the chloride content in the overhead receiver water is greater than 30 ppm, then caustic should be injected at the rate of 1–3 lbs per 1000 barrels of charge to reduce the chloride content to the range of 10–30 ppm. Salting out (i.e. the effect that occurs when adding salt to a solvent containing a solute that reduces the solubility of that solute) will occur below 10 ppm and severe corrosion will occur above 30 ppm. Another controlling factor on the overhead receiver water is pH which should be controlled between pH 5.5 and 6.5. Ammonia injection into the tower top section can be used as a control for this. In addition to electrical methods, desalting may also be achieved by using the concept of a packed column that facilitates the separation of the crude oil and brine through the agency of an adsorbent.

More specifically, controlling pH of the desalter wash water and optimizing the pressure differential in the mix valve are important aspects of the process. Naphthenic acids in the high-acid crude oil or in heavy crude oils (Chapter 1) can (or will) cause emulsion formation when high pH (i.e. highly alkaline) water is used, and an excessive pressure differential in the mix valve can exacerbate rag layer formation (a *stable* emulsion of oil, water, and solids). Re-evaluation of emulsion-breaking chemistries employed is also important. Also, high solid levels in heavy crude oils can promote rag layer formation and oil under-carry in the desalter. In addition to primary emulsion breakers, it may also be necessary to use wetting agents or asphaltene dispersants/stabilizers to better manage the desalters. Optimizing the mud washing process is also important since over-indulgence in the mud washing (i.e. too much mud washing) will maintain the desalter at an inefficient level, while too little mud washing allows solids to collect in the desalter which will eventually force a shutdown.

Finally, installation of a flasher unit for flashing the crude oil feed can frequently reduce corrosion in the principal distillation column. In the flashing operation, desalted crude is heat exchanged against other heat sources that are available to recover maximum heat before crude is charged to the heater, which ultimately supplies all the heat required for operation of the atmospheric distillation unit. Having the heater transfer temperature offset, the flow of fuel to the burners allows control of the heat input. The heater transfer temperature is merely a convenient control, and the actual temperature, which has no great significance, will vary from 320°C (610°F) to as high as 430°C (805°F), depending on the type of crude oil and the pressure at the bottom of the fractionating tower.

As presented before, dewatering followed by desalting of crude oil is a significant process to be carried out by the crude oil industry. It is a key process for upstream operation of crude distillation units for the removal of undesirable constituents including all types of mineral salts. The salt content of the crude oil is reduced in order to meet the desired specification.

Finally, the optimal performance of the entire process is achieved by taking samples at regular intervals and analyzing for salts and other objectional constituents or by mounting analyzers before and after the desalter (Nasehi et al., 2018).

2.4 PROCESS OPTIONS FOR HEAVY FEEDSTOCKS

Heavy crude oils are some of the most challenging crudes to desalt because of their oil properties, composition, and contaminants (Forero et al., 2001; Xu et al., 2007; Collins and Barletta, 2012; Zhao et al., 2015). The geographic location of the producing basin and, to some extent, the production method determine the degree of desalting difficulty due to variability in filterable solids, viscosity, composition, naphthenic acid content, and other contaminants. Even though many of these crudes are often cited arbitrarily as crude oils that have an API gravity between 10° and 20° API (Speight, 2014a, 2017), the desalting characteristics of these feedstocks are not always the same.

Also, by definition, heavy crude oil, extra heavy crude oil, and tar sand bitumen have a high concentration of asphaltene constituents which present problems when they exhibit incompatibility and from a separate phase (i.e. precipitate) from the heavy feedstock either in the desalter or the preheat train (Chapter 2). In fact, heavy feedstocks mixed with paraffinic condensates and other paraffinic type crude oils increase the likelihood of asphaltene precipitation and stable rag layer (a *stable* emulsion of oil, water, and solids) formation in the desalter.

The primary challenges in processing heavy crude oils are (1) poor flow properties due to high viscosity, (2) low API gravity (<20°), and (3) the presence of higher amounts of impurities than in conventional crude oils (Chapter 1) (Speight, 2014a, 2015, 2017; Babalola and Susa, 2019). With treatment options that can increase the flowability of the feedstocks, a variety of technologies have been employed to improve heavy oil flowability in pipelines. Due to high viscosity and hence high pressure drop, pumping cost is extremely high and so it is energy intensive; alternative solutions to higher pumping power include preheating of the heavy crude oil by heating of the pipes, dilution with light (low-boiling, low-density) hydrocarbon fluids, oil-in-water emulsification, and partial upgrading (Babalola and Susa, 2019). Flowability enhancement technologies for transporting heavy oil and for dewatering and desalting operations are of particular importance in the current context. However, caution is advised when diluting the viscous feedstock with light (low-boiling, low-density) hydrocarbon fluids because of the high potential for the precipitation of asphaltene constituents.

Because of the properties of the heavy feedstocks, larger than typical, desalters are required to desalt heavy crude oils. Most crude units designed for light or even moderately heavy crudes require additional desalter volume to satisfactorily desalt heavy crude oil. Due to the difficulties associated with desalting heavy crude oil, it is necessary to pay special attention to other critical desalter parameters. The overall performance will depend on desalter size, as well as attention to a series of other variables which are (1) operating temperature, (2) amount and quality of the water, (3) water and oil mixing, (4) mud washing to remove solids, (5) brine cooling heat exchanger design, (6) chemical treatment, (7) desalter design, and (8) transformer size.

In the desalting process, the wash water to oil ratio is 3–4 for light oils of API >30° and 7°–10° for light oils of API <30°, and pH of 6 is ideal for the wash water and may be adjusted using caustic soda or acid (Babalola and Susu, 2019). For the case of heavy oils, however, with an API gravity <20° but >10°, much higher water volume ratios will be required.

For heavy crude oils and extra heavy crude oils, however, there may be need to include a two- or three-stage desalting unit that may very well serve all the purposes needed to achieve about 90% removal of contaminants from the heavy crude. Heavy crude oils have higher proportions and wider ranges of contaminants such that a one-stage desalting process alone will not be adequate to pretreat them for efficient refining. In a typical single-step desalting unit, the feedstock is pumped from storage through a heater, and freshwater, dosed with demulsifying agents, is added to the oil stream, which goes through a mixing valve for proper mixing. The stream then enters the gravity settling tank, where the water phase settles by gravity at the bottom of the tank, while the oil phase remains as a top layer.

The separation of the oil from the water phase in the gravity settling tank is often aided by installing two electrodes between which an electrostatic AC or DC field is created with a potential of between 12,000 and 35,000 V but these values are most likely going to increase significantly in the desalting of heavy oils. The electrostatic field promotes coalescence of water globules and enhances separation. Effluent water is sent to water treatment unit, while the desalted crude is drawn from the top of the settling tank to the refinery for processing.

For heavy crude oils, however, except for crude oils with exceptionally low amounts of contaminants, a two-stage desalting unit may be adequate to the task. As the crude is pumped from storage, part or all of it is first diverted to a special treatment unit as the special need of the particular crude may require.

The traditional dewatering and desalting process was developed for light (low-density) crude oils of relatively high API gravity and low viscosity, but with the proportional increase in available heavier crudes, modifications have become necessary for effectiveness and efficiency in the desalting process. Heavy oil, especially with high contaminants, may be desalted after it is passed through its own special purification unit to significantly reduce the particular contaminant in question. For example, heavy oil with a high sulfur content can be first passed through a microwave irradiation unit before desalting. Also, an adsorption-packed bed with appropriate zeolite grades can be used to selectively reduce metal ions for a heavy crude, which is contaminated with a specific heavy metal. Another possible option is a minor deasphalting unit using solvent extraction (Babalola and Susu, 2019).

From the special treatment, the freshwater with demulsifying agents is added before the back-mixing by the mixing valve and then the stream flows to the first gravity settling tank. With the obvious need for a second gravity settling tank, the oil is sent to it for further separation, while the effluent water from the first settling tank is sent to water treatment unit. The desalted oil from the second tank goes to the final step in the pretreatment section which is the hydrotreatment unit before it is sent to the refinery for processing, while the water from the bottom of the second tank is recycled. This proposed arrangement is expected to achieve excellent results for the pretreatment of heavy crudes. This special treatment unit may be bypassed if it is not required for a feedstock with minimal contaminants (Babalola and Susu, 2019).

Also, the crude oil preheat trains must have the flexibility to meet the desalter temperature required for optimum performance. In addition, the preheat train must have the flexibility to vary the temperature of the desalter to accommodate various blends. Thus, the optimum temperature of the desalter depends on the specific heavy crude oil or heavy crude oil blend, and it should be an operating variable and should not be a consequence of the exchanger network design. In fact, temperatures are on the order of 105°C–115°C (220°F–240°F) and 104°C–115°C to process heavy crude oils, and furthermore, higher temperatures are on the order of 140°C–145°C (280°F–290°F) for some heavy feedstocks. On the other hand, the optimum temperature may be as low as 115°C–130°C (240°F–260°F) to avoid excessive asphaltene precipitation and the increase in conductivity occurs at higher temperatures. When asphaltene constituents precipitate in the desalter, they collect at the oil/water interface and stabilize the emulsion.

Efficient feedstock and water mixing is essential and the mix valve must create enough shear to produce a small enough droplet size to allow the water to contact the oil, permitting contaminants to be dissolved in the water. The objective is to try and make wash water droplets the same or similar size as the brine droplets, so that when coalescence occurs the brine will be removed with the make-up water. If the mix valve pressure drop is too low, the oil and water will not mix properly. Hence, salt, filterable solids, and amines removal will be poor. Optimum mix valve pressure drop will vary, and it must be determined through adjustment and desalter performance monitoring.

The amounts of filterable solids in the feedstocks vary significantly depending on the specific crude source. These solids tend to stabilize the oil/water emulsion, leading to a large rag layer (a *stable* emulsion of oil, water, and solids). The desalter size is often a function of the emulsion layer resolution rather than the droplet coalescing. The filterable solids that accumulate in the bottom of the desalter must be removed through intermittent or continuous mud washing. If the solids are not removed, they reduce residence time leading to poor brine effluent quality. Mud washing is recommended in both the first- and second-stage desalters.

The more time crude oil and water can spend in the desalter separating, the better. However, this residence time is hindered whenever there is a buildup of solids and grease (often referred to as mud) in the bottom of the desalter. Mud occupies space in the desalter that the water and oil could be occupying, decreasing the residence time. Desalters typically have mud-wash headers that consist of a brine recycle system. Water is pumped out of the desalter and back in through small nozzles that pressure wash the bottom of the vessel. It is very important to have a mud washing routine, whether it is continuous or scheduled throughout the day.

A primary emulsion breaker is standard for desalting operations, and they are typically injected into the crude oil. For example, a water clarifier can help remove oil from effluent water before it is processed at the wastewater plant.

Adding caustic before, at, or after the desalter has a number of benefits, one of which is reducing the corrosion risk in the crude unit overhead. Salts that come in with crude oil include magnesium chloride, calcium chloride, and sodium chloride, and they hydrolyze at varying temperatures. Magnesium and calcium chloride that escape the desalter will hydrolyze at crude unit heater temperatures and create potential for corrosion. Caustic converts hydrochloric acid formed to sodium chloride, which will not hydrolyze at the temperature of the atmospheric distillation unit.

Caustic has also shown to be helpful in breaking difficult emulsions. By successfully breaking emulsions, less water (containing contaminants) is carried over into the crude unit, and therefore, fouling and corrosion potential is reduced. Oil under-carry to the wastewater plant is minimized. Also, the addition of caustic can aid in the removal of iron at the desalter – the iron is forced from the oil and is removed from the system in the effluent water which decreases the potential for the occurrence of fouling.

Amines that arrive with crude oil can be problematic. These typically originate from triazine-based hydrogen sulfide (H_2S) scavengers added in the upstream and midstream industry to mitigate safety and environmental hazards. If the amines and salts make it through the desalting process, they can form amine chloride ($RNH_3^+Cl^-$) salts that can be very corrosive. Acidifying the desalter can force the amines to partition to the water phase.

The ultimate objective of desalting is to have water, salt, and solid-free oil leaving the top of the desalter destined for the crude unit and oil-free, and contaminant-laden water exiting the bottom of the desalter to the wastewater plant.

The make-up water must be heated to the desalter temperature and brine cooled sufficiently to allow further effluent treating. Since the amount of solids removed during mud washing fouls conventional heat exchangers, it is not unusual for the make-up water rate to be limited by the amount of brine that can be pressured from the system.

Thus, each desalter should be specifically designed for the crude type or blend, based on operating conditions and process specifications. The most important factor in designing for heavy crude oil is control of the interface emulsion. Most desalters are sized based on gravity and viscosity, typically making the desalter too small to resolve the interface emulsion at a rate as fast as it is being created. This leads to (1) a decrease of the efficiency of the unit, (2) the production of oily effluent water, (3) an increase in the consumption of added chemicals, and (4) frequent use of interface draw-off. Also, since the desalting stage can be carried out in two stages aided by electrostatic fields, the properties of the heavy feedstock are improved for downstream refining (Babalola and Susa, 2019).

Due to the conductivity of these types of feedstocks, voltage gradient must also be optimum for steady-state operation. Lower voltages tend to provide more flexibility than conventional desalter secondary outputs. A range of voltage outputs is selected so that optimization for any particular crude can be realized.

Mud washing (or sediment removal) is also critical to good, long-term operation and should be closely evaluated to match the equipment design and water-treating plant limitations.

2.5 POTENTIAL FOR CORROSION AND FOULING OF EQUIPMENT

Produced crude oil is usually accompanied by hydrocarbon gases, hydrogen sulfide, carbon dioxide, and formation water that contains soluble salts giving rise to many types of corrosion in the crude oil and natural gas industries (Speight, 2014c). Also, the variation of solids content of individual crude oils solids – which originate in the formation from which the crude oil was produced – and, as a result, large variations in the level of filterable solids are common. These solids can stabilize desalter emulsions and desalter emulsions contain high concentrations of the same solids that are found in (raw) unprocessed crude oils. In addition, the presence of scale (produced by corrosion and introduced during transportation) is also detrimental to the desalting operation.

Before being transferred to the refinery, crude oil is stored in steel storage tanks which, depending on the working capacity of the field site, can be as high as 50 feet and as wide as 100 feet. The majority of these tanks are constructed to withstand varying degrees of temperature and pressure. However, corrosion can occur regardless of geometry, size, or function. Their exterior of the tanks is typically coated with a thermally insulating material. However, if that protective condition fails, the inner walls of the tanks may corrode, for example, when hydrogen sulfide (H_2S) gas inside the tank reacts to produce iron sulfide, it can act as an ignition source, and when water is allowed to accumulate at the bottom of storage tanks, it will cause corrosion.

Different types of corrosion occur in various parts of a crude oil refinery depending on the interaction between the crude oil constituents and the refinery equipment. Once corrosion occurs, the outcome can be major disruptions in refinery operations such as unplanned shutdowns, leaks, and product loss (Speight, 2014c; Obot et al., 2019; Al-Moubaraki and Obot, 2021). If the proper mitigation methods were not used, it is common to find general corrosion, localized pitting corrosion, hydrogen induced cracking, erosion–corrosion, microbiologically influenced corrosion, sulfide stress cracking, stress corrosion cracking (intergranular or transgranular), chloride stress corrosion cracking, corrosion fatigue, high temperature corrosion, hydrogen flaking, corrosion under insulation, metal dusting, carburization, and graphitization throughout the refinery equipment (Humooudi et al., 2017).

It is because of the potential for corrosion that the dewatering and desalting processes are necessary and, hence, play important roles in the pretreatment (in the current context) of crude oil and other refinery feedstocks.

Initially, separators are used to degas produced crude and to remove the bulk of formation water. To meet the water content specified by pipeline companies, dehydrators are used to remove much of the remaining formation water and a portion of emulsified water. While salt content of produced crude depends primarily on salt content of formation water, salt content of dehydrated crude depends on the content of the BS&W which originates from the reservoir formation (ASTM D4007, ASTM D7829). Conventional crude oil leaving the dehydrators typically contains 0.05%–0.2% v/v BS&W, medium crude 0.1%–0.4% v/v, and heavy crude 0.3%–3% v/v. These values may be sufficiently low to meet the usual shipping specifications of pipeline operators but dehydrators are often of necessity followed by field desalters before the crude can be shipped to a refinery.

Desalting of the crude oil is the first step in refining that has a direct effect on corrosion. Desalting is the means by which inorganic salts that cause fouling or hydrolyze and form corrosive acids are largely removed. By mixing and washing the crude oil with water, salts and solids transfer to the water phase which settles out in a tank. Often, chemicals are added in the form of demulsifiers to break the oil/water emulsion. The efficiency of the desalting operation is often directly related to the occurrence of corrosion in the distillation unit(s) as well as in other parts of the refinery.

In the process, an electrostatic field is induced to speed up the separation of oil and water. In this way, inorganic salts that could cause fouling or hydrolyze and form corrosive acids are largely removed. In addition, chemicals are often added in the form of demulsifiers to break the oil/water emulsion.

The sources of corrosion obstruction credited to components present in the crude oil are mostly for hydrogen chloride, organic chloride derivatives, and inorganic chloride derivatives. The brine includes salts such as sodium chloride (NaCl) and magnesium chloride ($MgCl_2$) which can produce hydrogen chloride by hydrolysis at temperatures on the order of 150°C to 205°C (300°F–400°F). After the hydrogen chloride is formed, it passes through a preheat exchanger, a furnace, and then to the distillation column. Thus, the presence of hydrogen chloride causes corrosion to the distillation equipment. In addition, passage of hydrogen chloride through the distillation column(s) can poison the catalysts in the downstream processes.

A high salt content of the feedstock to the desalter can invariably leave some salt in the "desalted" feedstock. When this occurs, there can be significant corrosion of the unit. In addition, some crude oils, such as the opportunity crude oils (Chapter 1), also contain difficult-to-remove organic chlorides and inorganic chlorides, which require special treating chemistry.

A portion of the salt leaving the desalter hydrolyzes to hydrogen chloride (HCl) in the atmospheric and vacuum heaters. The amount of hydrolysis depends on the heater temperature, the type of salt, and the presence of other compounds such as naphthenic acids contained in high-acid crude oils (Chapter 1). An option to counteract such effects is to inject caustic downstream of the desalter to convert chlorides to hydrolyze to the more stable sodium chloride (NaCl) thereby reducing the amount of hydrogen chloride in the crude overhead system.

A portion of the thermally stable sodium chloride that remains in the desalted crude oil, which does not hydrolyze in the atmospheric heater, may decompose to hydrogen chloride in the HCl in the vacuum heater. Hence, corrosion rates and fouling in the vacuum tower overhead system can be very high relative to the atmospheric tower overhead system. Corrosion in the vacuum column overhead system and salt laydown in the top of the vacuum column have become much more common with heavy crude oils.

Poor desalting generally leads to a very high corrosion rate in the atmospheric crude overhead system and in the vacuum column overhead systems, and there is the potential for the top of these distillation columns to experience corrosion and/or salting. In addition, the peripheral equipment such as top pump-arounds and product rundown systems have experienced fouling and corrosion. Piping, exchangers, ejector equipment, and drums have all been severely corroded. Rapid laydown of amine chloride salts (e.g., $RNH_3{}^+Cl^-$) occurs at the top of the atmospheric distillation and the internals in the top of vacuum distillation columns can be fouled by chloride salts. These chlorides can eventually make their way to downstream hydrotreating equipment and other consequences of poor desalting include severe exchanger fouling from poor filterable solids removal as well as (and in conjunction with) poor exchanger design that does not match the properties of the desalted feedstock.

In addition to chloride derivatives, many refinery feedstocks (especially the viscous feedstocks) contain sulfur which can be released through processing as hydrogen sulfide (H_2S) and, in addition to sulfur, many crude oils contain chemical constituents that can be quantified by the total acid number (TAN) of the feedstock (Speight, 2014a, b). This number is not specific to a particular acid but refers to all possible acidic components in the crude, and it is defined by the amount of potassium hydroxide required to neutralize the acids in one gram of oil. Typically found are organic naphthenic acids as well as inorganic acids (mineral acids) such as hydrogen sulfide (H_2S), hydrogen cyanide (HCN), and carbon dioxide (CO_2) which can make significant contributions to the corrosion of equipment. Even materials suitable for sour service do not escape damage under such an onslaught of aggressive compounds.

Also, chemicals such as caustic soda are introduced to neutralize acidic components. However, the uncontrolled use of caustic soda can have a detrimental effect through the formation of soap due to, for instance, the presence of fatty acid derivatives (e.g., $R\text{-}CO_2H$, where R is an organic group). The soap, once formed, stabilizes the oil–water mixture and obstructs the separation process. Also, mixing of crude oil and water can create an emulsion that is very difficult to break. In fact, the crude oil may arrive at the refinery as an emulsion due to the presence of water that had been used to maximize the oil extraction from the oil reservoir or water that might have occurred naturally in the reservoir (which can be connate water, formation water, or interstitial water). If the emulsions are too strong and prove impossible to break, the contaminants may end up in downstream processes, which may have serious consequences.

One process parameter that can play a vital role in both neutralizing acids and demulsification is the process pH. Careful monitoring of pH in the desalter water effluent allows for efficient dosing of caustic or acid which may result in reduced corrosion. The stability of the oil–water emulsion depends partly on pH and maintaining the pH of the mixture within a certain range helps the demulsifier chemicals in breaking the emulsion by interacting directly with the water droplets. The speed and quality of the separation process can thus be improved which leads to less water carryover, which in turn can result in a significant reduction in downstream corrosion.

Thus, desalting (1) reduces salt buildup and under-deposit corrosion in preheat exchangers; (2) corrosion in the flash zone, upper sections, and side-stream piping of the atmospheric column; and (3) corrosion in the upper sections of the atmospheric and vacuum columns. Desalting also decreases the amount of BS&W in the crude oil as well as the amount of suspended metal compounds sent to downstream units via reduced crude (the atmospheric residuum) and the vacuum residuum.

Inadequate desalting can cause fouling of heater tubes and heat exchangers throughout the refinery, which restricts product flow and heat transfer and leads to failures due to increased pressures and temperatures. Corrosion, which occurs due to the presence of hydrogen sulfide, hydrogen chloride, naphthenic (organic) acids, and other contaminants in the crude oil, also causes equipment failure. Neutralized salts (ammonium chlorides and sulfides), when moistened by condensed water, can also cause corrosion. In addition, where elevated operating temperatures are used when desalting sour crudes, hydrogen sulfide will be present. There is the possibility of exposure to ammonia, dry chemical demulsifiers, caustics, and/or acids during this operation.

Chemicals used in the desalting operation improve overall desalting efficiency, reduce water and solids carryover with desalted crude, and reduce oil carry-under with brine effluent. Most desalting chemicals are demulsifiers that help break up the tight emulsion formed by the mix valve and produce relatively clean phases of desalted crude and brine effluent. If necessary, demulsifiers can be custom formulated for high water removal rates from crude oils, but at the cost of poor solids wetting and likely oil carry-under with the brine discharge. When formulated for high solids wetting rates, brine quality often decreases and water carryover with desalted crude increases.

Also, chemicals such as caustic soda are introduced into the desalting operation to neutralize acidic components but uncontrolled use of caustic can have a detrimental effect. An excess of caustic can result in the formation of soap-type products due to the presence of fatty acids – the soap-like products stabilize the oil–water mixture and obstruct the separation process. Regular monitoring of the pH in the desalter water effluent allows for efficient dosing of caustic which may result in more efficient operation.

The characteristics of many heavy feedstocks include high solids levels, unstable asphaltene constituents, non-extractable chlorides, and considerable variability in one or more of these parameters for a given grade of crude oil. In particular, opportunity crudes typically have high levels of naphthenic acids, sulfur, and metals and require more intensive processing to yield high-quality products. These crudes are medium to heavy (15°–25° API) and vary considerably in properties, which relates to high (negative) impact on plant operations and equipment leading to elevated levels of downstream corrosion. Improved desalter operating procedures, improved contact, chemical and monitoring programs, and constraint analysis can offer ways to mitigate the negative effects from such crude oils (Forero et al., 2001; Humooudi et al., 2017).

The challenges associated with the production of crude oil from tight shale deposits (sometimes erroneously called *shale oil*) are a function of their compositional complexities and the varied geological formations where they are found. These crude oils have a low-molecular-weight envelope but are waxy and reside in oil-wet formations. These properties create difficulties associated with refining which include phenomena such as (1) scale formation, (2) salt deposition,

(3) paraffin wax deposits, (4) destabilized asphaltene constituents, (5) corrosion, and (6) bacterial growth in storage tanks.

In the case of oil from tight shale deposits, solids loading can be highly variable, leading to large shifts in solids removal performance. Sludge layers from the tank farm may cause severe upsets, including growth of stable emulsion bands and intermittent increases of oil in the brine water. Agglomerated asphaltene constituents can enter the desalter from storage tanks or can flocculate in the desalter, leading to oil in the effluent brine.

Crude oil from tight shale deposits often contains high concentrations of hydrogen sulfide that require treatment with scavengers due to safety purposes. Amine-based scavengers often decompose as the crude oil is preheated through the hot preheat train and furnace, forming amine fragments. Ethanolamine (monoethanolamine, MEA), one of the most commonly used amines, readily forms an amine chloride salt in the atmospheric tower. These salts deposit in the upper sections and, often, under-deposit corrosion is the major cause of failures in process systems because in the tower the under-salt corrosion rates can be 10–100 times faster than a general acidic attack. Mitigation strategies include (1) controlling chloride to minimize the chloride occurrence in the tower overhead and (2) acidifying the desalter brine to increase the removal of amines into the water phase.

Heavy crude oil (15°–25° API) is being used more often as refinery feedstocks. Inasmuch as there can be issues with conventional crudes in the desalter, anticipation of problems in the desalter from heavy feedstocks must be recognized and realized. These oils often contain high levels of filterable solids, unstable asphaltene constituents (i.e. constituents likely to form a separate phase), and/or difficult-to-remove chloride salts. Since the filterable solids, asphaltene constituents, and salts are concentrated in the low-volatile (bottoms) fractions, the more difficult feedstocks tend to be (1) heavy, conventionally produced crude oils, (2) extra heavy oil that has a low degree of mobility unless heated, and (3) tar sand bitumen diluted with lighter (lower density, lower boiling) hydrocarbon derivatives or synthetic crudes to meet pipeline gravity and viscosity specifications. However, there are also lower density crude oils (25°–28° API) that can be problematic owing to their high filterable solid contents and only increased vigilance can overcome the impact of these feedstocks on desalter operations as well as on refinery operations in general.

Asphaltene instability can cause desalter problems when lower boiling (lower density) liquid hydrocarbons are employed for dilution of the extra heavy crude oil or tar sand bitumen – the asphaltene constituents are known to precipitate under such conditions and also stabilize water-in-oil emulsions, perhaps due to their concentration from the oil phase to the oil/water interface. The result can be BS&W carryover into the desalted crude oil and the appearance of asphaltene particulates in the desalter effluent water – both of which can cause sludge accumulation in crude storage tanks as well as preheat fouling in heat exchanger fouling and pre-flash units as well as foaming in the atmospheric column and corrosion.

Non-aqueous extractable chloride derivatives or organic chloride derivatives that are not removed during desalting operations can be typically defined as organic or inorganic chlorides that are not dissolved in emulsified water, removed by desalters, or indicated in normal, extractable chloride measurement methods. The identity of the chlorides varies with the specific feedstock (ASTM D4929) but, nevertheless, the presence of these materials results in higher crude unit atmospheric column overhead system chloride loadings, increased overhead neutralizer demand, and higher overhead condensing, as well as a high potential for system corrosion.

Maintaining desalter performance while processing the challenging heavy feedstocks requires careful attention to equipment design and operating conditions such as (1) wash water rate, (2) mixing energy, (3) temperature, and (4) mud-wash practice.

Increasing the *wash water rate* generally increases desalter performance in terms of salt content and BS&W of the desalted crude and desalter effluent water quality. It is often assumed that higher percentages of wash water result in wetter crude oil but the opposite is true. When the wash water rate is increased, there are more water droplets which are closer together and coalescing of small droplets into larger ones is facilitated. Increasing the wash water rate also dilutes the concentration of stabilizing molecules at the oil/water interface, which reduces the coalescing of droplets.

In terms of *mixing energy*, contaminant removal from the crude requires direct contact of the wash water droplets with the contaminant. If there is not enough mixing, then not all of the contaminants will be removed. If there is too much mixing, then very small water droplets are formed, producing an emulsion which cannot be fully resolved in the desalter vessel. Over-mixing results in high BS&W in the desalted crude oil and in poor salt removal both of which lead to corrosion.

Temperature is a process variable that is not typically changed much in the desalter but there are cases where favorable exchanger configurations can be used to increase desalter temperature. There are also cases in which changing the crude slate will change the heat balance on the tower and affect the desalter temperature. In general, raising the temperature will improve the oil/water separation in the desalter because the viscosity of the hydrocarbon decreases as the temperature increases. There are also naturally occurring materials at the oil/water interface that stabilize the desalter emulsion but they are dissolved in the oil or water phase at higher temperature making the emulsion easier to break.

However, there are negative impacts to raising the desalter temperature: (1) asphaltene constituents can become unstable as the temperature is increased and precipitated asphaltene constituents collect at the oil–water interface to stabilize the emulsion and cause oil under-carry, and (2) water is more soluble in oil at higher temperatures, reducing the ability to dehydrate the crude oil and leading to high BS&W in the desalted crude. For various practical reasons, the upper limit of the desalter temperature is typically on the order of 155°C (310°F) – most desalters run at temperatures substantially below the 155°C (310°F) maximum but the optimum operating temperature for the crude feedstock to be processed should be determined.

Mud-wash practice refers to the removal of contaminants such as sand, clay, iron sulfide, iron oxide, and other solids which settle to the bottom of the desalting vessel and form a *mud* or *sludge*. Heavy crude oil typically contains more filterable solids than conventional crude oil and tends to generate higher volumes of mud in the desalter – this reduces the working volume of the desalter and water outlet can be partially blocked and, in desalters with inlet headers in the bottom of the vessel, the inlet manifold can be partially plugged.

The mud buildup will lead to increased oil in the effluent brine and poor desalter performance leading to passage of undesirable materials through the desalter and corrosion. It is critical for desalters processing heavy feedstocks to incorporate a mud-wash system to remove mud from the bottom of the desalter. Furthermore, depending upon the properties of the heavy feedstock, it may be essential to perform the mud-wash operation at least once per day – if the mud is allowed to accumulate for several weeks, it becomes compacted and cannot be easily removed from the desalter.

REFERENCES

Abdel-Aal, H.K., Aggour, M.A., and Fahim, M.A. 2015. *Petroleum and Gas Field Processing*, 2nd Edition. CRC Press, Taylor & Francis Publishers, Boca Raton, FL.

Abdel-Aal, H.K., Zohdy, K., and Abdelkreem, M. 2018. Waste management in crude oil processing: Crude oil dehydration and desalting. *International Journal of Waste Resources* 8: 1000326. https://www.researchgate.net/publication/324199958_Waste_Management_in_Crude_Oil_Processing_Crude_Oil_Dehydration_and_Desalting.

Al-Moubaraki, A.H., and Obot, I.B. 2021. Corrosion challenges in petroleum refinery operations: Sources, mechanisms, mitigation, and future outlook. *Journal of Saudi Chemical Society*, 25: 101370. https://www.researchgate.net/publication/355678568_Corrosion_Challenges_in_Petroleum_Refinery_Operations_Sources_Mechanisms_Mitigation_and_Future_Outlook.

ASTM D4007. 2021. Standard Test Method for Water and Sediment in Crude Oil by the Centrifuge Method (Laboratory Procedure). Annual Book of Standards, ASTM International, West Conshohocken, PA.

ASTM D4929. 2021. Standard Test Methods for Determination of Organic Chloride Content in Crude Oil. Annual Book of Standards, ASTM International, West Conshohocken, PA.

ASTM D7829. 2021. Standard Guide for Sediment and Water Determination in Crude Oil. Annual Book of Standards, ASTM International, West Conshohocken, PA.

Atta, A.M. 2013. Electric desalting and dewatering of crude oil emulsion based on Schiff base polymers as demulsifier. *International Journal of Electrochemical Science*, 8: 9474–9498.

Babalola, F.U., and Susu, A.A. 2019. Pre-treatment of heavy crude oils for refining. In: *Processing Heavy Crude Oils*. R.M. Gounder (Editor). InTechOpen. https://www.intechopen.com/chapters/69325.

Ben Mahmoud, M., and Aboujadeed, A.A. 2017. Compatibility assessment of crude oil blends using different methods. *Chemical Engineering Transactions*, 57: 1705–1710. https://www.aidic.it/cet/17/57/285.pdf.

Collins, T., and Barletta, T. 2012. Desalting heavy Canadian crudes. *Digital Refining*. https://www.digitalrefining.com/article/1000566/desalting-heavy-canadian-crudes#.Y176DdrMJPY2001.

Forero, J., Duque, J., Diaz, J., Nuñez, A., Guarin, F., and Carvajal, F. 2001. New contact system in crude oil desalting process. *CT&T – Ciencia Tecnología y Futuro*, 2(2): 81–91. http://www.scielo.org.co/pdf/ctyf/v2n2/v2n2a07.pdf.

Gary, J.G., Handwerk, G.E., and Kaiser, M.J. 2007. *Petroleum Refining: Technology and Economics*, 5th Edition. CRC Press, Taylor & Francis Group, Boca Raton, FL.

Hsu, C.S., and Robinson, P.R. (Editors) 2017. *Handbook of Petroleum Technology*. Springer, New York.

Humooudi, A., Hamoudi, M.R., and Ali, B.M. 2017. Corrosion mitigation in crude oil process by implementation of desalting unit in erbil refinery. *American Scientific Research Journal for Engineering, Technology, and Sciences (ASRJETS)*, 36(1): 224–241. https://core.ac.uk/download/pdf/235050309.pdf.

Johnson, O.D., and Affam, A.C. 2019. Petroleum sludge treatment and disposal: A review. *Environmental Engineering Research*, 24(2): 191–201.

Kim, Y.H., and Wasan, D.T. 1996. Effect of demulsifier partitioning on the destabilization of water-in-oil emulsions. *Industrial & Engineering Chemistry Research*, 35: 1141–1149.

Liu, G., Xu, X., and Gao, J. 2003. Study on the compatibility of asphaltic crude oil with the electric desalting demulsifiers. *Energy & Fuels*, 17: 543–548.

McDaniels, J., and Olowu, W. 2016. Removing Contaminants from Crude Oil. Digital Refining, PTQ, Q1. https://www.digitalrefining.com/article/1001246/removing-contaminants-from-crude-oil#.YmgTBdrMJPY.

Mohammed, R.A., Bailey, A.I., Luckham, P.F., and Taylor, S.E. 1994. Dewatering of crude oil emulsions 3. Emulsion resolution by chemical means. Colloids and surfaces A: *Physicochemical and Engineering Aspects*, 83: 261–271.

Murphy, G., Campbell, J., Bohent, M. 1992. Fouling in refinery heat exchangers. *Fouling Mechanisms, Proceedings, GRETh Seminar*, Grenoble, France, pp. 249–261.

Nasehi, S., Sarraf, M.J., Ilkhani, A., Mohammadmirzaie, M.A., and Fazaelipoor, M.H. 2018. Study of crude oil desalting process in refinery. *Journal of Biochemical Technology*, Special Issue (2): 29–33. https://jbiochemtech.com/storage/models/article/3SWt6rilVhczHGfuNtwHqVRQndU3NmdRBqQSx266JkdVe6zBVPzZgiZDnIzv/study-of-crude-oil-desalting-process-in-refinery.pdf.

Obot, I.B., Solomon, M.M., Umoren, S.A., Suleiman, R., Elanany, M., Alanazi, N.M., and Sorour, A.A. 2019. Progress in the development of sour corrosion inhibitors: Past, present, and future perspectives. *Journal of Industrial and Engineering Chemistry*, 79: 1–18.

Parkash, S. 2003. *Refining Processes Handbook*. Gulf Professional Publishing, Elsevier, Amsterdam.

Pereira, J., Velasquez, I., Blanco, R., Sanchez, M., Pernalete, C., and Canelón, C. 2015. Crude oil desalting process. In: *Advances in Petrochemicals*, V. Patel (Editor). IntechOpen, London. Chapter 4. https://www.intechopen.com/chapters/48963.

Speight, J.G. 2014a. *The Chemistry and Technology of Petroleum*, 5th Edition. CRC Press, Taylor & Francis Group, Boca Raton, FL.

Speight, J.G. 2014b. *High Acid Crudes*. Gulf Professional Publishing Company, Elsevier, Oxford.

Speight, J.G. 2014c. *Oil and Gas Corrosion Prevention*. Gulf Professional Publishing Company, Elsevier, Oxford.

Speight, J.G. 2017. *Handbook of Petroleum Refining*. CRC Press, Taylor & Francis Group, Boca Raton, FL.

Speight, J.G. 2021. *Refinery Feedstocks*. CRC Press, Taylor & Francis Group, Boca Raton, FL.

Xu, X.R., Tang, J.Y., Zhang, B.L., and Gao, J.S. 2007. Demulsification of extra heavy crude oil. *Petroleum Science and Technology*, 25(11): 1375–1390.

Zhao, H., Memon, A., Gao, J., Taylor, S.D., Sieben, D., Ratulowski, J., Aboudwarei, H., Pappas, J., and Creek, J. 2015. Heavy crude oil dewatering processes for fluid property measurements: Comparison, impact, and suitability of methods. Proceedings. Paper No. SPE-174465-MS. *SPE Canada Heavy Oil Technical Conference*, Calgary, Alberta, Canada, June 2015. https://onepetro.org/SPECHOC/proceedings-abstract/15CHOC/All-15CHOC/SPE-174465-MS/181726?redirectedFrom=PDF.

3 Feedstock Blending

3.1 INTRODUCTION

The study of the pretreatment processes (i.e. the dewatering and desalting processes as they affect the properties of refinery feedstocks) would not be complete without some attention to the phenomena of blending various crude oils that form the feedstock to the modern refinery. This also gives rise to the need to understand the instability and incompatibility in which a separate phase is formed during the use of the feedstock, especially if the feedstock to the refinery operations is blend of two or more crude oils (Saleh et al., 2005). Both phenomena can result in the formation of degradation products and other undesirable changes in the original properties of crude oil products. And it is the analytical methods that provide the data that point to the reason for problems in the refinery or for the failure of products to meet specifications and to perform as desired.

The modern refinery no longer has the option of accepting a single crude oil as the sole refinery feedstocks. Decreasing supplies of conventional crude oils has forced refineries to accept heavier crude oils as one of several feedstocks that, for convenience, are fed as a blend into the refinery system. Commonly, this is achieved by blending high-value light crude oils with heavy crude oils (unconventional crude oils and lower quality crude oils, in terms of distillate production) or by the purchase of ready-made blends. The lower quality crude oils include heavy crude oils from known locations, as well as opportunity crude oils and the high-acid crude oils (Chapter 1).

In addition, the high viscosity of many heavy crude oils is disadvantage for many refineries and blending heavy crude oils with light crude oils or other diluents is often required to give the heavy crude oils the flow properties that enable their effective use as a refinery feedstock. Also, heavy crude oils are hydrogen deficient and have high levels of contaminants such as sulfur, nitrogen, organic acids, vanadium, nickel, silica, and asphaltene constituents. Blending a lower quality feedstock with a conventional crude oil is an inevitable necessity in the modern refinery in order to produce a blend that bears optimal properties to be processed.

Refineries worldwide are constructed from an engineering point of view and from materials that enable the distillation of well-defined types of crude oils. These refineries were built based on the availability of certain types of crude oils in reservoirs that are relatively close to the refinery, the cost of certain crude oils on the market, and demand for predominantly light distillates for naphtha and kerosene from which gasoline and diesel fuel are produced.

Distillation of crude oil was mainly targeted to produce gasoline components, such as light and middle distillate. More recently, many refineries have experienced a higher demand for diesel fuel over the demand for gasoline. Thus, in the past, a refinery accepted a conventional (light) crude oil as the refinery feedstock; the modern refinery must be able to distil heavier crude oils to increase the amount of middle distillate and heavier distillates.

However, different crude oils are not distilled as separate feedstocks but as a blend which is produced either on-site at the refinery or off-site by a company that is specialized in the production of blends to the specifications required by the refinery. The chemical properties and physical properties are adjusted to produce a blend which can be processed in refinery equipment and will yield the desired distillates.

Thus, the selection of crude oils as refinery feedstocks is a major part of refinery operations. The constant change in the quality of crude oils (including heavy crude oils, extra heavy crude oils, and tar sand bitumen) as well as mixtures of these feedstocks is accompanied by the risk of equipment fouling along with the failure of the refiner to achieve the desired objectives. In fact, the incompatibility of different crude oils is a common problem in the crude oil industry and may lead to severe fouling of refinery units, such as the pre-distillation dewatering and desalting units as well as the

DOI: 10.1201/9781003184959-3

distillation heat exchanger train (Sayles and Routt, 2011; Álvarez et al., 2012). Thus, the study of the properties of feedstocks as they enter the refinery (and before any blending operations) would not be complete in the current context unless some attention is given to the potential for the incompatibility of the blended constituents and the stability–instability of the blend (Saleh et al., 2005; Rathore et al., 2011; Speight, 2014a; Rogel et al., 2018).

3.2 BLENDING

Blending of crude oils is a process of mixing two or more crude oils together and is done to improve the overall value or quality of the blend. In the refining industry, blending is an old art (that has evolved into an accurate science) and is designed so that various products can meet the specifications required as a saleable product. For example, gasoline and diesel fuel rate typically blends of several product streams as there is no single process that can produce specification gasoline or specification diesel fuel (Parkash, 2003; Gary et al., 2007; Speight, 2014a, 2017; Hsu and Robinson, 2017). The crude oil blending process is one of the most accurate processes for companies that blend crude oil insofar as each step of the process is monitored for accuracy to maintain a low-to-zero error rate.

In fact, product blending plays a key role in preparing the refinery products for the market to satisfy the product specifications and environmental regulations. The objective of product blending is to assign all available blend components to satisfy the product demand and the specifications of the product that are required for sales (Castro and Grossmann, 2014). Almost all refinery products are blended for the optimal use of all of the intermediate product streams for the most efficient and profitable conversion of crude oil to marketable products. For example, typical motor gasolines may consist of straight-run naphtha from distillation, the cracked product (from a fluid catalytic cracking unit, FCC), reformate, alkylate, isomerate, and polymerate in proportion to make the desired grades of gasoline and the necessary specifications (Parkash, 2003; Gary et al., 2007; Speight, 2014a, 2017; Hsu and Robinson, 2017).

Basic intermediate streams can be blended into different finished products. For example, naphthas can be blended into gasoline, or jet fuel streams, depending on the demand. Until the 1960s, the blending was performed in batch operations; but with the advances in the digital world (i.e. advances in computerization and the availability of the required equipment), online blending operations have replaced blending in batch processes. Keeping inventories of the blending stocks along with cost and physical data has increased the flexibility of and profits from online blending through optimization programs (Naji et al., 2021). In most cases, the components blend nonlinearly for a given property (such as vapor pressure, octane number, cetane number, viscosity, and pour point), and correlations and programming are required for reliable predictions of the specified properties in the blends leading to an index for the crude oil blend compatibility that can be determined by means of the physical parameter ratios of the crude oils that are the constituents of the blend (Kumar et al., 2018). In the modern refinery, the blending operation has received considerable attention with the demise of conventional crude oils and the increased use of more viscous feedstocks. Except in this case, the blending operation is applied to the crude oils entering the refinery as feedstocks for the production of products.

Typically, blending refinery feedstocks is the last step after the dewatering and desalting processes in which the optimal combination of components (among various crude oil streams) are mixed to produce the final feedstock for the distillation section of the refinery. The main purpose of blending various crude oil feedstocks prior to refining is to counteract the varying quality of crude oil sent to the refinery (Chapter 1) (Saleh et al., 2005; Lopez et al., 2013; Cerdá et al., 2018). However, blending is much more complicated than a simple mixing of components.

Blending of two or more crude oils allows the production of a feedstock blend which has upgraded their chemical and physical properties over those of some of the feedstocks in the blend and which is better suited to refinery processing than the poorer quality crude oils in the blend (Shahnovsky et al., 2014). The issue is to determine at what stage should the blending operation be carried out. Preferably, after the dewatering and desalting processes of each crude oil that will be used in the blend, the properties of the individual crude oil will be known (through careful application of the necessary test methods).

The most important quality characteristics for a refinery feedstock are (1) the density, usually expressed as the API gravity, (2) the total acid number, often referred as TAN, and (3) the sulfur content (Parkash, 2003; Gary et al., 2007; Speight, 2014a, 2017; Hsu and Robinson, 2017). The API ranges from light crudes (high API on the order of > 30°) to heavy crude oils (low API on the order of <20°). Sulfur is present in crude oils as hydrogen sulfide and as polysulfide derivatives. These sulfur-containing constituents will decompose (partially or fully) during distillation, while hydrogen sulfide is evolved as a gas. The sulfur content and other acidic components in crude oil, such as naphthenic acid derivatives (Chapter 1), are highly corrosive, and the sulfur-containing constituents are responsible for crude oil to be expressed as sour (high-sulfur) crude oil or sweet (low-sulfur) crude oil.

The high-acid crude oils (i.e. the high TAN crude oils) typically have (1) a lower amount of low-boiling constituents than the conventional light crude oils, (2) a high density, i.e. a low API gravity, (3) a high viscosity, (4) low solidification point insofar as they exist as viscous liquids or are near-solid materials in the reservoir or deposit, (5) a high nitrogen content, (6) a high asphaltene content, (7) a high mineral salt content, (8) a content of heavy metals, which are generally defined as any metallic chemical element that has a relatively high density and is toxic or poisonous at low concentrations and examples include mercury Hg, cadmium Cd, arsenic As, chromium Cr, thallium Tl, and lead Pb (Figure 3.1), and (9) a low yield of light oil distillates.

These properties have an adverse influence on oil separation in the desalter insofar as the process is more difficult than for conventional crude oils. These properties also cause these crude oils to be very corrosive and produce low-quality products. Commonly, these crude oils are of lower cost than conventional crude oil and utilization of these crude oils in any way possible is an attractive option for refiners (Shahnovsky et al., 2014).

Thus, the key to a successful blending operation is to ensure that the components of the blend are compatible and the blend will remain stable (without the formation of a second phase). For example, the formation of a second phase such as is caused by the precipitation of asphaltene constituents can lead (in the refinery) to the plugging of atmospheric distillation unit and the vacuum distillation unit. Also, the crude oil preheat train and the vacuum bottom heat exchangers can plug, and as a result, require chemical or mechanical cleaning; otherwise, throughput has to be reduced leading to a loss of production. Therefore, a detailed knowledge of factors that affect the composition and the physical and chemical structure of the crude oils is necessary (Ben Mahmoud and Aboujadeed, 2017). In addition, the careful selection of the crude oils that are to make up a blend based on compatibility considerations is a critical part of a successful refinery operation. The development and use of compatibility prediction processes that can effectively predict the performance of crude oil blends in a specific refinery is necessary. The best strategy for reducing fouling is to use basic knowledge to eliminate its formation (Saleh et al., 2005; Ben Mahmoud and Aboujadeed, 2017).

3.2.1 Process Description

Crude oil blending has several operational parameters, each of which contributes to the quality of the feedstock entering to the feedstock distillation section of the refinery and focuses on (1) the limitations of the atmospheric and vacuum distillation units to refine any type of crude oil, (2) the chemical and physical properties of the feedstock, and (3) the production of products according to the market demand. Thus, for blending the heavy crude oils, the opportunity crude oils and/or the high-acid crude oils may be required to reduce viscosity of the feedstock which improves flow properties. Furthermore, the ratio of a component in a blend is actually limited by the physical properties required for the production of a feedstock suitable for the distillation section of the refinery with the goal of producing high-value distillates.

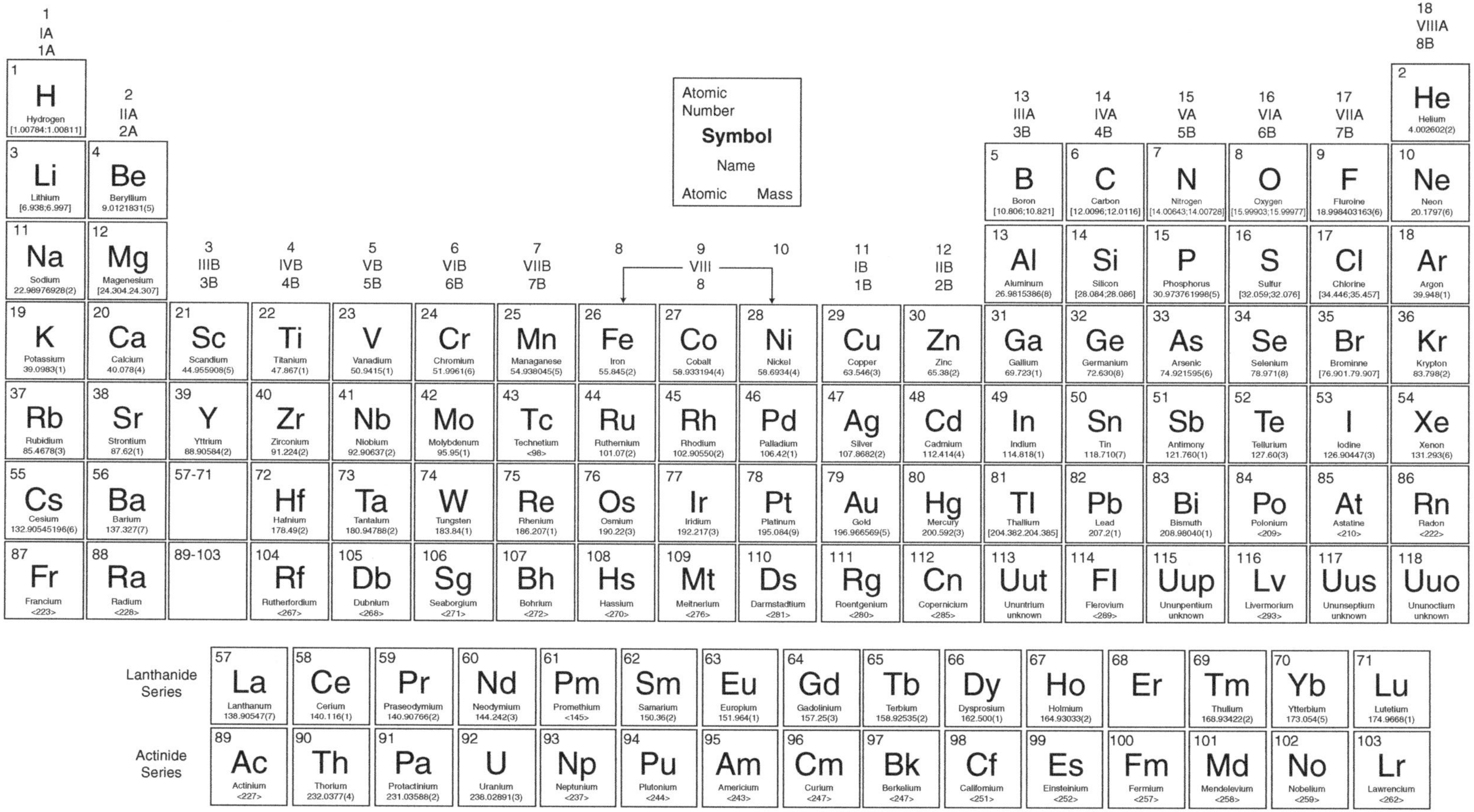

FIGURE 3.1 The periodic table of the elements.

It is also preferable that each crude oil that is destined to be a component of the blend should be tested for suitability (i.e. compatible with the other components of the blend) and blend stability (Section 3.5). For example, the path of the crude from recovery to distillation may be represented simply by the following path:

In addition to testing the individual company of the potential blend for suitability, once the blends are produced, a series of test methods can be applied to determine that usefulness of the blend in refinery operations (Section 3.5). From these test methods, the refinery may be able to develop a blending index (sometimes referred to as a blended index) based on the crude oils received (Rathore et al., 2011; Speight, 2014a, 2015).

Thus, a blending index combines two or more standard comparative indices which are useful in evaluating the performance of blends of crude oils and whether or not the constituents of the blend will be stable. Typically, the criterion for the compatibility of crude oils in a blend is that, for example, the volume average solubility blending number of the mixture is higher than the insolubility number of any component in the mixture.

This is, of course, dependent upon the quality of each component of the blend. Furthermore, the quality of each of the crude oils that constitute a refinery blend crude is highly dependent on the production location, the production, and the final blending decisions. Also, for the increased use of more viscous feedstocks, additional steps may be required to convert this material into something that can be transported (usually via pipeline) to the refinery. Typically that means blending the viscous crude with a diluent which after transportation is removed at the refinery. Indices such as the blending index: The blending index is a useful tool for the crude oil refining industry insofar as it allows the prediction of not only the compatibility of the constituents of the blend but also the properties of the blend. Indices such as the blending index make it possible to determine many more properties than simple linear blending insofar as there are properties related to nonlinear blending formulations that are not linear relationships and enquire information on the blends (such as the blending index) that are more useful.

In summary, there are major differences in the physical and chemical properties that make the more viscous heavier crude oils more difficult to process than the light, less viscous crude oils. Heavy crude oils are sour and more corrosive than more conventional crude oils. The higher viscosity, the fouling tendency, and different flow regimes make it more difficult to maintain stable feedstock rates (which are required for stable product yields and product quality) into the distillation unit (Saleh et al., 2005). In addition, the differences in boiling range between the conventional feedstocks and the more viscous feedstocks crudes require different process temperature requirements such as (1) preheating temperature, (2) distillation temperature, and (3) column overhead parameters. Finally, and not to be ignored, the more viscous feedstocks have a high content of asphaltene constituents and other contaminants (than the less viscous feedstocks) which contribute to poorer desalting performance (Shahnovsky et al., 2014).

3.2.2 EQUIPMENT

Different blending options exist to upgrade unconventional crude oils (such as heavy crude oil, opportunity crude oil, and high-acid crude oil) into blends thereby reducing the tendency of these crude oils to have an adverse effect on the crude oil distillation section of the refinery. In fact, there are two options for crude blending and they are (1) refineries and (2) blend producers/trading companies (Shahnovsky et al., 2014).

Blending crude oils into a single feedstock is an on-site refinery option to prepare the blend for consumption within the refinery. Efficient crude blending can be applied throughout the entire supply chain of crude oil, from (1) the recovery well to be suitable for transportation, (2) blending at the terminal, and (3) at the refinery. Thus:

Crude recovery → on-site dewatering/desalting → transportation → refinery tank farm
Refinery tank farm → dewatering–desalting → blending → distillation unit

The final blend supplied to the distillation unit may be a combination of several such activities.

The strategy of crude oil blending includes several parameters and each parameter contributes to the overall quality of the blend that is fed to the atmospheric distillation unit. For example, the blend may be suitable to match the engineering design (or limitations) of the atmospheric distillation unit. In addition, the blend must be suitable to accommodate the market demand for the products that the blend will produce. Also, any high-viscosity component of the blend will more likely affect the flow properties of blend and blending high-viscosity feedstocks with diluents or conventional crude oils may be required to reduce viscosity and to improve the flow properties. It is at this point that incompatibility and/or instability issues might arise and a suite of test methods should be performed before the blend is produced.

More specifically, in terms of process options, crude oil blending can be performed by two technologies which are (1) an in-tank blending process, also known as batch blending, and (2) an in-line blending process.

The batch blending process is an older process and was in operation before in-line blending process came into operations. In the batch blending process (the in-tank process), specific (calculated and measured) volumes of different crude oils, which are stored in separate tanks are loaded into a blending tank where they are mixed (by a mechanical stirring operation). Thus, specific volumes (determined by a pre-blending test program) of different types of crude oils stored in separate tanks are loaded into a blending tank where they are mixed by mechanical stirring until the homogeneous composition is achieved. Samples must be withdrawn to determine whether the blend is homogeneous and whether it conforms to its predetermined specification. This must not be haphazard process and must be monitored carefully – the next-step refining operations depend upon the quality of the blend! In the event of any discrepancy, adjustment of the composition (and properties) of the blend must be conducted. Thus, for monitoring purposes, samples must be withdrawn to determine whether or not the blend is homogeneous and whether or not the blend conforms to a predetermined specification. In the event of a failure of the blend to meet the necessary specifications, discrepancy, the process must be continued until specifications are met.

On the other hand, the in-line blending process is performed by simultaneously transferring different crude oils through an online static mixing device to the final blend storage tank. In-line blending is the controlled proportioning of two or more component streams to produce a final blended product of closely defined quality from the beginning to the end of the batch which permits the blended product to be used immediately for the prescribed purpose.

Using in-line blending stations, feedstocks for the atmospheric distillation unit with the desired properties are obtained by mixing flows of different types of crude oils using the right blending recipe. To operate the blending process efficiently and without error, online process analyzers are required to instantaneously measure the blend downstream and to feed the blending operators with the required quality details of the blend in production. This enables real-time and online correction during the blending process, providing the blend of predetermined properties and reduced the need for corrective reblending of an entire tank.

In-line blending has several advantages over the batch blending process, including (1) improved quality of the blended product and (2) reduced processing time which leads to a product that can be produced faster. In-line blending can also be supplied in many different configurations including barge units, trailer mounted units, or skid-mounted systems. Once the configuration is in place, the type of blending is often referred to as ratio blending which is dependent on accuracy through the use of flow meters and flow transmitters and the accuracy of the process.

Thus, the in-line blending system is an effective alternate process to the batch blending system and gives companies a better chance to process a wider range of refinable feedstocks. The blending system consists of flow meters, control valves, mixing, control, and analyzer systems. For a successful in-line blending system for crude oil, the in-line blender must perform at a high level of quality once integrated with the blender and the control system.

However, when using the in-line blending system, the actual mixing of the crude oil streams together is critical. The mixing of an in-line blend has to be accurate and accomplished dynamically when the components come together and the process must be continuous. The in-line blending process almost always has inadequate nature turbulence and that is what helps to blend the product throughout the flow. There are mixing devices that help to combine the streams accurately and consistently throughout the process, one of which is the static mixer element. in this mixer, the elements will divide and turn the fluid, which increases the rate of energy dissipation through pressure loss. The loss that occurs is proportionate to other parts of the process, especially small mixtures. An option to the static mixer system is the system (sometime referred to as the jet-mix system) which employs a powered loop to produce the necessary energy to blend the components. This jet-mix system is particularly useful when an analyzer such as a viscometer is installed in the blend header.

Trimming can be used to assist in the accurate monitoring of the blending system. Using this approach, a trim set point (such as the viscosity) is a part of the overall blend process and that is used to monitor (or trim) the volumetric ratios of the blend components. This part of the process can give an increased level of accuracy in the blended product. However, great care must be used when applying such a monitoring device to a mix or it could have adverse effects on the accuracy of the blending system. As part of the monitoring process, consideration should be given to (1) installing the analyzer in the mix section of the blend, (2) maintaining the blend system efficient throughout the process, and (3) protecting the analyzer from physical effects such as noise and/or vibration. In addition, sampling of the blend throughout the process is necessary in order to determine the quality of the blended product.

In large refineries, several atmospheric distillation units may be available to process a variety of crude oils that have been retained in dedicated storage tanks. The designed schedule for the blending operations must not only select the cluster of tanks allocated to each atmospheric distillation unit but also determine the scheduling of the blending operations which will provide the appropriate feedstocks for every distillation unit. The content of trace elements in the proposed blending components and the temperature boiling point curve are the properties normally controlled to set the feedstock quality (Cerdá et al., 2018). However, other parameters may also be necessary to determine the reliability (i.e. the compatibility of the blend components and the stability of the blend) of the blend (Section 3.5).

The predetermined flow ratio between different crude oils is used to provide a blend of the required specifications. This process enables online correction of the quality of the blend by changing the ratio between different crude oils. The blended mixture is produced instantaneously and no stirred blending tanks are required. In addition, in order to operate the in-line blending process efficiently, online process analyzers are required to instantaneously measure the blend downstream and to produce the required quality details of the blend in production which allows real-time and online corrections during the process to provide a blend of the required specifications.

3.3 INCOMPATIBILITY AND INSTABILITY OF BLENDS

In the current context, the meaning of the term incompatibility is found when it is applied to crude oil recovery and crude oil refining. Incompatibility during refining can occur in a variety of processes, either by intent (such as in the deasphalting process) or inadvertently when the separation is detrimental to the process. Thus, separation of solids occurs whenever the solvent characteristics of the liquid phase are no longer adequate to maintain polar and/or high molecular weight material in solution. Examples of such occurrences are (1) asphaltene separation which occurs when the paraffinic nature of the liquid medium increases, (2) wax separation which occurs when the temperature drops or the aromaticity of the liquid medium increases, (3) sludge and/or sediment formation in a reactor which occurs when the solvent characteristics of the liquid medium change so that asphaltic or wax materials separate, coke formation which occurs at high temperatures and commences when the solvent power of the liquid phase is not sufficient to maintain the coke precursors in solution, and (4) sludge/sediment formation in fuel products which occurs because of the interplay of several chemical and physical factors (Mushrush and Speight, 1995; Speight, 2014a).

Instability/incompatibility occurs when a product is formed in the formation or well pipe (recovery) or in a reactor (refining) and the product is incompatible with (immiscible with or insoluble in) the original crude oil or its products. Such an example is the formation and deposition of wax and other solids during recovery or the formation of coke precursors and even of coke during many thermal and catalytic operations. Coke formation is considered to be an initial phase separation of an insoluble, solid, coke precursor prior to coke formation proper. In the case of crude oils, sediments and deposits are closely related to sludge, at least as far as compositions are concerned. The major difference appears to be in the character of the material.

There is also the suggestion (often, but not always, real) that a sediment and/or a deposit can originate from the inorganic constituents of crude oil. For example, the sediment may be formed from the inherent components of the crude oil (i.e. the metalloporphyrin constituents) or from the ingestion of contaminants by the crude oil during the recovery operations or during the initial processing operations. For example, crude oil is known to assimilate iron and other metal contaminants from contact with pipelines and pumps during the transportation of the crude oil to the refinery. This emphasizes the need for an efficient dewatering/desalting system at the refinery even though the crude oil may have undergone preliminary dewatering and desalting at the wellhead to meet the specifications for entry of the crude oil into the pipeline system.

Sediments can also be formed from organic materials but the usual inference is that these materials are usually formed from inorganic materials. The inorganic materials can be salt, sand, rust, and other contaminants that are insoluble in the crude oil and which settle to the bottom of the storage vessel. For example, gum typically forms by the way of a hydroperoxide intermediate that induces polymerization of olefins. The intermediates are usually soluble in the liquid medium. However, gum that has undergone extensive oxidation reactions tends to be higher in molecular weight and much less soluble. In fact, the high molecular weight sediments that form in fuels are usually the direct result of autoxidation reactions. Active oxygen species involved include both molecular oxygen and hydroperoxides. These reaction proceed by a free radical mechanism and the solids produced tend to have increased incorporation of heteroatom and are thus also more polar so they are increasingly less soluble in the fuel.

The most significant and undesirable instability change in fuel liquids is the formation of solids, termed filterable sediment. Filterable sediments can plug nozzles, filters, coke heat exchanger surfaces, and otherwise degrade engine performance. These solids are the result of free radical autoxidation reactions. Although slight thermal degradation occurs in non-oxidizing atmospheres, the presence of oxygen or active oxygen species, such as hydroperoxides, will greatly accelerate oxidative degradation as well as significantly lower the temperature at which undesirable products are formed. Solid deposits that form as the result of short-term high temperature reactions share many similar chemical characteristics with filterable sediment that form in storage.

The soluble sludge/sediment precursors that form (or are present) in a crude oil may have a molecular mass in the several hundred range. For this soluble precursor to reach a molecular weight sufficient to precipitate (or to phase-separate), one of two additional reactions must occur. Either the molecular weight must increase drastically as a result of condensation reactions leading to the higher molecular weight species or the polarity of the precursor must increase (without necessarily increasing the molecular weight) by incorporation of additional oxygen, sulfur, or nitrogen functional groups. Additionally, the polarity may increase because of the removal of non-polar hydrocarbon moieties from the polar core, as it occurs during cracking reactions. In all three cases, insoluble material will form and separate from the liquid medium.

Additives are chemical compounds intended to improve some specific properties of fuels or other crude oil products. Different additives, even when added for identical purposes, may be incompatible with each other, for example, react and form new compounds. Consequently, a blend of two or more fuels, containing different additives, may form a system in which the additives react with each other and so deprive the blend of their beneficial effect.

The chemistry and physics of incompatibility can, to some extent, be elucidated (Por, 1992; Power and Mathys, 1992; Mushrush and Speight, 1995) but many unknowns remain. In addition to the chemical aspects, there are also aspects such as the attractive force differences: (1) specific interactions between like/unlike molecules, such as hydrogen bonding and electron donor–acceptor phenomena, (2) field interactions such as dispersion forces and dipole–dipole interactions, and (3) any effects imposed on the system by the size and shape of the interacting molecular species.

Such interactions are not always easy to define and, thus, the measurement of incompatibility and instability has involved visual observations, solubility tests hot filtration sediment (HFS), and gum formation. However, such methods are often considered to be after-the-fact methods insofar as they did not offer much in the way of predictability. In refinery processes (Parkash, 2003; Gary et al., 2007; Speight, 2014a, 2017; Hsu and Robinson, 2017), predictability is not just a luxury, it is a necessity. The same principle must be applied to the measurement of incompatibility and instability. Therefore, methods are continually being sought to aid in achieving this goal.

In addition to the gravimetric methods, there have also been many attempts to use crude oil and/ or product characteristics and their relation to the sludge and deposit formation tendencies. In some cases, a modicum of predictability is the outcome but, in many cases, the data appear as preferred ranges and are subject to personal interpretation. Therefore, caution is advised.

The purpose of this chapter is to document the concept of incompatibility (and stability) of feedstocks in the refinery and how this event can might affect the dewatering–desalting and distillation processes. In addition, some of the more prominent methods are used for analyzing refinery feedstocks and determining the potential for incompatibility and instability of crude oil feedstocks and crude oil products. The choice of the method is subject to the composition and properties of the feedstock and process parameters as well as desired product. No preference will be shown, and none will be given, to any individual method. Also, it is the choice of the individual experimentalist to choose the method on the basis of the type of feedstock, the immediate needs, and the projected utilization of the data.

Crude oil blending is a common-place event in many refineries and blending of two or more crude oil allows the production of a feedstock blend which has upgraded their chemical and physical properties over those of some of the feedstocks in the blend and which is better suited to refinery processing than the poorer quality crude oils in the blend. However, blending is not as simple as it may seem to many observers and if two or more crude oils are blended in the wrong proportions or even the wrong order, the asphaltene constituents can precipitate as a separate phase either immediately or after a period of time (often referred to as the induction period). Once the separate phase forms, redispersion of the asphaltene constituents is difficult, if not impossible, because the natural asphaltene–resin association has been broken (Speight 1994, 2014a, 2021). Thus, it may be necessary for a refinery-associated laboratory not only to test each crude oil for compatibility with the remainder of the feedstocks to be included in the blend but also to develop a blending index.

Incompatibility can occur in the refinery when, for example, there is (1) a phase separation, (2) the precipitation of asphaltene constituents when a paraffinic crude oil is blended with a viscous crude oil, (3) the precipitation of asphaltene constituents when a paraffinic crude oil product is blended with a viscous crude oil, (4) the blend is heated in pipes leading to a reactor, and also (5) through the formation of degradation products and other undesirable changes in the original properties of crude oil products.

For example, opportunity crude oils and the high-acid crude oils (Chapter 1) have attracted the attention of refiners; but although blends of various crude oils are often used in refining processes, this practice has some unwanted consequences in terms of fouling in the preheat trains and heat exchangers. These problems may be caused by the precipitation/separation of asphaltene constituents as a separate solid phase. Also, salts, sediment, and corrosion products can arise from the impurities in the opportunity crude oils. Therefore, a detailed knowledge of the factors that affect the composition and physical–chemical structure of crude oils is necessary. When such phenomena occur, which may not be occurring immediately at the time of the blend but can occur after an induction period, it is often referred to as instability of the blends (Rathore et al., 2011; Speight, 2014a, 2017; Rogel et al., 2018).

In some cases, the terms incompatibility and instability are used separately. For example, the term incompatibility might refer to a blend of two or more feedstocks that, after blending, forms two or more separate phases with or without an induction period. On the other hand, the term instability might refer to a blend of two or more feedstocks that, after blending, appears to be stable but after the passage of a given time period (the induction period) forms two or more separate phases.

Briefly, and by the way of explanation, an induction period in chemical (or physical) reactions is an initial stage of the reaction (or a slow stage of a reaction) in which the reaction does not appear to occur after which the reaction may accelerate, even leading to (in some reactions) an explosion. Typically, the end of the induction period marks the end of the period of low (or no) chemical or physical activity and the onset of the reaction.

For example, using the coking process as an example, there is an induction period during which coke is not formed but the chemistry of coke formation is already stated and coke depletion is manifested after a period to time that is dependent upon the feedstock composition and the temperature of the process (Magaril and Aksenova, 1967, 1970a, b, 1972; Magaril and Ramazaeva, 1969; Magaril et al., 1970, 1971; Wiehe, 1993; Wiehe and Kennedy, 2000; Speight, 2014a, 2017). In another case, difficult to handle at that stage without shutting down the unit during the deasphalting procedure, there is an induction volume (or induction period) during which time the blend may appear to be stable and the amount of the lighter crude oil precipitant does not cause any asphaltene constituents to separate at the time of mixing. Again, the appearance of the asphaltene phase is dependent upon the feedstock composition and the temperature of the process (Speight, 2014a, 2015).

An another example of the issues that can arise in terms of the blended feedstocks is that high-acid (high TAN) crude oils are characterized by high density and viscosity, low solidification point, high nitrogen content, high asphaltene content, high salts and high heavy metals content as well as a low yield of low-boiling distillates. This can confer a high potential for incompatibility with the more conventional crude oils. If this occurs in the desalter, it will use more conventional crude oils. These properties also bestow on these crude oils a high potential for the corrosion of refinery equipment.

Moreover, instability and incompatibility through the deposition of (semi-solid) sludge or (solid) sediment is real and the challenge in mitigating instability and incompatibility is to eliminate or modify the prime chemical reactions in the formation of incompatible products during the processing of feedstocks containing resin constituents and asphaltene constituents, particularly those reactions in which the insoluble lower molecular weight products (frequently referred to as carbenes and carboids) (Figure 3.2) are formed (Speight, 2014a, b, 2017). In fact, the asphaltene constituents of feedstocks have been a source of problems in the crude oil industry for decades (Speight, 1994, 2014a, 2015).

Blending two or more crude oils can significantly change the overall concentrations of the SARA constituents (Figure 3.2) at a molecular level by disturbing the delicate balance of the crude oil system which could well result in the precipitation of the asphaltene fraction or a portion of that fraction. Thus, the deposition of asphaltene constituents is the consequence of instability of the crude oil (i.e. in this context, a feedstock blend of different feedstocks or a blend of a feedstock with a crude oil product produced from the feedstock) – an example of such a product is fuel oil. The asphaltene constituents are stabilized by resin constituents and remain in the crude oil as a result of this stabilization. Asphaltene dispersants are substitutes for the natural resin constituents and are able to keep the asphaltene constituents dispersed to prevent flocculation/aggregation and phase separation (Speight, 1994). Dispersants will also clean up sludge in the fuel system and they have the ability to adhere to surface of materials that are insoluble in the oil and convert them into stable colloidal suspensions.

Complexity and the means by which the product is evaluated (ASTM, 2021) have made the industry unique among industries. But product complexity has also brought to the fore issues such as

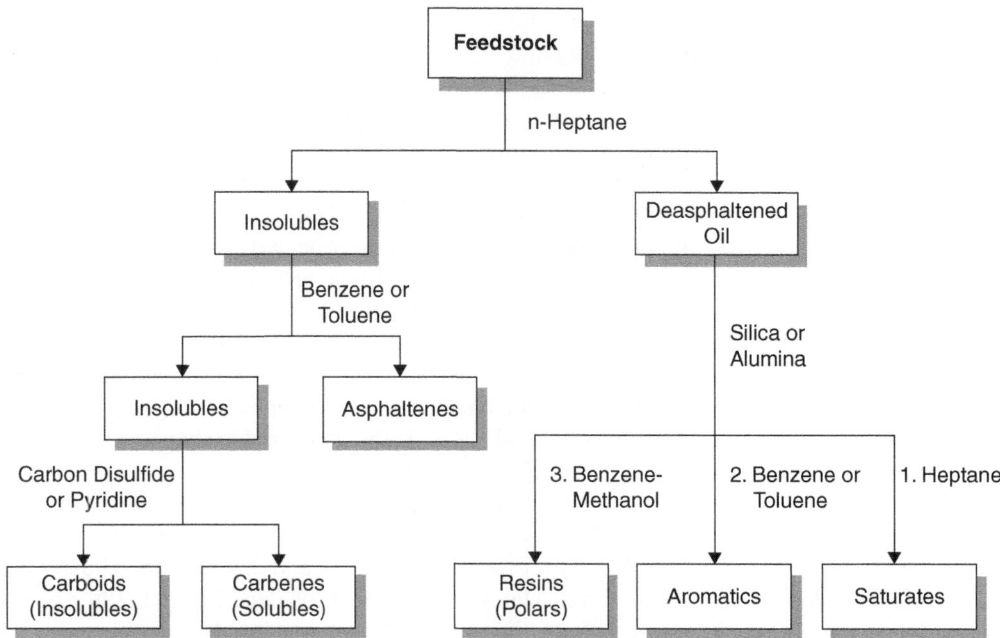

FIGURE 3.2 General fractionation scheme and nomenclature of petroleum fraction; carbenes and carboids are thermally generated product fractions.

incompatibility and instability. Product complexity becomes even more disadvantageous when various fractions from different types of crude oil are blended or are allowed to remain under conditions of storage (prior to use) and a distinct phase separates from the bulk product. Typically, the individual components of a feedstock blend have a high, medium, or low fouling potential. The blends of those that are incompatible however typically show an increased instability and may be the cause of the fouling problems encountered during processing. The adverse implications of this for refining the fractions to saleable products increase (Batts and Fathoni, 1991; Por, 1992; Mushrush and Speight, 1995; Saleh et al., 2005; Speight, 2014a). Therefore, it is appropriate here to define some of the terms that are used in the liquid fuels field so that their use later in the text will be more apparent and will also alleviate some potential for misunderstanding.

In general, fuel instability and fuel incompatibility can be related to the heteroatom-containing compounds (i.e. nitrogen-, oxygen-, and sulfur-containing compounds) that are present. The degree of unsaturation of the fuel (i.e. the level of olefinic species) also plays a role in determining instability/incompatibility. In fact, recent investigations have also implicated catalytic levels of various oxidized intermediates and acids as especially deleterious for middle distillate fuels.

Incompatibility in crude oil (i.e. in the context of refinery feedstocks) can have many meanings. The most obvious example of incompatibility is the inability of two or more crude oils to mix and form a separate phase. Instability typically refers to the formation of a separate phase at some time after the blending operation and is defined in terms of the formation of filterable and non-filterable sludge (i.e. a sediment, a deposit, and/or a gum), an increased peroxide level, and the formation of color bodies. Color bodies in and of themselves do not predict instability but the chemical (or physical) reactions that initiate the formation of color body formation can be closely linked to heteroatom-containing (i.e. nitrogen-, oxygen-, and sulfur-containing) functional groups and their respective chemistry.

Incompatibility and instability in crude oil refineries affects the operation of crude oil pre-distillation heat exchanger trains. Although feedstock instability/incompatibility has been the subject of study for decades, there is a lack of fundamental understanding of the chemical transformations leading to instability/incompatibility. The premise (and reality) is that the analysis of foulant deposits may provide a key to understanding organic and inorganic instability/incompatibility in these heat exchangers. However, before progressing further, there are several terms that need definition for a better understating of incompatibility and instability.

The term incompatibility refers to the formation of a precipitate (or sediment) or separate phase when two liquids are mixed. The term instability is often used in reference to the formation of color, sediment, or gum in the liquid over a period of time and is usually due to chemical reactions, such as oxidation, and is chemical rather than physical. This term may be used to contrast the formation of a precipitate in the near term (almost immediately). In addition, self-incompatible feedstocks are those containing insoluble asphaltenes that were not formed by mixing incompatible components. As stated above, examples of incompatible blends, in the current context, that require caution are the following: (1) a paraffinic crude oil and a viscous crude oil, (2) a paraffinic crude oil product is blended with a viscous crude oil, (3) when the blend is heated in pipes leading to a reactor, and (4) also through the formation of degradation products and other undesirable changes in the original properties of crude oil products.

The phenomenon of instability is often referred to as incompatibility, and more commonly known as sludge formation, and sediment formation, or deposit formation. In crude oil and its products, instability often manifests itself in various ways (Tables 3.1 and 3.2) (Stavinoha and Henry, 1981; Hardy and Wechter, 1990; Power and Mathys, 1992; Mushrush and Speight, 1995; Speight, 2014a). Hence, there are different ways of defining each of these terms but the terms are often used interchangeably.

Gum formation occurs when soluble organic material is produced by blending, whereas sediment is the insoluble organic material. Storage stability (or storage instability) (ASTM D381) is a term used to describe the ability of the liquid to remain in storage over extended periods of time without appreciable deterioration as measured by gum formation and/or the formation sediment. Thermal stability is also defined as the ability of the liquid to withstand relatively high temperatures for short periods of time without the formation of sediment (i.e. carbonaceous deposits and/or coke). Thermal oxidative stability is the ability of the liquid to withstand relatively high temperatures for short periods of time in the presence of oxidation and without the formation of sediment or deterioration of properties, and there is standard equipment for various oxidation tests (ASTM D4871). Stability is also the ability of the liquid to withstand long periods at temperatures up to 100°C (212°F) without degradation.

Existent gum is the name given to the non-volatile residue present in the fuel as received for test (ASTM D381). In this test, the sample is evaporated from a beaker maintained at a temperature of

---

TABLE 3.1

Examples of Properties Related to Instability in Crude oil and Crude oil Products

Property	Comments
	Separates from oil when gases are dissolved
	Separates when heavy crude oil is blended with light crude oil
	Thermal alteration can cause phase separation
Heteroatom constituents	Provide polarity to crude oil constituents
	Preferential reaction with oxygen
	May be soluble in water
Aromatic constituents	May be incompatible with paraffinic medium
Non-asphaltene constituents	Phase separation of polar species

TABLE 3.2

Examples of the Influence of Asphaltene Constituents in Feedstocks on Various Processes

Process	Issues
Oil recovery	Wellbore plugging and pipeline deposition
Dewatering and desalting	Equipment plugging during/after blending
Distillation	Equipment plugging during/after blending
Visbreaking process	Degraded asphaltene constituents are more aromatic (loss of aliphatic chains) and less soluble and from deposits in the reactor
Cracking processes	Degraded asphaltene constituents are more aromatic (loss of aliphatic chains) and less soluble and deposit on the catalyst
Emulsion formation	Emulsions may be formed. Asphaltene constituents are highly polar and surface active and responsible for the undesired stabilization of emulsions
Preheating	Preheating of a feedstocks prior to entering a fraction combustion encourages reaction and precipitation of the reacted asphaltene constituents leading to premature coking
Combustion	A high content (>6% w/w) of asphaltene constituents in the feedstock can result in ignition delay and poor combustion further leading to boiler fouling, poor heat transfer, stack solid emission, and corrosion problems
Blending of feedstocks	Blending and the resulting change of the properties of the liquid medium during mixing can cause destabilization of asphaltene (and reacted asphaltene) constituents
Storage	In the case of visbroken products, sedimentation and plugging can occur due to oxidation of the asphaltene constituents

160°C–166°C (320°F–331°F) with the aid of a similarly heated jet of air. This material is distinguished from the potential gum that is obtained by aging the sample at an elevated temperature.

Dry sludge is defined as the material separated from crude oil and crude oil products by filtration and which is insoluble in heptane. Existent dry sludge is the dry sludge in the original sample as received and is distinguished from the accelerated dry sludge obtained after aging the sample by chemical addition or heat. The existent dry sludge is distinguished from the potential dry sludge that is obtained by aging the sample at an elevated temperature. It is operationally defined as the material separated from the bulk of a crude oil or crude oil product by filtration and which is insoluble in heptane. The test is used as an indicator of process operability and as a measure of potential downstream instability/incompatibility.

Instability/incompatibility is a complex phenomenon which follows different mechanisms involving factors such as crude oil character (type, composition), crude blending, temperature, fluid velocity, and deposit properties (Young et al., 2009). In order to understand crude oil behavior leading to instability/incompatibility, there is a need to investigate crude oil characterization as well as crude oil blending to determine compatibility or incompatibility of the constituents of the blends which will lead to a better understanding of the mechanism of instability/incompatibility as well as potential methods of instability/incompatibility mitigation.

It is worthy to note here that opportunity crude oils (Chapter 1) and the blends with other crude oils play an important role in increasing the refinery but the risks are high because they are usually laden with contaminants such as destabilized asphaltene constituents and high metal content. These contaminants can cause stable oil–water emulsion problems, heat exchanger fouling, and coking in furnace tubes, leading to high maintenance costs and equipment losses. Furthermore, incompatible crude oil blends can result in the flocculation and deposition of asphaltene constituents as a preliminary step in the formation of (unwanted) coke.

In fact, opportunity crude oils have attracted the attention of oil companies looking to increase their gross refinery margin, and although blends of various crude oils are often used in refining processes, this practice has several constraints in terms of logistics, such as the non-availability of sufficient numbers of storage tanks and vessels. It also has some unwanted consequences in terms of fouling in the preheat trains and heat exchangers, and coking in the pipe still furnace tubes. These problems may be caused by the precipitation of asphaltene constituents, oxidative polymerization, and the components of coke formation in the oil. Salts, sediment, and corrosion products can arise from the impurities (Speight, 2014a, b). The problems associated with the flocculation and deposition of asphaltene constituents can further increase the cost of oil recovery processes. Therefore, a detailed knowledge of the factors that affect the composition and physicochemical structure of crude oils is necessary.

In fact, as a result of the growing need for refineries to accept the opportunity crude oils, high-acid crude oils, and the viscous feedstocks (heavy crude oil, extra heavy crude oil, and tar sand bitumen), there is the need to reduce the effect of these feedstocks on refinery equipment. Blending is often cited as an option for dealing with such feedstocks and is an effective method for lowering the concentration in the feedstock of any objectional constituents. However, every feedstock varies in composition and properties, resulting in varying levels of compatibility in crude oil blends. Some blend components are inherently incompatible, owing to the insolubility of asphaltene constituents in crude oil that are more paraffinic. Nevertheless, blending does not mitigate the problems of the incompatible components but reduces the concentration of the incompatible constituents to an acceptable level which is determined by the downstream processing of the blend. In addition, blending may cause additional problems during refining if the crude being blended contains high levels of sulfur compounds or other corrosive agents.

In general, instability/incompatibility mechanisms are classified into five categories: (1) reaction instability/incompatibility, which can occur in reactors during crude oil processing, (2) particulate instability/incompatibility, which is caused by the presence of particulate matter such as inorganic materials in crude oil, (3) corrosion instability/incompatibility, which is caused by the presence of corroded metals flakes in crude oil, (4) crystallization instability/incompatibility, which is typical of the presence of wax constituents in crude oil, and (5) biological instability/incompatibility, which is caused by the accumulation of microorganisms, plants, and algae on surfaces. Furthermore, while instability/incompatibility can be conveniently divided into the above five categories, in practice, it may be much more complicated and caused by interaction of two or more of these mechanisms.

Feedstock composition (which is even more complex with the use of feedstock blends for transportation, storage, and in the refinery) is the most common option studied to determine the instability/incompatibility potential of a feedstock (Gary et al., 2007; Speight, 2014a). Evaluation of crude oil and the means by which products are evaluated are the object of a series of standard test methods to determine feedstock complexity leading to an estimation of the potential for instability/incompatibility (Speight, 2015). Feedstock and product composition becomes even more important when feedstocks (particularly different types of crude oils) or products are blended or are allowed to remain in storage (prior to use) and a distinct phase separates from the bulk liquid (Batts and Fathoni, 1991; Por, 1992; Mushrush and Speight, 1995; Speight, 2014a).

Thus, feedstock constituents cause instability/incompatibility when they are incompatible resulting in the formation of sludge or sediment under the conditions of the process. If the sludge or sediment is marginally soluble in a blended feedstock mix, there is the possibility that the sludge/sediment can be separated by filtration or by extraction (ASTM D4310) – in the refinery guard, reactors are commonly used to remove any such solid matter from the feedstock before entry into the reactor proper (Speight, 2014a). In addition, the sludge/sediment constituents typically increase the viscosity of the feedstock. The viscosity change might be due to the separation of paraffins (wax deposition) as might occur when a paraffin-based feedstock is allowed to cool in a low temperature environment.

It is generally recognized and accepted that that the crude oil and heavy oil system (which has been extended to include extra heavy oil and tar sand bitumen) is a colloid-type system comprising fractions of saturates, aromatics, resin constituents, and asphaltene constituents (Speight, 1994, 2014a).

The asphaltene constituents are stabilized by the resin constituents but they can be naturally or artificially precipitated if the molecular association with the resin constituents is disturbed.

As a result, deposition of the asphaltene constituents can occur in different parts of the crude oil system including (1) the reservoir, (2) the well tubing, and (3) the surface flowlines. The deposition depends on the changes in flow conditions, pressure, temperature, and oil composition. Such occurrences will decrease well and reservoir productivity, and correction requires frequent chemical treatments for the removal of the asphaltene constituents. Destabilization (i.e. flocculation) of the asphaltene constituents depends on breaking up the balance of attraction forces between the associated resin molecule constituents and the asphaltene constituents (Koots and Speight, 1975; Branco et al., 2001; Stark and Asomaning, 2003; Speight, 2014a).

Thus, feedstocks and feedstock constituents are incompatible when sludge, semi-solid, or solid particles (for convenience here, these are termed secondary products to distinguish them from the actual crude oil product) are formed during and after blending. This phenomenon usually occurs prior to use. If the secondary products are marginally soluble in the blended crude oil product, their use might detract from solubility of the secondary products, and they will appear as sludge or sediment that can be separated by filtration or by extraction (ASTM D4310). When the secondary products are truly insoluble, they separate and settle out as a semi-solid or solid phase floating in the fuel or are deposited on the walls and floors of containers. In addition, secondary products usually increase the viscosity of the crude oil product. Standing at low temperatures will also cause a viscosity change in certain fuels and lubricants. Usually, the viscosity change might be due to separation of paraffins as might occur when diesel fuel and similar engines are allowed to cool and stand unused overnight in low temperature climates.

In attempts to define and understand the nature of instability/incompatibility by asphaltene constituents, many models for asphaltene deposition, but very few, if any, consider the complexity of the asphaltene fraction and instead choose to use average parameters therefore leading to frequent references to models that need to be fine tuned. Furthermore, the effect of resins in the feedstock has been largely ignored in such models and there has been a general failure of the model proponents to recognize that the resin constituents also play a major role in crude oil chemistry and physics insofar as the resin constituents stabilize (peptize, disperse) the asphaltene constituents (Koots and Speight, 1975; Hammami and Freworn, 1998; Carnahan et al., 1999; Hammami et al., 2000; Gawrys et al., 2003).

In fact, the complexity and variation in the properties of sub-fractions of the resin fraction and the asphaltene fraction (Speight, 1994, 2014a) detracts from the concept of any form of meaningful average structure for these fractions. Thus, the mechanism of instability/incompatibility by these constituents is a multi-path approach where each constituent (chemical-type) must be considered on the basis of individual chemistry and chemical properties.

However, instability/incompatibility is not only a refinery phenomenon but it can also be manifested at the wellhead when the crude oil is prepared for shipping along with the hydrocarbon gases. The presence of these gases, which are soluble in the crude oil, causes a disturbance of the crude oil system that can result in irreversible flocculation of asphaltene constituents which can severely reduce the permeability of the reservoir, cause formation damage, and can also plug up the wellbore and tubing. This phenomenon is largely ascribed to different extents of compressibility of the low-boiling feedstock constituents and the higher boiling constituents (resin and asphaltene constituents) of the under-saturated crude. In fact, the relative volume fraction of the lower-boiling constituents within the liquid phase increases as the pressure of the under-saturated reservoir fluid approaches its bubble point. Thus, any factor such as changes in pressure, temperature, or composition that disrupt this adsorption equilibrium and precipitation can cause asphaltene separation and deposition. Precipitation and deposition may occur during primary production, during the displacement of reservoir oil by carbon dioxide, hydrocarbon gas, or water and gas application (Khanifar et al., 2011).

In the modern refinery where the refinery feedstock slate often is composed of several crude oils that are introduced into the refinery as a blend, the instability/incompatibility of crude oil (and of crude oil products) is manifested in the formation of sludge, sediment, and general darkening in color of the liquid (ASTM D1500). Sludge (or sediment) formation takes one of the following forms: (1) material dissolved in the liquid, (2) precipitated material, and (3) material emulsified in the liquid. Under favorable conditions, sludge or sediment will dissolve in the crude oil or product with the potential of increasing the viscosity. Sludge or sediment, which is not soluble in the crude oil (ASTM D96, ASTM D473, ASTM D1796, ASTM D2273, ASTM D4007, ASTM D4807, ASTM D4870), may either settle at the bottom of the storage tanks or remain in the crude oil as an emulsion. In most of the cases, the smaller part of the sludge/sediment will settle satisfactorily and the larger part will stay in the crude oil as emulsions. In any case, there is a need of breaking the emulsion, whether it is a water-in-oil emulsion or the sludge itself, which has to be separated into the oily phase and the aqueous phase. The oily phase can be then processed with the crude oil and the aqueous phase can be drained out of the system.

Phase separation can be accomplished either by the use of suitable surface active agents allowing for sufficient settling time or by the use of a high voltage electric field for breaking such emulsions after admixing water at a rate of about 5% and at a temperature of about 100°C (212°F).

Emulsion breaking, regardless of whether the emulsion is due to crude oil–sludge emulsions, crude oil–water emulsions, or breaking the sludge itself into organic (oily) and inorganic components, is of major importance from operational as well as commercial aspects. With some heavy fuel oil products and heavy crude oils, phase separation difficulties often arise (Por, 1992; Mushrush and Speight, 1995; Speight, 2014a). Also, some crude oil emulsions may be stabilized by naturally occurring substances in the crude oil. Many of these polar particles accumulate at the oil–water interface, with the polar groups directed toward the water and the hydrocarbon groups toward the oil. A stable interfacial skin may be so formed; particles of clay or similar impurities, as well as wax crystals present in the oil, may be embedded in this skin and make the emulsion very difficult to break (Schramm, 1992).

Chemical and electrical methods for sludge removal and for water removal, often combined with chemical additives, have to be used for breaking such emulsions. Each emulsion has its own structure and characteristics: water-in-oil emulsions where the oil is the major component or oil-in-water emulsions where the water is the major component. The chemical and physical nature of the components of the emulsion plays a major role in their susceptibility to various surface active agents used for breaking them.

Therefore, appropriate emulsion breaking agents have to be chosen very carefully, usually with the help of previous laboratory evaluations. Water- or oil-soluble demulsifiers, the latter being often non-ionic surface active alkylene oxide adducts, are used for this purpose. But, as had been said in the foregoing, the most suitable demulsifier has to be chosen for each case from a large number of such substances in the market by a prior laboratory evaluation.

3.4 FACTORS AFFECTING INCOMPATIBILITY OF BLENDS

There is a need for a series of test methods that will assist in various attempts to predict the incompatibility and stability of crude oil blends. In fact, it is a productive and accurate route to use a series of test methods that can assist in determining the following parameters: (1) the crude oil in the blend that is compatible/incompatible with the other crude oils in the blend, (2) the ratio of the crude oils in the blend at which incompatibility is manifested, (3) whether or not there an induction period before incompatibility or instability occurs as manifested by the formation of a separate phase, and (4) whether or not there is a temperature range in which incompatibility or instability of the blend occurs.

On the other hand, multiple methods exist for measuring the specific properties of refinery feedstocks – most refiners utilize standard methods according to internationally accepted measurement techniques. However, the refiner must decide on the method that is utilized in their facility or use methods that are dictated by crude oil properties and by product specifications.

The standard test methods (ASTM, 2019) applied to an investigation of the properties of and feedstock are presented in the form of a feedstock assay (crude oil assay, crude assay), which provides a summary of its properties for various boiling point cuts that exist in the feedstock, in addition to the quantity of each boiling point fraction. The assay provides information that can be used to decide the crude slate and product slate for a refinery. The information can also be used to estimate the properties of blended feedstock based on the properties of each crude oil in the blend. However, this does not take into account any interactions between the components of blend and the assay is a snapshot in time because of the potential for crude oil properties from a reservoir to vary with the age of the reservoir. Thus, the validation of any feedstock assay and the actual properties compared to assumed or predicted properties are critical components of the feedstock selection process – the assumption that a specific company (i.e. crude oil) of a blend will have a fixed and reproducible set of properties is erroneous.

Furthermore, the ability of any analytical laboratory to accurately measure a property is strongly influenced by the following factors: (1) sampling and a thorough mixing of the feedstock to overcome any tendency to undergo stratification during storage, (2) sample preparation and handling, (3) availability and accuracy of the measurement equipment, and (4) adherence to a necessary analytical procedure (Speight, 2014a, 2015).

Briefly, stratification can occur in a blend when the components of the blend vary considerably in density, with the higher density component tending to settle to the bottom of the storage tank.

Following these guidelines should allow refinery personnel to make critical decisions based on accurate measurements of properties of feedstocks. In addition, any given method will have a level of repeatability and reproducibility (Speight, 2015). Therefore, when the results of laboratory testing are reported for a specific property, it is preferable that the users of the data recognize that the property is reported within the context of the repeatability and reproducibility levels.

The methods presented below are the most common methods used for the analysis of feedstocks as a means of determining the potential for instability/incompatibility of a feedstock blend. However, one test method producing one set of data is unlikely to provide the complete answer to the suitability of the crude oil for refining. The data from several test methods must be used in conversion to determine whether or not the crude oil is suitable for the proposed refining sequence.

3.4.1 Acidity

The acidity of crude oil or crude oil products is usually measured in terms of the acid number which is defined as the number of milli-equivalents per gram of alkali required to neutralize the acidity of the crude oil sample (ASTM D664, ASTM D974, ASTM D3242).

Acidity due to the presence of inorganic constituents is not expected to be present in crude oils, but organic acidity might be found. Acidic character is composed of contributions from strong organic acids and other organic acids. Typically, the total acidity of crude oils is in the range of 0.1–0.5 mg potassium hydroxide per gram, although higher values are not exceptional. Values above 0.15 mg potassium hydroxide per gram are considered to be significantly high. Crude oils of higher acidity may exhibit a tendency of instability. The acid imparting agents in crude oils are naphthenic acids and hydrosulfides (thiols, mercaptans, R-SH). Acidity can also form by bacterial action insofar as some species of aerobic bacteria can produce organic acids from organic nutrients. On the other hand, anaerobic sulfate-reducing bacteria can generate hydrogen sulfide, which, in turn, can be converted to sulfuric acid (by bacterial action).

The term naphthenic acid has roots in the somewhat archaic term "naphthene" (cycloaliphatic but non-aromatic) used to classify hydrocarbons. It was originally used to describe the complex mixture of crude oil-based acid derivatives when the analytical methods available in the early 1900s could identify only a few naphthene-type components with accuracy. Currently, the term naphthenic acid is used in a more generic sense to refer to all of the carboxylic acids present in crude oil, whether cyclic, acyclic, or aromatic compounds, and carboxylic acids containing heteroatoms such as N and

S. The naphthenic acid fraction is a mixture of several cyclopentyl and cyclohexyl carboxylic acids with molecular weight on the order of 120–700.

Example of a cyclopentyl component of the naphthenic acid fraction:

The main constituents of the fraction are carboxylic acids with a carbon backbone of 9–20 carbons, although acids containing up to 50 carbon atoms have been identified in heavy crude oil (Qian and Robbins, 2001). Although commercial naphthenic acids often contain a majority of cycloaliphatic acids, multiple studies have shown that they also contain straight chain and branched aliphatic acids and aromatic acids; some naphthenic acids contain >50% combined aliphatic and aromatic acids (Clemente and Fedorak, 2005; Qian and Robbins, 2001). Naphthenic acids are represented by a general formula $C_nH_{2n-z}O_2$, where n indicates the carbon number and z specifies a homologous series. The z is equal to 0 for saturated, acyclic acids and increases to two in monocyclic naphthenic acids, to four in bicyclic naphthenic acids, to six in tricyclic acids, and to eight in tetracyclic acids.

Free hydrogen sulfide is often present in crude oils, a concentration of up to 10 ppm being acceptable in spite of its toxic nature. However, higher hydrogen sulfide concentrations are sometimes present, 20 ppm, posing serious safety hazards. Additional amounts of hydrogen sulfide can form during the crude oil processing, when hydrogen reacts with some organic sulfur compounds converting them to hydrogen sulfide. In this case, it is referred to as potential hydrogen sulfide, contrary to free hydrogen sulfide.

Naphthenates are the salts of naphthenic acids, analogous to the corresponding acetates, which are better defined and have the formula M(naphthenate)$_2$ or are basic oxide derivatives with the formula M$_3$O(naphthenate)$_6$. The naphthenate derivatives are highly soluble in organic media. Industrially useful naphthenates include those of aluminum, magnesium, calcium, barium, cobalt, copper, lead, manganese, nickel, vanadium, and zinc (Clemente and Fedorak, 2005).

3.4.2 ASPHALTENE CONTENT

The asphaltene fraction (Figure 3.2) is a dark brown to black friable solids that have no definite melting point and usually intumesce on heating with decomposition to leave a carbonaceous residue. The fraction is obtained from crude oil by the addition of a non-polar solvent (such as a liquid hydrocarbon). Liquids used for this purpose are n-pentane and n-heptane (Table 3.3) (Speight, 1994, 2014a, 2021). Usually, the asphaltene fraction is removed by filtration through paper but more recently a membrane method has come into use (ASTM D4055). Liquid propane is used commercially in processing crude oil residua (Parkash, 2003; Gary et al., 2007; Speight, 2014a, 2017;

TABLE 3.3

Standard Methods for Asphaltene Precipitation[a]

Method	Precipitant	Volume Precipitant per g of Sample (mL)
ASTM D893	*n*-pentane	10
ASTM D2006	*n*-pentane	50
ASTM D2007	*n*-pentane	10
ASTM D3279	*n*-heptane	100
ASTM D4124	*n*-heptane	100

[a] ASTM Annual Book of Standards, 1980–2021.

Hsu and Robinson, 2017) – the asphaltene constituents are soluble in liquids such as benzene, toluene, pyridine, carbon disulfide, and carbon tetrachloride.

Asphaltene constituents are best known for the problems they cause as solid deposits that obstruct flow in the crude oil production systems as well as the formation of coke during processing. A better understanding of the effect of asphaltene constituents and resin constituents is key to preventing the formation of deposits during production and refining and mitigating the deleterious effects of these feedstock constituents (Speight, 1994, 2014a, b, 2015, 2017). In addition, the resin constituents of crude oil will also be considered since these constituents are regarded as those materials soluble in n-pentane or n-heptane (i.e. whichever hydrocarbon is used for the separation of asphaltene constituents) but insoluble in liquid propane and which maintain the asphaltene constituents in the crude oil.

In terms of composition, it is generally recognized that crude oil and heavy feedstocks are composed of four major fractions (saturate constituents, aromatic constituents, resin constituents, and asphaltene constituents) (Figure 3.2) that differ from one another sufficiently in solubility and adsorptive character that the separation can be achieved by application of relevant methods (Speight, 2014a, 2015). Indeed, although these four fractions are chemically complex, the methods of separation have undergone several modifications to such an extent that the evolution of the separation techniques is a study in itself (Speight, 2014a, 2015). And because the fractions are in a balanced relationship in crude oil, the chemical and physical character of these constituents needs reference in this chapter.

Thus, the asphaltene fraction is particularly important because as the proportion of this fraction increases, there is (1) concomitant increase in thermal coke yield and (2) an increase in hydrogen demand as well as catalyst deactivation. The constituents of the asphaltene fraction to form coke quite readily is of particular interest in terms of the compatibility/incompatibility of the coke precursors (Speight, 1994).

The effect of the asphaltene constituents and the micelle structure, and the state of dispersion also merit some attention. The degree of dispersion of asphaltene constituents is higher in the more naphthenic/aromatic crude oils because of higher solvency of naphthene constituents and aromatic constituents over paraffin constituents. This phenomenon also acts in favor of the dissolution of any sludge that may form thereby tending to decrease sludge deposition. However, an increase in crude oil often accompanies sludge dissolution.

The higher the asphaltene content of a feedstock, the greater the tendency of the feedstock to form a separate phase, especially when blended with other non-compatible stocks.

The resin constituents are closely related to the asphaltene constituents, and for the purposes of this text, the term resin generally implies material that has been eluted from various solid adsorbents (Koots and Speight, 1975; Speight, 2001, 2014a, 2015). Thus, after the asphaltene constituents are precipitated, adsorbents are added to the n-pentane or n-heptane solutions of the resin constituents and oils, the process by which the resin constituents are adsorbed and subsequently recovered by the use of a more polar solvent, while the oils remain in solution. The term maltenes (sometimes called petrolenes) indicates a mixture of the resin constituents and oil constituents obtained in the filtrates from the asphaltene precipitation (Speight, 2014a, b, 2015, 2017).

3.4.3 Density/Specific Gravity

In the earlier years of the crude oil industry, density and specific gravity (with the API gravity) were the principal specifications for feedstocks and refinery products. They were used to give an estimate of the most desirable product, i.e. kerosene, in crude oil. At the present time, a series of standard tests exists for determining density and specific gravity (Parkash, 2003; Gary et al., 2007; Speight, 2014a, 2015, 2017; Hsu and Robinson, 2017).

There is no direct relation between the density and the specific gravity of crude oils to their sludge forming tendencies, but crude oil having a higher density (thus, a lower API gravity) is generally more susceptible to sludge formation, presumably because of the higher content of the polar/asphaltic constituents.

3.4.4 Elemental Composition

The ultimate analysis (elemental composition) of crude oil and its products is not reported to the same extent as for coal (Speight, 2013). Nevertheless, there are ASTM procedures (ASTM, 2012) for the ultimate analysis of crude oil and crude oil products but many such methods may have been designed for other materials (Speight, 2015).

Of the data that are available, the proportions of the elements in crude oil vary only slightly over narrow limits: carbon 83.0%–87.0% w/w, hydrogen 10.0%–14.0% w/w, nitrogen 0.10%–2.0% w/w, oxygen 0.05%–1.5% w/w, and sulfur 0.05%–6.0% w/w (Speight, 2014a, 2015). And yet, there is a wide variation in physical properties from the lighter more mobile crude oils at one extreme to the extra heavy crude oil and tar sand bitumen at the other extreme (Chapter 1). In terms of the incompatibility and instability of crude oil and crude oil products, the heteroatom content appears to represent the greatest influence. In fact, it is not only the sulfur and nitrogen content of crude oil that are important parameters in respect of the processing methods which have to be used in order to produce fuels with the required sulfur concentrations but also the type of sulfur and nitrogen species in the oil. There could well be a relation between nitrogen and sulfur content and crude oil (or product) stability; higher nitrogen and sulfur crude oils are suspect of higher sludge forming tendencies.

In general, the reaction sequence for sediment formation can be envisaged as being dependent upon the most reactive of the various heteroatomic species that are present in fuels. The worst-case scenario would consist of a high olefin fuel with both high indole concentration and a catalytic trace of sulfonic acid species. This reaction matrix would lead to rapid degradation. However, just as there is no one specific distillate product, there is also no one mechanism of degradation. In fact, the mechanism and the functional groups involved will give a general but not specific mode of incompatibility. The key reaction in many incompatibility processes is the generation of the hydroperoxide species from dissolved oxygen. Once the hydroperoxide concentration starts to increase, macromolecular incompatibility precursors form in the fuel. Acid- or base-catalyzed condensation reactions then rapidly increase the polarity, the incorporation of heteroatoms, and the molecular weight.

When various feedstocks are blended at the refinery, incompatibility can be explained by the onset of acid–base-catalyzed condensation reactions of various organo-nitrogen compounds in the individual blending stocks. These are usually very rapid reactions with practically no observed induction time period.

3.4.5 Metals Content

The majority of crude oils contain metallic constituents that are often determined as combustion ash (ASTM D482). This is particularly so for the heavier feedstocks. These constituents, of which nickel and vanadium are the principal metals, are very influential in regard to feedstock behavior in processing operations.

The metal (inorganic) constituents of crude oil or a liquid fuel arise from either the inorganic constituents present in the crude oil originally or those picked up by the crude oil during storage. The former are mostly metallic substances like vanadium, nickel, sodium, iron, and silica and the latter may be contaminants such as sand, dust, and corrosion products (Speight, 2014a, b).

Incompatibility leading to deposition of the metals (in any form) on to the catalyst leads to catalyst deactivation whether it is by physical blockage of the pores or destruction of reactive sites. In the present context, the metals must first be removed if erroneously high carbon residue data are to be avoided. Alternatively, they can be estimated as ash by complete burning of the coke after carbon residue determination.

Metals content above 200 ppm are considered to be significant, but the variations are very large. The higher the ash content, the higher is the tendency of the crude oil to form sludge or sediment.

3.4.6 POUR POINT

The pour point defines the cold properties of crude oils and crude oil products, i.e. the minimal temperature at which they still retain their fluidity (ASTM D97). Therefore, pour point also indicates the characteristics of crude oils: the higher the pour point, the more paraffinic is the oil and vice versa. Higher pour point crude oils are waxy, and therefore, they tend to form wax-like materials that enhance sludge formation.

To determine the pour point (ASTM D97, ASTM D5327, ASTM D5853, ASTM D5949, ASTM D5950, ASTM D5985, IP 15, IP 219, IP 441), the sample is contained in a glass test tube fitted with a thermometer and immersed in one of three baths containing coolants. The sample is dehydrated and filtered at a temperature 25°C (45°F) higher than the anticipated cloud point. It is then placed in a test tube and cooled progressively in coolants held at −1°C to +2°C (30°F–35°F), −18°C to −20°C (−4°F to 0°F), and −32°C to −35°C (−26°F to −31°F), respectively. The sample is inspected for cloudiness at temperature intervals of 1°C (2°F). If conditions or oil properties are such that reduced temperatures are required to determine the pour point, alternate tests are available that accommodate the various types of samples.

3.4.7 VISCOSITY

The viscosity of a feedstock varies with the origin and type of the crude oil and also with the character of the chemical constituents, particularly the polar functions where intermolecular interactions can occur. For example, there is a gradation of viscosity between conventional crude oil, heavy oil, and bitumen (Speight, 2014a, 2015). Viscosity is a measure of fluidity properties and consistency at specific temperatures. Heavier crude oil, i.e. crude oil having lower API gravity, typically has higher viscosity. Increases of viscosity during storage indicate either an evaporation of volatile components or formation of degradation products dissolving in the crude oil.

3.4.8 VOLATILITY

Crude oil can be subdivided by distillation into a variety of fractions of different boiling ranges or cut points (Speight, 2014a, 2015). In fact, distillation was, and still is, the method for feedstock evaluation for various refinery options. Indeed, volatility is one of the major tests for crude oil products and it is inevitable that the majority of all products will, at some stage of their history, be tested for volatility characteristics (Speight, 2014a, 2015).

The very nature of the distillation process by which residua are produced (Parkash, 2003; Gary et al., 2007; Speight, 2014a, 2017; Hsu and Robinson, 2017), i.e. removal of distillate without thermal decomposition, dictates that the majority of the heteroatoms which are predominantly in the higher molecular weight fractions will be concentrated in the higher boiling products and the residuum. Thus, the inherent nature of the crude oil and the means by which it is refined can seriously influence the stability and incompatibility of the products (Mushrush and Speight, 1995; Speight, 2014a, 2017). In fact, whether in a blend or not, the more viscous feedstocks tend to form more sludge during storage compared to light crude oils.

3.4.9 WATER CONTENT, SALT CONTENT, AND BOTTOM SEDIMENT AND WATER

Water content (ASTM D4006, ASTM D4007, ASTM D4377, ASTM D4928) salt content (ASTM D3230), and bottom sediment and water (BS&W; ASTM D96, ASTM D1796, ASTM D4007) indicate the concentrations of aqueous contaminants, present in the crude either originally or picked up by the crude during handling and storage. Water and salt content of crude oils produced at the field can be very high, forming sometimes its major part. The salty water is usually separated at the field, usually by settling and draining, and surface active agents electrical emulsion breakers (desalters) are sometimes employed. The water and salt contents of crude oil supplied to the buyers are function of the production field. Water content below 0.5%, salt content up to 20 pounds per 1000 barrels, and bottom sediment and water up to 0.5% are considered to be satisfactory.

Although the centrifuge methods are still employed (ASTM D96, ASTM D1796, ASTM D2709 and ASTM D4007), many laboratories prefer the Dean and Stark adaptor (ASTM D95). The apparatus consists of a round-bottom flask of capacity 50 mL connected to a Liebig condenser by a receiving tube of capacity 25 mL, graduated in 0.1 mL. A weighed amount, corresponding to approximately 100 mL of oil, is placed in the flask with 25 nil of dry toluene. The flask is heated gently until the 25 mL of toluene have distilled into the graduated tube. The water distilled with the toluene separates to the bottom of the tube where the volume is recorded as ml, or the weight as mg, or per cent.

The Karl Fischer titration method (ASTM D1744), the Karl Fischer titration method (ASTM D377), and the colorimetric Karl Fischer titration method (ASTM D4298) still find wide application in many laboratories for the determination of water in liquid fuels, specifically the water content of aviation fuels.

The higher the bottom sediment and water content, the higher sludge and deposit formation rates can be expected in the stored crude oil.

In summary, the asphaltene constituents (and the resin constituents) can cause major problems in refineries through phase separation of insoluble products and through unanticipated coke formation. The thermal decomposition of these constituents has received some attention with the objective of examining the products and the potential of these products to form coke (Speight, 2014a, b, 2017). The data contradicted earlier theories that coke formation was predominantly polymerization reaction and, in fact, the initial stages of coke formation involved the separation of lower molecular weight insoluble products from which coke was produced by further reaction of these products. There is an induction period before coke begins to form that seems to be triggered by phase separation of reacted asphaltene. The phase separation is likely triggered by changes in the molecular structure of the resin constituents or changes in the molecular structure of the asphaltene constituents or both, and the products are isolated as insoluble carbene or carboids (Figure 3.2). In addition, the organic nitrogen originally in the resin and asphaltene constituents invariably undergoes thermal reaction to concentrate in the non-volatile coke and is considered likely that carbon–carbon bonds or carbon–hydrogen bonds in a heterocyclic nitrogen ring system may be susceptible to thermal decomposition as an initial events in the thermal decomposition process (Speight, 2014a, 2017). When denuded of the attendant hydrocarbon moieties, the nitrogen heterocyclic systems and various polynuclear aromatic systems are undoubtedly insoluble (or non-dispersible) in the surrounding hydrocarbon medium. The next step is gradual carbonization of these entities to form coke (thermal instability/incompatibility).

3.5   DETERMINING INCOMPATIBILITY OF BLENDS

Instability/incompatibility of feedstocks in crude oil refineries affects the operation of crude oil pre-distillation heat exchanger trains. Although feedstock incompatibility and instability have been the subject of study for decades, there is a lack of fundamental understanding of the chemical transformations leading to incompatibility and instability. The premise (and reality) is that analysis of deposits (sludge, sediment) arising from incompatibility and instability that will provide a key to understanding incompatibility and instability.

In terms of refinery operations starting with the dewatering/desalting operations, the effect of the blending of crude oils on the potential of the blend to discharge a separate phase should be investigated in order to determine the extent to which compatibility of the crude oil feedstocks can influence the formation of the separate phase. Testing should be performed for each of the crude oils in the blend as well as a series of blends. After testing, fouling any such deposits should be recovered and analyzed. In addition, the effect of temperature changes on the stability of the blends should also be investigated from which a correlation can be developed that is an indicator of the potential instability of the blend constituents (Rathore et al., 2011; Speight, 2014; Rogel et al., 2018).

In summary (from the previous section), the instability/incompatibility of crude oil products is a precursor to either the formation of degradation products or the occurrence of undesirable changes in the properties of the blend. Individually, the components of a product may be stable, and in compliance with specifications, their blend may exhibit poor stability properties, making them unfit for use. It is necessary to apply a suite of standard test methods (ASTM, D2021), including a recently developed titration method (Rogel et al., 2018), that can provide the data that point to the reason for the instability of the blends.

Use of standard test methods is one way of predicting in compatibility (or incompatibility) of refinery feedstocks and allows the refiner to predict the performance of blends of crude oils before they are processed.

Thus, incompatibility/compatibility and instability/stability of crude oils and crude oil products can be estimated using several tests (Table 3.4) (Schermer et al., 2004; Speight, 2014a, b; Mendoza de la Cruz et al., 2015; Ben Mahmoud and Aboujadeed, 2017). One test, or property, that is somewhat abstract in its application but which is becoming more meaningful popular is the solubility parameter (Speight, 2014a, 2015), which allows estimations to be made of the ability of liquids to become miscible on the basis of miscibility of model compound types where the solubility parameter can be measured or calculated. Although the solubility parameter is often difficult to define when complex mixtures are involved, there has been some progress. For example, crude oil fractions have been assigned a similar solubility parameter to that of the solvent used in the separation. Whichever method is the best estimate may be immaterial as long as the data are used to the most appropriate benefit and allow some measure of predictability.

Bottle tests constitute the predominant test method and the test conditions have varied in volume, type of glass or metal, vented and unvented containers, and type of bottle closure. Other procedures have involved stirred reactor vessels under air pressure or oxygen pressure, and small volumes of fuel employing a cover slip for solid deposition.

There are several accelerated fuel stability tests that can be represented as a time–temperature matrix (Goetzinger et al., 1983; Hazlett, 1992). A graphical representation shows that the majority of the stability tests depicted fall close to the solid line, which represents a doubling of test time for each 10°C (18°F) change in temperature. The line extrapolates to approximately 1 year of storage under ambient conditions. Temperatures at 100°C (212°F) or higher present special chemical problems.

TABLE 3.4

Examples of Standard Test Methods Suitable for the Estimation of Stability/Instability and Compatibility/Incompatibility in Feedstocks[a]

Test method	Observation
ASTM D5	Penetration
ASTM D113	Ductility
ASTM D445	Viscosity
ASTM D1500	Color
ASTM D1754	Change in physical properties
ASTM D3381	Change in physical properties
ASTM D4742	Oxygen stability
ASTM D7112	Optical detection of solid phase

[a] Listed numerically and not by any preference.

An analogous test, the thin film oven test (TFOT) (ASTM D1754) is used to indicate the rate of change of various physical properties such as penetration (ASTM D5), viscosity (ASTM D2170), and ductility (ASTM D113) after a film of asphalt or bitumen has been heated in an oven for five hours at 163°C (325°F) on a rotating plate. A similar test is available for the stability of engine oil by thin film oxygen uptake (TFOUT) (ASTM D4742). This test establishes the effects of heat and air based on changes incurred in the above physical properties measured before and after the oven test. The allowed rate of changes in the relevant bitumen properties after the exposure of the tested sample to the oven test is specified in the relevant specifications (ASTM D3381).

Attractive as they seem to be, any tests that involve accelerated oxidation of the sample must be used with caution and consideration of the chemistry. Depending on the constituents of the sample, it is quite possible that the higher temperature and extreme conditions (oxygen under pressure) may not be truly representative of the deterioration of the sample under storage conditions. The higher temperature and the oxygen under pressure might change the chemistry of the system and produce products that would not be produced under ambient storage conditions.

It is also worthy of note that fractionation of crude oil and its products may also give some indication of instability. There are many schemes by which crude oil and related materials might be fractionated (Speight, 2014a, 2015), and although it is not the intent to repeat the details of these in this chapter, a brief overview is necessary since fractional composition can play a role in stability and incompatibility phenomena.

Crude oil can be fractionated into four broad fractions by a variety of techniques although the most common procedure involves precipitation of the asphaltene fraction and the use of adsorbents to fractionate the deasphalted (deasphaltened) oil. The fractions are named for convenience and the assumption that fractionation occurs by specific compound type is not quite true.

In general terms, study of the composition of the incompatible materials often involves determination of the distribution of the organic functional groups by selective fractionation that is analogous to the deasphalting procedure and subsequent fractionation of the maltene (non-asphaltene) constituents: (1) heptane soluble materials: often called maltenes or petrolenes in crude oil work, (2) heptane insoluble material, benzene (or toluene) soluble material, often referred to as the asphaltene fraction, and (3) toluene-insoluble material, often referred to as carbenes (pyridine soluble constituents) and carboids (pyridine insoluble constituents) when the material under investigation is a thermal product.

Carbon disulfide and tetrahydrofuran have been used in place of pyridine. The former (carbon disulfide), although having an obnoxious odor and therefore not much different from pyridine, is easier to remove because of the higher volatility. The latter (tetrahydrofuran) is not as well established in crude oil science as it is in the coal-liquid related research. Thus, it is more than likely that the crude oil researcher will use carbon disulfide or, pyridine, or some suitable alternate solvent. It may also be necessary to substitute cyclohexane as an additional step for treatment of the heptane insoluble materials prior to treatment with benzene (or toluene). The use of quinoline has been suggested in place of pyridine but this solvent presents issues associated with its high boiling point.

Whichever solvent separation scheme is employed, there should be ample description of the procedure so that the work can be repeated not only in the same laboratory by a different researcher but also in different laboratories by various researchers. Thus, for any particular feedstock and solvent separation scheme, the work should be repeatable and reproducible within the limits of experimental error.

Fractionation procedures allow a before-and-after inspection of any feedstock or product and can give an indication of the means by which refining or use changes the composition of the feedstock. In addition, fractionation also allows studies to be made of the interrelations between various fractions. For example, the most interesting phenomenon (in the present context) to evolve from the fractionation studies is the relationship between the asphaltene constituents and the resin constituents.

In crude oil, the asphaltene constituents and the resin constituents have strong interactions to the extent that the asphaltene constituents are immiscible (insoluble/incompatible) with the remaining constituents in the absence of resins (Koots and Speight, 1975; Speight, 1994, 2014a). And there appears to be points of structural similarity between the asphaltene constituents and the resin constituents of a feedstock thereby setting the stage for a more than is generally appreciated complex relationship but confirming the hypothesis that crude oil is a continuum of chemical species, including the asphaltene constituents (Speight, 1994, 2014a).

This sets the stage for the incompatibility of the asphaltene constituents in any operation in which the asphaltene or resin constituents are physically or chemically altered. Disturbance of the asphaltene–resin relationships can be the stimulation by which, for example, some or all of the asphaltene constituents form a separate insoluble phase leading to such phenomena as coke formation (in thermal processes) or asphalt instability during use.

There is a series of characterization indices that can also present indications of whether or not a crude oil product is stable or unstable. For example, the characterization factor indicates the chemical character of the crude oil and has been used to indicate whether a crude oil was paraffinic in nature or whether it was a naphthenic/aromatic crude oil.

The characterization factor (sometimes referred to as the Watson characterization factor) is a relationship between boiling point and specific gravity:

$$K = T_b^{1/3}/d$$

where T_b is the cubic average boiling point, degrees Rankine (°F + 460) and d is the specific gravity at 15.6°C (60°F).

The characterization factor was originally devised to illustrate the characteristics of various feedstocks. Highly paraffinic oils have K = 12.5–13.0, while naphthenic oils have K = 10.5–12.5. In addition, if the characterization factor is above 12, the liquid fuel or product might, because of its paraffinic nature, be expected to form waxy deposits during storage.

The viscosity-gravity constant (vgc) was one of the early indices proposed to classify crude oil on the basis of composition. It is particularly valuable for indicating a predominantly paraffinic or naphthenic composition. The constant is based on the differences between the density and the specific gravity for various hydrocarbon species:

$$vgc = \left[10d - 1.0752 \log(v - 380) \right] / \left[10 - \log(v - 38) \right]$$

where d is the specific gravity and v is the Saybolt viscosity at 38°C (100°F). For viscous crude oils (and viscous products) where the viscosity is difficult to measure at low temperature, the viscosity at 99°C (210°F) can be used:

$$vgc = \left[d - 0.24 - 0.022 \log(v - 35.5) \right] / 0.755$$

In both cases, the lower index number is indicative of a more paraffinic sample. For example, a paraffinic sample may have a vgc on the order of 0.840, while the corresponding naphthenic sample may have an index on the order of 0.876.

The obvious disadvantage is the closeness of the indices, almost analogous to comparing crude oil character by specific gravity only where most crude oils fall into the range d = 0.800–1.000. The API gravity expanded this scale from 5 to 60 thereby adding more meaning to the use of specific gravity data.

In a similar manner, the correlation index which is based on a plot of the specific gravity (d) versus the reciprocal of the boiling point (K) in °K (°K = degrees Kelvin = °C + 273) for pure hydrocarbons adds another dimension to the numbers:

$$\text{Correlation index}(CI) = 473.7d - 456.8 + 48640/K$$

In the case of a crude oil fraction, K is the average boiling point determined by the standard distillation method.

The line described by the constants of the individual members of the normal paraffin series is given a value of the correlation index (CI) of zero and a parallel line passing through the point for benzene is given a value of the correlation index of 100. Values between 0 and 15 indicate a predominance of paraffin hydrocarbon derivatives in the sample and values from 15 to 20 indicate predominance either of naphthenes or of mixtures of paraffin derivatives/naphthene derivatives/aromatic derivatives, while an index value above 50 indicates a predominance of aromatics in the fraction.

Finally, there are various criteria for describing (or estimating) the stability of the asphaltene constituents in crude oils and crude oil blends. One example is a colloidal instability index (represented as CII) which is based on amount (% w/w) of saturates, aromatics, resins, and asphaltenes (SARA; Figure 3.2). Thus:

$$CII = (saturates = asphaltenes) / (aromatics + resins)$$

Thus, the colloidal instability index is expressed by using the mass ratio of the sum of asphaltenes and flocculants (saturates) to the sum of resins and aromatics in a crude oil. Generally, blends that have a $CII > 2$ tend to be incompatible or unstable which leads to the precipitation of an asphaltene fraction.

By the way of definition, the SARA fractionation involves prior separation of the asphaltene fraction from the crude oil by the addition of excess liquid hydrocarbon such as n-pentane or n-heptane (ASTM D4124) after which the saturates fraction, the aromatics fraction, and the resins fraction are extracted from the deasphalted oil (sometime referred to as the deasphaltened oil) using a liquid chromatographic method (ASTM D2007).

Another method uses an optical microscope to determine flocculation points for oils as well as evidence of asphaltene precipitation. In this method, various mixtures of the solvent are prepared by blending toluene and n-heptane in known proportions to determine solubility or insolubility for each crude oil. In order to prepare compatible blends of oils, the solubility parameter on a reduced n-heptane–toluene scale is called insolubility number (IN). In addition, the tests measure the solubility parameter of the oil on a n-heptane–toluene scale that is the solubility blending number (SBN). The determination of the INs and the SBN is key parameters in the prediction of flocculation when dealing with incompatible crude oil blends.

The criterion for the compatibility of any blend is that the volume average SBN is greater than the maximum INs of any component in the blend (i.e. $SBN > IN$). In this relationship, the INs represents the degree of asphaltene insolubility, while the SBN is a measure of the ability of oil to "dissolve" asphaltene constituents. The SBN can be calculated from the following equation:

$$SBN_{blend} = (V_1 SBN_1 + V_2 SBN_2 + V_2 SBN_3 \cdots + V_n SNBN_n) / (V_1 + V_2 + V_3 \ldots V_n)$$

From this calculation, the compatibility criterion for oil mixtures can be defined as follows:

$$SBN_{mix} > IN_{max}$$

Another test, the heptane dilution test, involves the determination of the maximum volume of n-heptane (HD) that can be added to a given volume of oil without precipitating asphaltenes. Any insoluble asphaltene constituents can be detected by observing a drop of the mixture between a glass slide and a cover slip under an optical microscope at 100–200 magnification (ASTM D7061).

Yet another test is the toluene equivalence test in which the minimum percent of toluene in mixture with n-heptane to dissolve asphaltenes is determined at a concentration of 1 g of oil and 5 mL of toluene–n-heptane mixture (test liquid). The volume percent of toluene in the test liquid is plotted versus 100 times the volume ratio of oil to test liquid. The SBN for the crude oil is calculated:

$$SBN = IN[1 + 100 / HD]$$

In this equation, HD is the abscissa intercept or was taken from the heptane dilution test. For self-compatible oils, the line slope would be negative, and for self-incompatible oils, the x-axis intercepts are infinite, and the SBN is equal to the INs and to the percentage of toluene in the test liquid for any volume ratio of oil to solvent mixture. The higher SBN favors higher stability of the asphaltene constituents and, hence, higher stability of the blend.

3.6 POTENTIAL FOR CORROSION AND FOULING OF EQUIPMENT

At any point in the life history of a crude oil, corrosion and fouling can occur and have an adverse effect on production equipment, transportation equipment, and refining equipment. Hence, there is a need for dewatering and desalting (which may be performed to some extent occurs at the well-head) at the refinery. In addition, although the crude oils and the blends play an important role in increasing refinery efficiency, the risks are high because there are several factors that determine the formation of deposits in the pipelines and pipes of heat exchangers, which includes the presence of contaminant such as inorganic salts, sediments, asphaltenes, waxes, stable emulsions or inorganic solids, and a high metal content (Murphy et al., 1992; Watkinson, 1992; Asomaning and Watkinson, 2000). These contaminants can cause stable oil–water emulsion problems leading to high maintenance cost and equipment losses.

Just as crude oil, which is a mixture of hydrocarbon constituents, is not corrosive, there are some impurities and components often found in crude oil that could cause corrosion in pipelines and refinery equipment such as the atmospheric distillation column, overhead lines, heat exchangers, and condensers. In fact, if the constituents that can cause corrosion are not removed during the dewatering and desalting processes, the corrosivity of the constituents of one crude oil in the bland can be manifested in the behavior of the blend.

Inefficient dewatering and desalting processes of the individual crude oil destined for incorporation into a blend can lead to blends of crude oils that are unstable mixtures which precipitate species such as asphaltene and result in rapid corrosion and fouling.

In most cases, water containing chloride salts (often referred to as brackish water) such as magnesium chloride ($MgCl_2$), calcium chloride ($CaCl_2$), and sodium chloride (NaCl) are drawn from crude oil wells along with the crude oil. The concentration of these salts in the crude oil depends on the oil field from which the crude is extracted, but it is usually present within the range of 3–300 pounds per barrel.

During preheating, if the crude oil reaches temperatures of more than 120°C (250°F), these chloride salts break down to HCl. Thus:

$$CaCl_2 + H_2O = CaO + 2HCl$$

There is a similar reaction for magnesium but sodium chloride is more stable and is less easily hydrolyzed. By increasing the preheating temperature up to 380°C (715°F), most of the magnesium chloride and calcium chloride will undergo hydrolysis.

Hydrogen chloride (HCl) in a gaseous state is not dangerous in terms of corrosion; but when the gas is cooled to temperatures lower than the dew point of water (which is the temperature and pressure at which the first drop of water vapor condenses into a liquid), it reacts with moisture (condensing water) to produce hydrochloric acid, which is an extremely corrosive substance. Once the hydrogen chloride reacts with steel, it will react with iron chloride to reproduce hydrogen chloride which causes corrosion of steel:

$$CaCl_2 + H_2O \rightarrow CaO + 2\,HCl$$

$$HCl + Fe \rightarrow FeCl_2 + H_2$$

$$FeCl_2 + H_2S \rightarrow FeS + 2\,HCl$$

One of the ways to mitigate the effects of hydrochloric acid is by adding ammonium (NH_3) as a basic material to neutralize the hydrogen chloride by reaction to produce ammonium chloride (NH_4Cl). This substance is highly hygroscopic (readily takes up and retains moisture) and can even react with water vapor (steam). The water containing ammonium chloride is very corrosive for copper-based alloys, such as brass and bronze.

In the current context, the method that is used to reduce this type of corrosion involves rinsing the crude oil with water and sending it to a desalting vessel to remove brackish water. Despite all of this, a small concentration of remaining chloride salts in crude oil is enough to cause a failure in the upstream units.

Naphthenic acids that are present in crude oil and hence can be transferred to the blend can cause severe corrosion in certain circumstances. This kind of corrosion usually occurs at temperatures between 230°C and 400°C (445°F and 750°F). Naphthenic acid corrosion typically occurs in refinery distillation equipment such as furnace tubes, transfer lines, vacuum columns, and side cut piping. NAC rarely happens in fluid catalytic units because the temperature at these units is more than 400°C (750°F), which can decompose naphthenic acids.

The following reaction illustrates the interaction between naphthenic acids and steel. The product of this reaction is hydrogen and a complex of iron–organic acid, which is soluble in crude oil:

$$Fe + 2\,RCOOH \rightarrow Fe(RCOO)_2 + H_2$$

In the presence of sulfide derivatives in crude oil, there is a further reaction to produce iron sulfide (FeS). Thus:

$$Fe(RCOO)_2 + H_2S = FeS + 2\,RCOOH$$

Iron sulfide (ferrous sulfide, FeS) is insoluble in water and oil and can form a protective layer on steel at low shear stress of fluid, therefore protecting it from further corrosion. As a result, the presence of sulfides in crude oil might decrease the rate of NAC, especially at low temperatures. On the other hand, the reproduced naphthenic acid keeps the corrosion happening.

Naphthenic acid corrosion is considered to be a localized corrosion and is seen in areas where fluid velocity is high and organic acid vapors are present. The lack of corrosion product in the corroded area is another feature of naphthenic acid.

While the dewatering and desalting processes may reduce the amount of naphthenic acid in a crude oil destined for a blend, one of the most common ways to reduce naphthenic acid corrosion in crude oil refining systems is by blending a high-acid (high TAN) crude oil with a low total acid number (low TAN) crude oil. In this condition, the overall total acid number value will be reduced to a much lower, less harmful range. The blending process for a new source of crude oil should be done with caution because, as mentioned in the above, adequate concentration of a certain kind of naphthenic acid in a crude oil with low total acid number can cause a high rate of naphthenic acid corrosion.

Crude oils usually contain sulfide derivatives that can cause corrosion at high temperatures – a process known as sulfidation – which occurs in many refineries. The amount of total sulfur in a crude oil depends on the type of oil field and it varies from 0.05% to 14%. Of course, sulfur values as low as 0.2% are enough to create sulfidation corrosion in plain steels and low alloy steels. These kinds of steels are often used in several parts of refinery units.

Most of the sulfur in crude oil is in the form of organic sulfide derivatives, such as mercaptan derivatives (RSH), alkyd sulfide derivatives (RSR), and thiophene derivatives along with trace amounts of elemental sulfur and hydrogen sulfide (H_2S). But not all kinds of sulfur compounds are corrosive; only a fraction of the sulfur compounds (the so-called active sulfur derivatives, such as elemental sulfur) can react with metallic compounds to create sulfidation corrosion. Despite that, in the presence of hydrogen gas, most of the organic sulfides (which are often categorized as inactive sulfide constituents) decompose to hydrogen sulfide which is an active sulfur compound that can lead to sulfidation.

Sulfidation happens at temperatures higher than 230°C (445°F) and the rate of the reaction accelerates when the temperature is raised to 480°C (895°F). At temperatures higher than 370°C (700°F), hydrogen sulfide H_2S decomposes into elemental sulfur, which is the most aggressive sulfur compound. In fact, the sulfidation rate reaches its maximum at around 400°C (750°F).

In addition to corrosion, inefficient dewatering and desalting processes can lead to blends of crude oils that are unstable mixtures which precipitate species such as asphaltene and result in rapid fouling (Wilson and Polley, 2001; Deshannavr et al., 2010). The crude oil incompatibility and the resulting precipitation of asphaltene constituents on blending of crude oils can cause significant fouling and coking in crude preheat train. In general, fouling mechanisms can be classified into five categories which are, alphabetically (1) biological fouling, (2) chemical reaction fouling, (3) corrosion fouling, (4) crystallization fouling, and (5) particulate fouling.

An important factor which influences the fouling is crude blending which can result in the formation of unstable mixtures which precipitate molecular species (such as asphaltene constituents) and result in rapid fouling. The crude oil incompatibility and the precipitation of asphaltene constituents as a result of crude oil blending can cause significant fouling and coking in the crude oil preheat train.

In crude oil blends, the occurrence of one or more chemical reactions between the organic constituents of the blend or the deposition of particulate matter (i.e. solid organic matter such as asphaltene constituents) plays an important role. In addition, there may be an induction period (sometimes referred to as an initiation period or as a delay period, which may last any time from few seconds to several hours) before the formation of a separate phase occurs.

REFERENCES

Abraham, H. 1945. *Asphalts and Allied Substances*, 5th Edition. Van Nostrand Inc., New York. Volume I, p. 1.

Álvarez, P., Menendez, J.L., Burrueco, C., Rostani, K., and Millan, M. 2012. Determination of crude oil incompatibility regions by ellipsometry. *Fuel Processing Technology*, 96: 16–21.

Asomaning, S., and Watkinson, A. P., 2000, Petroleum stability and heteroatom species effects in fouling of heat exchangers by asphaltenes. *Heat Transfer Engineering*, 21: 10–16.

ASTM. 2021. Annual Book of Standards. American Society for Testing and Materials, West Conshohocken, PA.

ASTM D96. 2021. Standard Test Method for Water and Sediment in Crude Oil by Centrifuge Method (Field Procedure). Annual Book of Standards. American Society for Testing and Materials, West Conshohocken, PA.

ASTM D445. 2021. Standard Test Method for Kinematic Viscosity of Transparent and Opaque Liquids (and Calculation of Dynamic Viscosity). Annual Book of Standards. American Society for Testing and Materials, West Conshohocken, PA.

ASTM D473. 2021. Standard Test Method for Sediment in Crude Oils and Fuel Oils by the Extraction Method. Annual Book of Standards. American Society for Testing and Materials, West Conshohocken, PA.

ASTM D1500. 2021. Standard Test Method for ASTM Color of Petroleum Products (ASTM Color Scale). Annual Book of Standards. American Society for Testing and Materials, West Conshohocken, PA.

ASTM D1796. 2021. Standard Test Method for Water and Sediment in Fuel Oils by the Centrifuge Method (Laboratory Procedure). Annual Book of Standards. American Society for Testing and Materials, West Conshohocken, PA.

ASTM D2007. 2021. Standard Test Method for Characteristic Groups in Rubber Extender and Processing Oils and Other Petroleum-Derived Oils by the Clay-Gel Absorption Chromatographic Method. Annual Book of Standards. American Society for Testing and Materials, West Conshohocken, PA.

ASTM D3279. 2021. Standard Test Method for n-Heptane Insolubles. Annual Book of Standards. American Society for Testing and Materials, West Conshohocken, PA.

ASTM D4007. 2021. Standard Test Method for Water and Sediment in Crude Oil by the Centrifuge Method (Laboratory Procedure). Annual Book of Standards. American Society for Testing and Materials, West Conshohocken, PA.

ASTM D4124. 2021. Standard Test Method for Separation of Asphalt into Four Fractions. Annual Book of Standards. American Society for Testing and Materials, West Conshohocken, PA.

ASTM D4310. 2021. Standard Test Method for Determination of Sludging and Corrosion Tendencies of Inhibited Mineral Oils. Annual Book of Standards. American Society for Testing and Materials, West Conshohocken, PA.

ASTM D4807. 2021. Standard Test Method for Sediment in Crude Oil by Membrane Filtration. Annual Book of Standards. American Society for Testing and Materials, West Conshohocken, PA.

ASTM D4871. 2021. Standard Guide for Universal Oxidation/Thermal Stability Test Apparatus.

ASTM D7061. 2021. Standard Test Method for Measuring n-Heptane Induced Phase Separation of Asphaltene-Containing Heavy Fuel Oils as Separability Number by an Optical Scanning Device. Annual Book of Standards. American Society for Testing and Materials, West Conshohocken, PA.

ASTM D7112. 2021. Standard Test Method for Determining Stability and Compatibility of Heavy Fuel Oils and Crude Oils by Heavy Fuel Oil Stability Analyzer (Optical Detection). Annual Book of Standards. American Society for Testing and Materials, West Conshohocken, PA.

Batts, B.D., and Fathoni, A.Z. 1991. A literature review on fuel stability studies with particular emphasis on diesel oil. *Energy & Fuels*, 5: 2–21.

Ben Mahmoud, M., and Aboujadeed, A., 2017. Compatibility assessment of crude oil blends using different methods. *Chemical Engineering Transactions*, 57: 1705–1710.

Branco, V.A.M., Mansoori, G.A., De Almeida Xavier, L.C., Park, S.J., and Manafi, H. 2001. Asphaltene flocculation and collapse from petroleum fluids. *Journal of Petroleum Science and Engineering*, 32: 217–230.

Carnahan, N.J., Salager, R. and Davila, A. 1999. Properties of resins extracted from Boscan crude oil and their effect on the stability of asphaltene constituents in Boscan and Hamaca crude oils. *Energy & Fuels*, 13: 309.

Castro, P.M., and Grossmann, I.E. 2014. Global optimal scheduling of crude oil blending operations with RTN continuous-time and multiparametric disaggregation. *Industrial & Engineering Chemistry Research*, 53(39): 15127–15145.

Cerdá, J., Pautasso, P.C., and Cafar, D.C. 2018. Optimization approaches for efficient crude blending in large oil refineries. *Industrial & Engineering Chemistry Research*, 57(25): 8484–8501.

Clemente, J.S., and Fedorak, P.M. 2005. A review of the occurrence, analyses, toxicity, and biodegradation of naphthenic acids. *Chemosphere*, 60(5): 585–600.

Deshannavr, J.B., Rafeen M.S., Ramasamy, M., and Subbarao, D. 2010. Crude oil fouling: A review. *Journal of Applied Sciences*, 10(24): 3167–3174. https://docsdrive.com/pdfs/ansinet/jas/ 2010/3167-3174.pdf.

Gary, J.G., Handwerk, G.E., and Kaiser, M.J. 2007. *Petroleum Refining: Technology and Economics*, 5th Edition. CRC Press, Taylor & Francis Group, Boca Raton, FL.

Gawrys, K., Spiecker, L., Matthew, P., and Kilpatrick, P.K. 2003. The role of asphaltene solubility and chemical composition on asphaltene aggregation. *Petroleum Science and Technology*, 21(3,4): 461–489.

Goetzinger, J.W., Thompson, C.J., and Brinkman, D.W. 1983. A review of storage stability characteristics of hydrocarbon fuels. US. Department of Energy, Report No. DOE/BETC/IC-83-3.

Hardy, D.R., and Wechter, M.A. 1990. Insoluble sediment formation in middle-distillate diesel fuel: The role of soluble macromolecular oxidatively reactive species. *Energy & Fuels*, 4: 270–274.

Hammami, A., and Freworn, K.A. 1998. Asphaltic crude oil characterization: An experimental investigation of the effect of resins on the stability of asphaltene constituents. *Petroleum Science and Technology*, 16(3,4): 227–249.

Hammami, A., Phelps, H., and Little, T.M. 2000. Asphaltene precipitation from live oils: An experimental investigation of the onset conditions and reversibility. *Energy Fuels*, 14, 14–18.

Hazlett, R.N. 1992. *Thermal Oxidation Stability of Aviation Turbine Fuels.* Monograph No. 1. American Society for Testing and Materials, Philadelphia, PA.

Hsu, C.S., and Robinson, P.R. (Editors). 2017. *Handbook of Petroleum Technology.* Springer International Publishing AG, Cham,

Khanifar, A., Demiral, B., and Darman, N. 2011. Modeling of asphaltene precipitation and deposition during WAG application. *Proceedings of International Petroleum Technology Conference.* Bangkok, Thailand, November 15–17. International Petroleum Technology Conference (IPTC), Richardson, Texas.

Koots, J.A., and Speight, J.G. 1975. The relation of petroleum resins to asphaltenes. *Fuel*, 54: 179.

Kumar, R., Voolapalli, K.A., and Upadhyayula, S. 2018. Prediction of crude oil blends compatibility and blend optimization for increasing heavy oil processing. *Fuel Processing Technology*, 177: 307–327.

Lopez, D.C., Mahecha, C.A., Arellano-Garcia, H., and Wozny, G. 2013. Optimization model of crude oil distillation units for optimal blending and operating conditions. *Industrial & Engineering Chemistry Research*, 52(36): 2993–13005.

Magaril, R.Z., and Akensova, E.I. 1967. Mechanism of coke formation during the cracking of petroleum Tars. *Izvestia Vyssh Ucheb. Zaved. Neft Gas*, 10(11): 134–136.

Magaril, R.Z., and Akensova, E.I. 1968. Study of the mechanism of coke formation in the cracking of petroleum resins. *International Chemical Engineering*, 8(4): 727–729.

Magaril, R.Z., and Aksenova, E.I. 1970a. Mechanism of coke formation in the thermal decompositon of asphaltenes. *Khimiya i Tekhnologiya Topliv i Masel*, 15(7): 22–24.

Magaril, R.Z., and Aksenova, E.I. 1970b. Kinetics and mechanism of coking asphaltenes. *Khim. Izvestia Vyssh. Ucheb. Zaved. Neft Gaz.*, 13(5): 47–53.

Magaril, R.Z., and Aksenova, E.I. 1972. Coking kinetics and mechanism of asphaltenes. *Khim. Kim Tekhnol. Tr. Tyumen Ind. Inst.*, 169–172.

Magaril, R.Z., and Ramazaeva, L.F. 1969. Study of carbon formation in the thermal decomposition of asphaltenes in solution. *Izvestia Vyssh. Ucheb. Zaved. Neft Gaz.*, 12(1): 61–64.

Magaril, R.Z., Ramazaeva, L.F., and Aksenova, E.I. 1970. Kinetics of coke formation in the thermal processing of petroleum. *Khimiya i Tekhnologiya Topliv i Masel*, 15(3): 15–16.

Magaril, R.Z., Ramazaeva, L.F., and Aksenova, E.I. 1971. Kinetics of the formation of coke in the thermal processing of crude oil. *International Chemical Engineering*, 11(2): 250–251.

Mendoza de la Cruz, J.L., Cedillo-Ramirez, J.C., Aguirre-Gutiérrez, A.d.J., Garcia-Sánchez, F., and Aquino-Olivos, M.A. 2015. Incompatibility determination of crude oil blends from experimental viscosity and density data. *Energy & Fuels*, 29(2): 480–487.

Murphy, G., Campbell, J., and Bohent, M., 1992, Fouling in Refinery Heat Exchangers (Eds), Fouling Mechanisms. *Proceedings of GRETH Seminar*, Grenoble, France, pp. 249–261.

Mushrush, G.W., and Speight, J.G. 1995. *Petroleum Products: Instability and Incompatibility.* Taylor & Francis Publishers, Washington, DC.

Naji, F.A., Ateeq, A.A., and Al-Mayyahi, M.A. 2021. Optimization of blending operation for the iraqi oils. *Journal of Physics: Conference Series*, 1773: 012037. https://www.researchgate.net/publication/349655511_Optimization_of_blending_operation_for_the_Iraqi_oils.

Parkash, S. 2003. *Refining Processes Handbook.* Gulf Professional Publishing, Elsevier, Amsterdam.

Por, N. 1992. *Stability Properties of Petroleum Products.* Israel Institute of Petroleum and Energy, Tel Aviv.

Power, A.J., and Mathys, G.I. 1992. Characterization of distillate fuel sediment molecules: Functional group derivatization. *Fuel*, 71: 903–908.

Qian, K. and Robbins, W.K. 2001. Resolution and identification of elemental compositions for more than 3000 crude acids in heavy petroleum by negative-ion micro-electrospray high-field fourier transform ion cyclotron resonance mass spectrometry. *Energy & Fuels*, 15: 1505–1511.

Rathore, V., Brahma, R., Thorat, T.S., Rao, P.V.C., and Choudary, N.V. 2011. Assessment of crude oil blends. *Digital Refining. PTQ*, Q4: 1–6. https://www.digitalrefining.com/article/1000381/assessment-of-crude-oil-blends#.YlmcBNrMJPY.

Rogel, E., Hench, K., Miao, T., Lee, E., and Dickakian, G. 2018. Evaluation of the compatibility of crude oil blends and its impact on fouling. *Energy Fuels*, 32(9): 9233–9242.

Saleh, Z.S., Sheikholeslami, R., and Watkinson, A.P. 2005. Blending effects on fouling of four crude oils. In: *Proceedings of ECI Symposium Series, Volume RP2. 6th International Conference on Heat Exchanger Fouling and Cleaning - Challenges and Opportunities*, H. Müller-Steinhagen. M. Malayeri, and A.P. Watkinson (Editors), Engineering Conferences International, Kloster Irsee, Germany, June 5–10. https://dc.engconfintl.org/cgi/viewcontent.cgi?referer=https://www.google.com/&httpsredir=1&article=1003&context=heatexchanger2005.

Sayles, S., and Routt, D.M. 2011. *Unconventional Crude Oil Selection and Compatibility.* Digital Refining, Croydon. https://cdn.digitalrefining.com/data/articles/file/488759177.pdf.

Schermer, W.E.M., Melein, P.M.J., and Van Den Berg, F.G.A. 2004. Simple techniques for evaluation of crude oil compatibility. *Petroleum Science and Technology,* 22(7): 1045–1054.

Schramm, L.L. (Editor). 1992. *Emulsions: Fundamentals and Applications in the Petroleum Industry.* Advances in Chemistry Series No. 231. American Chemical Society, Washington, DC.

Shahnovsky, G., Cohen, T., and McMurray, R. 2014. *Advanced Solutions for Efficient Crude Blending.* Digital Refining. https://cdn.digitalrefining.com/data/articles/file/1612762447.pdf.

Speight, J.G. 1994. Chemical and physical studies of petroleum asphaltenes. In *Asphaltenes and Asphalts. I. Developments in Petroleum Science, 40,* T.F. Yen and G.V. Chilingarian (Editors). Elsevier, Amsterdam, Chapter 2, pp. 7–65.

Speight, J.G. 2013. *The Chemistry and Technology of Coal,* 3rd Edition. CRC Press, Taylor & Francis Group, Boca Raton, FL.

Speight, J.G. 2014a. *The Chemistry and Technology of Petroleum,* 5th Edition. CRC Press, Taylor & Francis Group, Boca Raton, FL.

Speight, J.G. 2014b. *Oil and Gas Corrosion Prevention.* Gulf Professional Publishing Company, Elsevier, Oxford.

Speight, J.G. 2015. *Handbook of Petroleum Product Analysis,* 2nd Edition. John Wiley & Sons Inc., Hoboken, NJ.

Speight, J.G. 2017. *Handbook of Petroleum Refining.* CRC Press, Taylor & Francis Group, Boca Raton, FL.

Speight, J.G. 2021. *Refinery Feedstocks.* CRC Press, Taylor & Francis Group, Boca Raton, FL.

Stark, J.L., and Asomaning, S. 2003. Crude oil blending effects on asphaltene stability in refinery fouling. *Petroleum Science and Technology,* 21(3–4): 569–579.

Stavinoha, L.L., and Henry, C.P. (Editors). 1981. *Distillate Fuel Stability and Cleanliness.* Special Technical Publication No. 751. American Society for Testing and Materials, Philadelphia, PA.

Watkinson A.P, 1992, Chemical reaction fouling of organic fluids. *Chemical Engineering and Technology,* 15: 82–90.

Wiehe, I.A. 1993. A phase-separation kinetic model for coke formation. *Industrial & Engineering Chemistry Research,* 32: 2447–2454.

Wiehe, I.A., and Kennedy, R.J. 2000. The oil compatibility model and crude oil incompatibility. *Energy & Fuels,* 14(1): 56–59.

Wilson, D.I., and Polley, G.T. 2001. Mitigation of refinery preheat train fouling by nested optimization. *Advances in Refinery Fouling Mitigation Session,* 46: 287–294.

4 Distillation

4.1 INTRODUCTION

Crude oil in the unrefined state (Chapter 1) is of limited value and of limited use and refining, which is a series of steps by which the crude oil is converted into saleable products, and it is required to produce the products in the amounts dictated by the market (Parkash, 2003; Gary et al., 2007; Speight, 2014a, 2017; Hsu and Robinson, 2017). In fact, a refinery is essentially a group of manufacturing plants that perform a variety of functions (Figure 4.1) with a range of products produced and methods used to make those products to provide a balanced operation (Speight, 2019, 2020, 2021).

Refinery processes must be selected and products manufactured to give a balanced operation: that is, crude oil must be converted into products according to the rate of sale of each. Crude oil refining is a very recent science and many innovations have evolved during the 20th century (Parkash, 2003; Gary et al., 2007; Eser and Riazi, 2013; Speight, 2014a, 2017; Hsu and Robinson, 2017). As the 20th century evolved, distillation techniques in refineries became more sophisticated to handle a wider variety of crude oils to produce marketable products or feedstocks for other refinery units. However, it became apparent that the distillation units in the refineries were incapable of producing specific product fractions (also referred to as "cuts" based on different temperatures at which they evaporate and condense – cut points). In order to accommodate this type of product demand, refineries have, in the latter half of this century, incorporated azeotropic distillation and extractive distillation in their operations.

All compounds have definite boiling temperatures but a mixture of chemically dissimilar compounds will sometimes cause one or both components to boil at a temperature other than what is expected. A mixture that boils at a temperature lower than the boiling point of any of the components is an azeotropic mixture. When it is desired to separate components with very similar boiling points, the addition of a non-indigenous component will form an azeotropic mixture with one of the components of the mixture thereby lowering the boiling point by the formation of an azeotrope and facilitate separation by distillation.

The separation of these components with similar volatility may become economically feasible if an entrainer can be found that effectively changes the relative volatility (Long, 1995). It is also desirable that the entrainer be reasonably cheap, stable, nontoxic, and readily recoverable from the components. In practice, it is probably this last-named criterion that limits severely the application of extractive and azeotropic distillation. The majority of successful processes are those in which the entrainer and one of the components separate into two liquid phases on cooling if direct recovery by distillation is not feasible. A further restriction in the selection of an azeotropic entrainer is that the boiling point of the entrainer be in the range of 10°C–40ºC (18°F–72°F) below that of the components.

The crude oil refinery of the 21st century is a much more complex operation than those refineries of 100–120 years ago. Early refineries were distillation units, perhaps with ancillary units to remove objectionable odors from various product streams. The refinery of the 1930s was more complex but was essentially a distillation unit but at this time cracking and coking units were starting to appear in the scheme of refinery operations. These units were not what is used in the current refineries as cracking and coking units but were the forerunners of modern units. Also at this time, asphalt was becoming a recognized crude oil product. Finally, the current refineries are a result of major evolutionary trends and are highly complex operations. Most of the evolutionary adjustments to refineries have occurred during the decades since the commencement of World War II. In the crude oil industry, as in many other industries, supply and demand are key factors in efficient and economic operation, and innovation is also a key.

DOI: 10.1201/9781003184959-4

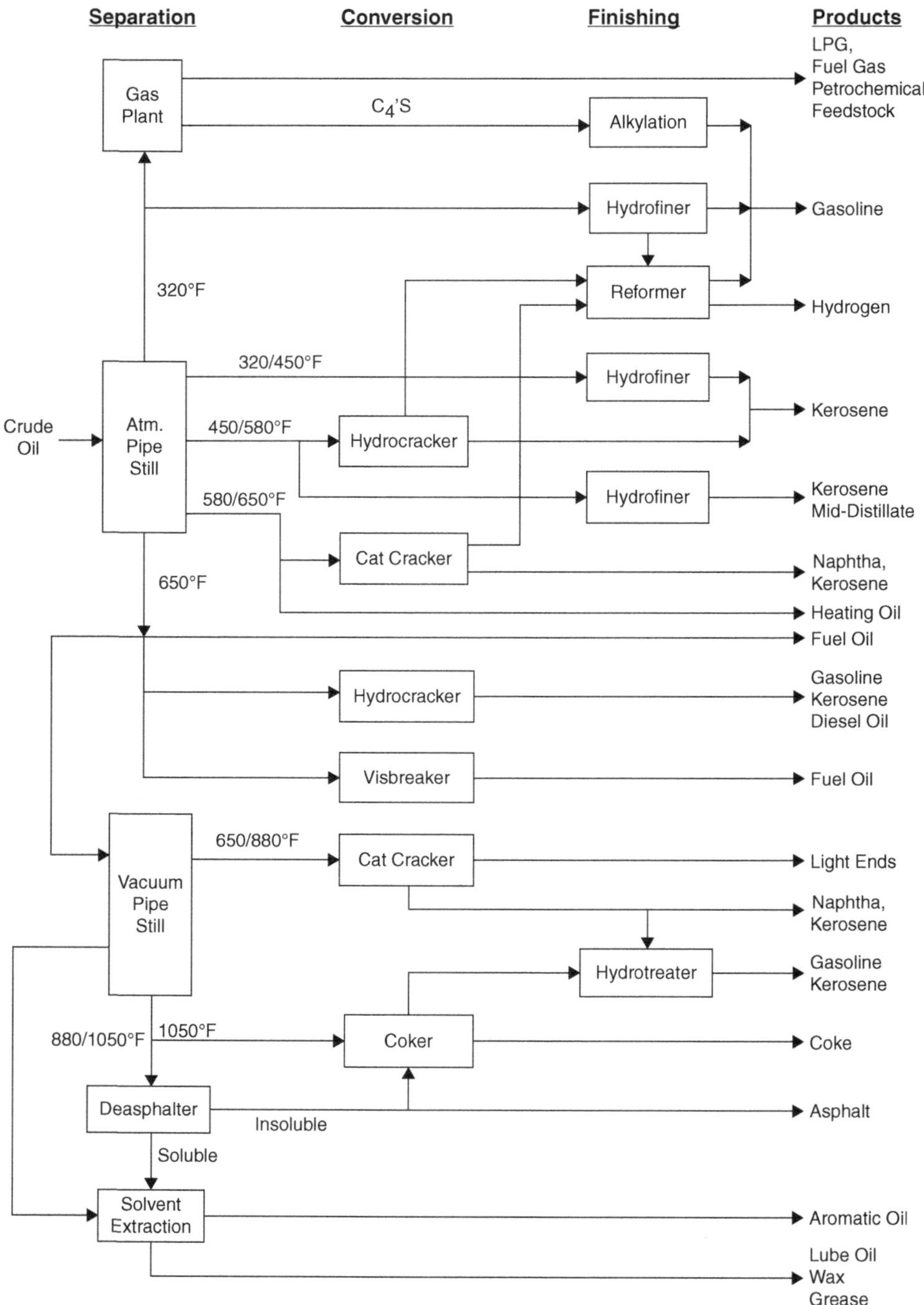

FIGURE 4.1 General schematic of a refinery. (Speight, J.G. 2017. *Handbook of Petroleum Refining*, CRC Press, Taylor & Francis Group, Boca Raton, FL. Figure 7.1, p. 251.)

It is not the objective of this chapter to dispel distillation as a refining process but to place the distillation processes in the sequence of processes that are used to produce various saleable products from crude oil. It is the purpose of this chapter to illustrate the initial processes that are applied to a feedstock in a refinery. The first processes (desalting and dewatering) are focused on the cleanup of the feedstock, particularly the removal of the troublesome brine constituents (Chapter 2).

This is followed by distillation to remove the volatile constituents with the concurrent production of a residuum that can be used as a cracking (coking) feedstock or as a precursor to asphalt. Current methods of bitumen processing involve direct use of the bitumen as feedstock for delayed or fluid coking (Speight, 2000, 2014a). Other methods of feedstock treatment that involve the concept of volatility are also included here even though some of the methods (such as stripping and rerunning; Chapter 4) might also be used for product purification.

4.2 FEEDSTOCK EVALUATION

While there are many test methods that can be used to evaluate crude oil, the focus of the evaluation of a crude oil feedstock for the distillation process is the volatility of the feedstock or, in these cases of the blends of several crude oils today, the volatility of each member of the blend should be assessed.

Briefly, the volatility of a liquid is defined as the tendency of liquid to vaporize, that is, to change from a liquid to a vapor or gaseous state.

The vaporizing tendencies of crude oil and crude oil products are the basis for the general characterization of liquid fuels, such as liquefied petroleum gas, natural gasoline, motor and aviation gasoline, naphtha, kerosene, gas oil, diesel fuel, and fuel oil (ASTM D2715). A test (ASTM D6) also exists for determining the loss of material when crude oil and asphaltic compounds are heated. Another test (ASTM D20) is a method for the distillation of road tars that might also be applied to estimate the volatility of high molecular weight residues.

For some purposes, it is necessary to have information on the initial stage of vaporization. To supply this need, standard test methods for the flash point, the fire point, the vapor pressure, and the tendency for a product to evaporate are available (ASTM, 2021). The data from the early stages of several distillation methods are also useful. For other uses, it is important to know the tendency of a product to vaporize partially or completely, and in some cases, to know if small quantities of high-boiling components are present. For such purposes, chief reliance is placed on the distillation methods.

As an early part of characterization studies, a correlation was observed between the quality of crude oil products and their hydrogen content since gasoline, kerosene, diesel fuel, and lubricating oil are made up of hydrocarbon constituents containing high proportions of hydrogen. Thus, it is not surprising that tests to determine the volatility of crude oil and crude oil products were among the first to be defined. Indeed, volatility is one of the major tests for crude oil products and it is inevitable that all products will, at some stage of their history, be tested for volatility characteristics.

Distillation involves the general procedure of vaporizing the crude oil liquid in a suitable flask either at atmospheric pressure (ASTM D86, ASTM D2892) or at reduced pressure (ASTM D1160). In this process, heated crude oil or heated blends of crude oils are introduced into the distillation unit and volatile constituents of the mix (referred to as overhead) are separated from non-volatile constituents (referred to as bottoms or residuum). The two prime methods of distillation are (1) distillation at atmospheric pressure and (2) distillation at reduced pressure, both of which bring about a controlled and efficient degree of feedstock fractionation (separation) (Figure 4.2). Collectively, both the atmospheric distillation unit (Figure 4.3) and the vacuum distillation unit (Figure 4.4) are often referred to as the distillation section of the refinery.

One of the main properties of crude oil that serves to indicate the comparative ease with which the material can be refined is the volatility (Speight, 2015). Investigation of the volatility of crude oil is usually carried out under standard conditions, thereby allowing comparisons to be made between data obtained from various laboratories. Thus, nondestructive distillation data (US Bureau of Mines method) show that, not surprisingly, bitumen is a higher boiling material than the more conventional crude oils (Speight, 2014a, 2017). There is usually little, or no, naphtha (sometimes arbitrarily referred to as the gasoline fraction) in bitumen and the majority of the distillate falls in the gas oil-lubrication distillate range (>260°C, >500°F). More than 50% of each bitumen is

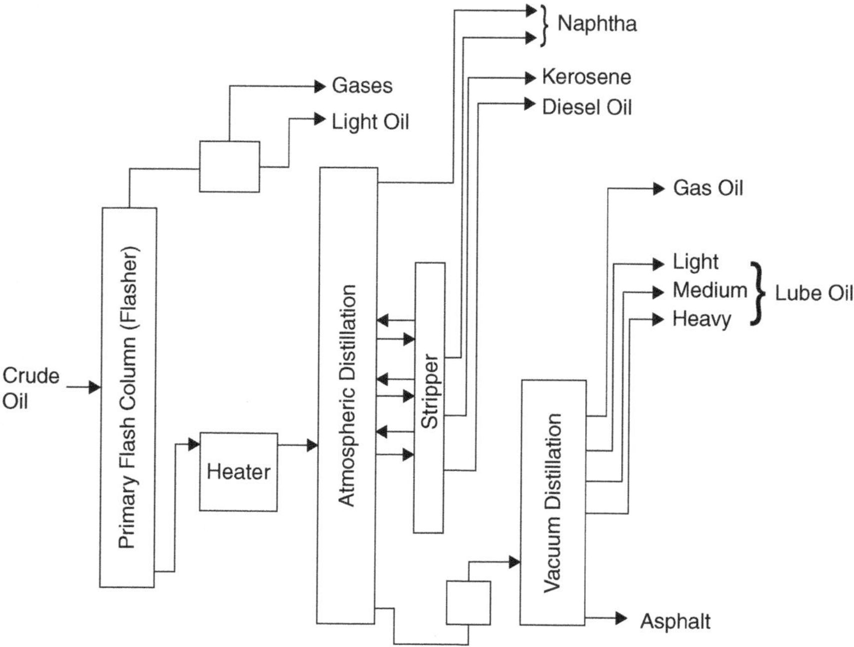

FIGURE 4.2 Representation of a refinery distillation section. (Speight, J.G. 2017. *Handbook of Petroleum Refining*, CRC Press, Taylor & Francis Group, Boca Raton, FL. Figure 7.12, p. 264.)

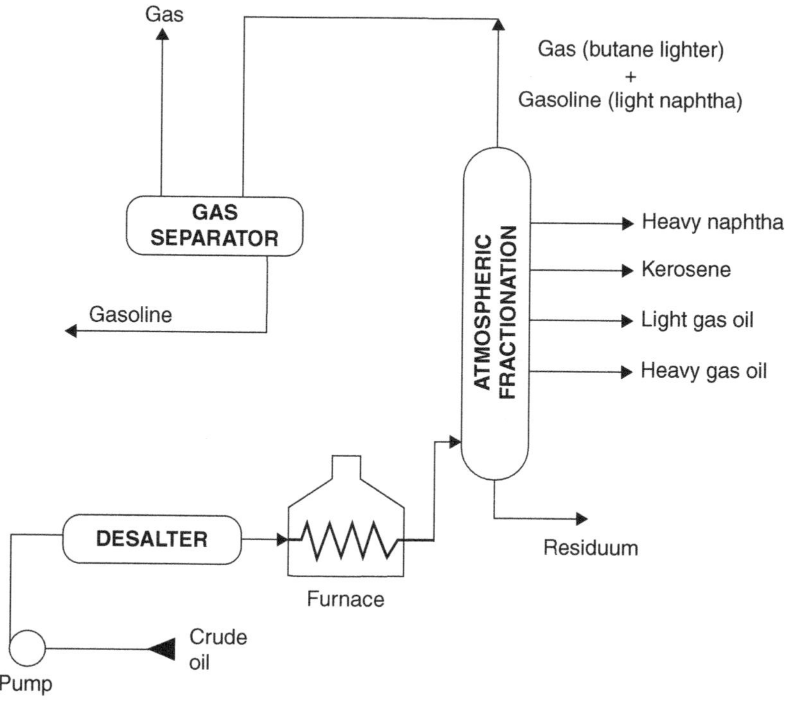

FIGURE 4.3 Representation of an atmospheric distillation unit showing position of desalter. (OSHA Technical Manual, Section IV, Chapter 2: Petroleum Refining Processes. http://www.osha.gov/dts/osta/otm/otm_iv/otm_iv_2.html.)

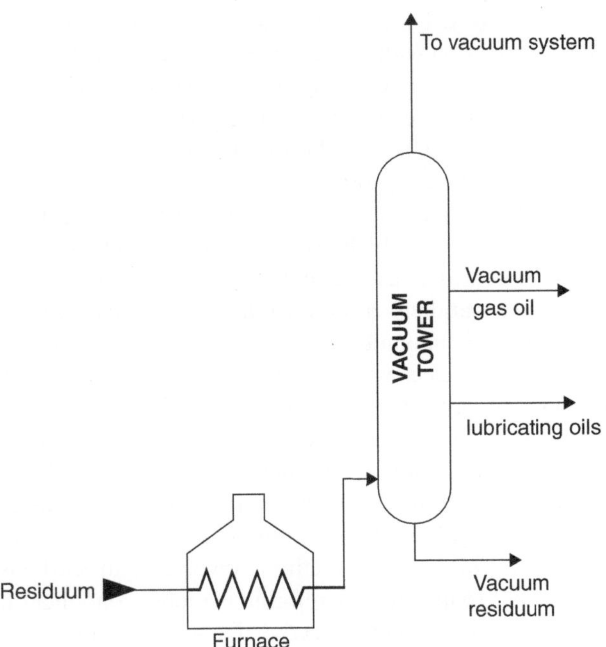

FIGURE 4.4 Representation of a vacuum distillation unit. (OSHA Technical Manual, Section IV, Chapter 2: Petroleum Refining Processes. http://www.osha.gov/dts/osta/otm/otm_iv/otm_iv_2.html.)

non-distillable under the test conditions, and the yield of the non-volatile material corresponds very closely to the amount of the asphaltic constituents (asphaltene constituents plus resin constituents) in each feedstock. In fact, detailed fractionation of the sample might be of secondary importance. Thus, it must be recognized that the general shape of a one-plate distillation curve is often adequate for making engineering calculations, correlating with other physical properties, and predicting the product slate.

There is also another method that is increasing in popularity for application to a variety of feedstocks and that is the method commonly known as simulated distillation (ASTM D2887) (Carbognani et al., 2012). The method has been well researched in terms of method development and application (Romanowski and Thomas, 1985; MacAllister and DeRuiter, 1987; Schwartz et al., 1987; Neer and Deo, 1995). The benefits of the technique include good comparisons with other ASTM distillation data as well as the application to higher boiling fractions of crude oil. In fact, data output includes the provision of the corresponding Engler profile (ASTM D86) as well as the prediction of other properties such as vapor pressure and flash point. When it is necessary to monitor product properties, as is often the case during refining operations, such data provide a valuable aid to process control and online product testing.

For a more detailed distillation analysis of feedstocks and products, a low-resolution, temperature-programmed gas chromatographic analysis has been developed to simulate the time-consuming true boiling point distillation (ASTM D5307). The method relies on the general observation that hydrocarbons are eluted from a non-polar adsorbent in the order of their boiling points. The regularity of the elution order of the hydrocarbon components allows the retention times to be equated to distillation temperatures and the term simulated distillation (or SimDis) by gas chromatography is used throughout the industry to refer to this technique.

Simulated distillation by gas chromatography is often applied in the crude oil industry to obtain true boiling point data for distillates and crude oils (Speight, 2015). Two standardized methods (ASTM D2887 and ASTM D3710) are available for the boiling point determination of crude oil fractions and gasoline, respectively. The ASTM D2887 method utilizes non-polar, packed gas chromatographic columns in conjunction with flame ionization detection. The upper limit of the boiling range covered by this method is approximately 540°C (1000°F) atmospheric equivalent boiling point. Recent efforts in which high-temperature gas chromatography was used have focused on extending the scope of the ASTM D2887 method for higher boiling crude oil materials to 800°C (1470°F) atmospheric equivalent boiling point.

Finally, a comment on the expected changes in the refinery feedstock slate that will arise because of the depletion of crude oil resources which brings serious challenges to the crude oil economy is facing serious challenges. Biorefining has gradually become a new method used to produce energy and chemical products, which parallels crude oil refining. Biomass is used as a raw material to produce some middle platform compounds by thermochemical, chemical, or biological methods, which are then processed into bio-fuels or chemicals. Biorefining offers the possibility of achieving sustainable production of bioenergy and biomaterials.

Biorefining or the production of fuels and chemicals (Speight, 2014a) is a new concept in industrial manufacturing, which will become a new manufacturing technology paradigm, becoming more and more important because of its use of alternative resources and its environmental benefits. Compared with crude oil refining, biorefining is unique for the following reasons: (1) biorefinery feedstocks are carbohydrates, while crude oil refining involves hydrocarbons without oxygen molecules and lacking functional groups, thus biorefining is more economical from the point of view of atom utilization and, therefore, it is cost-effective due to the avoidance of the high-cost oxidation process; (2) compared with the crude oil-derived fuels, biodiesel obtained from biorefining has better oxidation stability and higher energy density, which can reduce carbon deposits in the engine, so that operating costs are only half of those of the conventional diesel apparatus; and (3) in the crude oil refining process, the resource-rich region is separated from the end market, which means long-distance transport is required, while biomass resources are usually found near to the market.

With the advances of technology, a number of bio-based products that are able to replace or partially replace crude oil-based products – such as ethanol, lactic acid, citric acid, biodiesel, and 1,3-propanediol – have arisen. These products can be divided into many groups according to carbohydrates platform, oil platform, and thermochemical platform. Although there are a variety of products from biorefining, many of them are not needed in large quantities. Therefore, it would be uneconomical to prepare all of these products directly from the basic raw materials. Hence, it is necessary to develop many platform products, such as succinic acid, fumaric acid, or malic acid, which can then be economically converted into various chemicals (Speight, 2019).

4.3 DISTILLATION

By way of introduction, distillation is the process by which crude oil is separated into a variety of bulk fractions based on the boiling points or volatility of the constituents of the crude oil. More volatile constituents (constituents with low-boiling points) tend to vaporize more readily than heavy (high boiling) constituents and this forms the basis of separation through the distillation process.

The crude oil distillation unit separates raw crude oil into fractions of useful products (sometimes termed bulk products), which are then processed further at other units in the refinery (Parkash, 2003; Gary et al., 2007; Speight, 2014a, 2017; Hsu and Robinson, 2017). The crude oil distillation system comprises (1) a heat exchanger network, (2) a pre-flash drum, (3) an atmospheric distillation unit, and (4) a vacuum distillation unit. Briefly, a distillation column has several perforated trays or packings that allow vapors to rise to the top of the column. High temperatures are required to separate light (lower boiling) constituents (such as naphtha and kerosene) from heavier (higher boiling) constituents, such as gas oil fractions. In the vacuum distillation unit, reduced pressure is required to prevent thermal cracking when further distilling the crude from the atmospheric tower at high temperature.

The principle of vacuum distillation is similar to that of atmospheric distillation, with the exception of using a larger diameter column to mitigate the effects of pressure variations in the unit. Some vacuum distillation units differ from the atmospheric distillation units insofar as random packing and demister pads are used instead of trays. The vacuum heater tubes, vacuum tower transfer lines, lining, trays, packing, oil pumps, vacuum tower bottoms line, and heat exchangers are all vulnerable to corrosion by naphthenic acid derivatives corrosion.

In fact, despite an efficient desalting operation, corrosion agents can still be passed over from the desalter, causing serious corrosion problems in the distillation section (i.e. the atmospheric unit and the vacuum unit) to the units (Fajobi et al., 2019). The desalter brine and aqueous condensate from the overhead reflux drums of the atmospheric fractioning column as well as the water from the vacuum distillation unit may contain some hydrogen sulfide. Steam is injected into the distillation unit and condenses in the upper part of the unit to improve fractionation.

Typically, distillation involves heating the crude oil feedstock (or a blend of crude oils) to produce a series of condensed products. More specifically, the object of the process target of distillation is to remove the lower boiling constituents component from the feedstock. In the refinery, the crude oil passes through several processes before becoming a finished product of which the pretreatment stages (i.e. dewatering, desalting, and distillation) are essential.

In the refinery distillation column, light (low boiling) constituents are removed from the top of the column, and the heavy (high-boiling) constituents of the mixture are removed from the lower level of the distillation column. For a crude oil that is typically a mixture of thousands of hydrocarbon derivatives, some very low-boiling constituents (such as ethane, propane, and butane) only appear in the top fraction, while extremely high-boiling and non-volatile constituents only appear in the bottom of the column. The lowest boiling constituents that emerge from the top of the distillation column (referred to as overhead distillate or full-range naphtha) are sent to the light-ends unit for further separation into liquefied petroleum gas and naphtha. The side streams separated in the atmospheric distillation column give fractions that include the straight-run products such as kerosene, light gas oil, and heavy gas oil. The residue from the atmospheric distillation column is sent to the vacuum distillation unit and produces two side streams which are light vacuum gas oil (LVGO) and heavy vacuum gas oil (HVGO), as well as the vacuum residue from the bottom of the vacuum unit. All of these distillation products are subjected to subsequent processing to produce various fuels as well as a variety of non-fuel products.

On a historic basis, distillation was the first method by which crude oil was refined. The original technique involved a batch operation in which the still was a cast-iron vessel mounted on brickwork over a fire and the volatile materials were passed through a pipe or gooseneck which led from the top of the still to a condenser. The latter was a coil of pipe (worm) immersed in a tank of running water.

In the early stages of refinery development, when illuminating and lubricating oils were the main products, distillation was the major, and often only, refinery process (Stephens and Spencer, 1956). At that time, naphtha (sometimes, in the modern sense, incorrectly referred to as gasoline) was a minor product, but as the demand for gasoline increased, conversion processes were developed because distillation could no longer supply the necessary quantities.

As the technologies for refining evolved into the 21st century, refineries became much more complex but distillation remained the prime means by which crude oil is refined. Indeed, the distillation section of a modern refinery (Figure 4.2) is the most flexible section in the refinery since conditions can be adjusted to process a wide range of refinery feedstocks from the lighter crude oils to the heavier more viscous crude oils. However, the maximum permissible temperature (in the vaporizing furnace or heater) to which the feedstock can be subjected is 350°C (660°F). Thermal decomposition occurs above this temperature which, if it occurs within a distillation unit, can lead to coke deposition in the heater pipes or in the tower itself with the resulting failure of the unit. The contained use of atmospheric and vacuum distillation has been a major part of refinery operations during this century and will undoubtedly continue to be employed throughout the remainder of the century as the primary refining operation.

The distillation of crude oil to produce a series of bulk fractions that can be sold as products or used as feedstocks for downstream refinery processes is well established. In the process, the atmospheric oil crude distillation unit must produce three objectives which are (1) initial separation of the crude oil for sales or for downstream processing, (2) heat recovery to reduce investment required and to reduce energy costs, and (3) feedstock preparation for the downstream units which includes the vacuum distillation unit. These objectives can only be achieved with any degree of certainty if the preceding units that accomplish dewatering and desalting have operated successfully.

Thus, it is possible to obtain products ranging from gases which are taken off at the top of the distillation column to a non-volatile residue or reduced crude (the atmospheric residuum, also referred to as residuum or, bottoms), with correspondingly lighter materials at intermediate points. The reduced crude may then be processed by vacuum distillation or by steam distillation in order to separate the high-boiling lubricating oil fractions without the danger of decomposition, which occurs at high (>350°C, >660°F) temperatures. Atmospheric distillation may be terminated with a lower boiling fraction (cut) if vacuum or steam distillation yields a better quality product or if the process appears to be economically more favorable. Not all crude oils yield the same distillation products and there may be variations by several degrees in the boiling ranges of the fractions as specified by different companies and the nature of the crude oil dictates the processes that may be required for refining (Speight, 2014a, 2017).

However, the success of the refinery distillation units depends very much on the success of the dewatering and desalting operations (Chapter 2). For example, salt dissolved in water (brine) enters the crude stream as a contaminant during the production or transportation of oil to refineries. If salt is not removed from crude oil, serious damage can result, especially in the heater tubes, due to corrosion caused by the presence of Cl. Salt in the crude oil feedstock also causes reduction in heat transfer rates in heat exchangers and furnaces.

Heating a crude oil causes the more volatile, lower boiling components to vaporize and then condense in the worm to form naphtha. As the distillation progresses, the higher boiling components are vaporized and then condensed to produce kerosene, which was the major crude oil product in the late 19th century and the early 20th century. When all of the possible kerosene had been obtained, the material remaining in the still was discarded. The still was then refilled with crude oil and the operation repeated.

The capacity of the distillation units (often referred to as stills) at that time was usually several barrels of crude oil and it often required three or more days to distill (run) a batch of crude oil. The simple distillation as practiced in the 1860s and 1870s was notoriously inefficient. The kerosene was more often than not contaminated by naphtha, which distilled during the early stages, or by heavy oil, which distilled from the residue during the final stages of the process. The naphtha generally made the kerosene so flammable that explosions occurred when it ignited. On the other hand, the presence of heavier oil adversely affected the excellent burning properties of the kerosene and created a great deal of smoke. This condition could be corrected by redistilling (rerunning) the kerosene during which process the more volatile fraction (front-end) was recovered as additional naphtha, while the kerosene residue (tail) remaining in the still was discarded.

In the 1880s, continuous distillation was introduced in refineries. The method employed a number of stills coupled together in a row and each still was heated separately and was hotter than the preceding one. The stills were arranged so that oil flowed by gravity from the first to the last. Crude oil in the first still was heated so that a light naphtha fraction was distilled from it before the crude oil flowed into the second still, where a higher temperature caused the distillation of a heavier naphtha fraction. The residue then flowed to the third still where an even higher temperature caused kerosene to distill. The oil thus progressed through the battery to the last still, where destructive distillation (thermal decomposition; cracking) was carried out to produce more kerosene. The residue from the last still was removed continuously for processing into lubricating oils or for use as fuel oil.

In the early 1900s, a method of partial (or selective) condensation was developed to allow a more exact separation of crude oil fractions. A partial condenser was inserted between the still and the conventional water-cooled condenser. The lower section of the tower was packed with stones and insulated with brick so that the heavier less volatile material entering the tower condensed and drained back into the still. Non-condensed material passed onto another section where more of the less volatile material was condensed on air-cooled tubes and the condensate was withdrawn as a crude oil fraction. The non-condensable (overhead) material from the air-cooled section entered a second tower that also contained air-cooled tubes and often produced a second fraction. The volatile material remaining at this stage was then condensed in a water-cooled condenser to yield a third fraction. The van Dyke tower is essentially one of the first stages in a series of improvements that ultimately led to the distillation units found in modern refineries, which separate crude oil fractions by fractional distillation.

Thus, in early refineries, distillation was the primary means by which products were separated from crude oil. As the technologies for refining evolved into the twentieth century, refineries became much more complex but distillation remained the prime means by which crude oil is refined. Indeed, the distillation section of a modern refinery is the most flexible unit in the refinery since conditions can be adjusted to process a wide range of refinery feedstocks from the lighter (low-density) crude oils to the heavier, more viscous crude oils. Generally, the maximum permissible temperature (in the vaporizing furnace or heater) to which the feedstock can be subjected is 350°C (660°F). The rate of thermal decomposition increases markedly above this temperature, although higher temperatures (up to approximately 395°C, 745°F) are part of the specifications for some distillation units – serious cracking does not occur at these higher temperatures but is subject to the properties of the crude oil feedstock and the residence time of the feedstock in the hot zone. If unplanned cracking occurs within a distillation unit, coke deposition can occur in the heater pipes or in the tower itself, resulting in failure of the distillation unit.

Distillation has remained a major refinery process and it is a process to which crude oil that enters the refinery is subjected. A multitude of separations are accomplished by distillation, but its most important and primary function in the refinery is its use for the separation of crude oil into component fractions (Chapter 1). Thus, it is possible to obtain products ranging from gaseous materials taken off the top of the distillation column to a non-volatile atmospheric residuum (bottoms, reduced crude) with correspondingly lower boiling materials (gas, gasoline, naphtha, kerosene, and gas oil) taken off at intermediate points (Parkash, 2003; Gary et al., 2007; Speight, 2014a, 2017; Hsu and Robinson, 2017).

The reduced crude may then be processed by vacuum or steam distillation to separate the high-boiling lubricating oil fractions without the danger of decomposition, which occurs at high (>350°C, 660°F) temperatures, and the amount of decomposition is subject to the residence time of the feedstock in the distillation unit. Indeed, atmospheric distillation may be terminated with a lower boiling fraction (boiling cut) if it is thought that vacuum or steam distillation will yield a better quality product or if the process appears to be economically more favorable (Parkash, 2003; Gary et al., 2007; Speight, 2014a, 2017; Hsu and Robinson, 2017).

In the modern sense, distillation was the first method by which crude oil was refined (Speight 2014a). As crude oil refining evolved, distillation became a formidable means by which various products were separated. Further evolution saw the development of topping or skimming or hydroskimming refineries and conversion refineries (Parkash, 2003; Gary et al., 2007; Speight, 2014a, 2017) named for the manner in which crude oil was treated in each case. And many of these configurations exist in the world of modern refining. However, of all the units in a refinery, the distillation section comprising the atmospheric unit and the vacuum unit is required to have the greatest flexibility in terms of variable quality of feedstock and range of product yields. This flexibility is somewhat reduced because of the tendency to omit the distillation section when heavy oil, extra heavy oils, and tar sand bitumen enter the refinery (Parkash, 2003; Gary et al., 2007; Speight, 2014a, 2017; Hsu and Robinson, 2017). Thus, refinery configurations can be adapted to the properties of the feedstocks which may dictate no distillation or a simple removal of any volatile constituents.

Generally, the maximum permissible temperature of the feedstock in the vaporizing furnace is the factor limiting the range of products in a single-stage (atmospheric) column. Thermal decomposition or cracking of the constituents begins as the temperature of the oil approaches 350°C (660°F) and the rate increases markedly above this temperature. However, the decomposition is time-dependent and temperatures on the order of 395°C (745°F) may be employed provided the residence time of the feedstocks in the hot zone does not cause thermal decomposition of the constituents. Thermal decomposition is generally regarded as being undesirable because the coke-like material produced tends to be deposited on the tubes with consequent formation of hot spots and eventual failure of the affected tubes. In the processing of lubricating oil stocks, an equally important consideration in the avoidance of these high temperatures is the deleterious effect on the lubricating properties. However, there are occasions when cracking distillation might be regarded as beneficial and the still temperature will be adjusted accordingly. In such a case, the products will be named accordingly using the prefix cracked, e.g., cracked residuum, in which case the term pitch is used incorrectly. By definition, pitch is a coal-derived product.

Based on chemical characteristics, a very approximate estimation of the potential for thermal decomposition of various feedstocks can be made using the Watson characterization factor (Speight, 2014a), K_w:

$$K_w = T_b^{1/3}/\text{sp gr}$$

where T_b is the mean average boiling point in degrees Rankine (490 plus degrees Fahrenheit) and sp gr is the specific gravity. The factor ranges from approximately 10 for paraffinic crude oil to approximately 15 for highly aromatic crude oil. On the assumption that the components of paraffinic crude oil are more thermally labile than the components of aromatic crude oil, it might be supposed that a relationship between the characterization factor and the temperature is viable. However, the relationship is so broad that it may not be sufficiently accurate to help the refiner. There are occasions when cracking distillation might be regarded as beneficial and the still temperature will be adjusted accordingly. In such a case, the products will be named accordingly using the prefix cracked, e.g., cracked residuum, in which case the arbitrary term pitch would be used.

The simplest refinery configuration, called a topping refinery, is designed to remove volatile constituents from a feedstock under simple conditions. It consists of tankage, a distillation unit, recovery facilities for gases and light hydrocarbons, and the necessary utility systems (steam, power, and water treatment plants). Topping refineries produce large quantities of unfinished oils and are highly dependent on local markets, but the addition of hydrotreating and reforming units to this basic configuration results in a more flexible hydroskimming refinery, which can also produce desulfurized distillate fuels and high-octane gasoline (Chapter 1) (Parkash, 2003; Hsu and Robinson, 2006; Gary et al., 2007; Speight, 2014a, 2017).

The most versatile configuration in the refinery is known as the conversion refinery which incorporates all of the basic building blocks found in both the topping and hydroskimming refineries, as well as features gas oil conversion plants such as catalytic cracking and hydrocracking units, olefin conversion plants such as alkylation or polymerization units, and frequently, coking units for sharply reducing or eliminating the production of residual fuels (Parkash, 2003; Hsu and Robinson, 2006; Gary et al., 2007; Speight, 2014a, 2017). Modern conversion refineries may produce two-thirds of their output as unleaded gasoline, with the balance distributed between high-quality jet fuel, liquefied petroleum gas (LPG), low-sulfur diesel fuel, and a small quantity of crude oil coke. Many such refineries also incorporate solvent extraction processes for manufacturing lubricants and petrochemical units with which to recover high-purity propylene, benzene, toluene, and xylenes for further processing into polymers.

A multitude of separations are accomplished by distillation, but its most important and primary function in the refinery is its use of the distillation tower and the temperature gradients therein

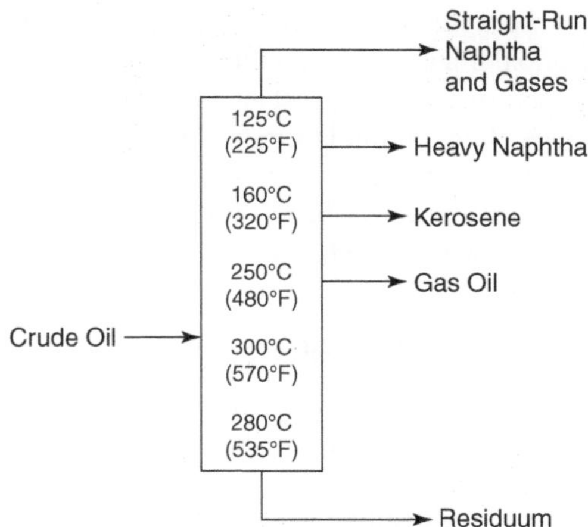

FIGURE 4.5 Representation of the temperature profiles in an atmospheric distillation tower. (N.B.: These may vary somewhat from refinery to refinery.)

(Figure 4.5) for the separation of crude oil into fractions that consist of varying amounts of different components (Speight, 2000, 2014a, 2015, 2017). Thus, it is possible to obtain products ranging from gaseous materials taken off the top of the distillation column to a non-volatile atmospheric residuum (atmospheric bottoms, reduced crude) with correspondingly lower boiling materials (gas, gasoline, naphtha, kerosene, and gas oil) taken off at intermediate points with each crude oil providing different amounts of various fractions (Diwekar, 1995; Jones, 1995; Speight, 2000, 2014a, 2017; Parkash, 2003; Hsu and Robinson, 2006; Gary et al., 2007).

The reduced crude may then be processed by vacuum or steam distillation to separate the high-boiling lubricating oil fractions without the danger of decomposition, which occurs at high (>350°C, 660°F) temperatures (Speight, 2000, 2014a). Indeed, atmospheric distillation may be terminated with a lower boiling fraction (boiling cut) if it is thought that vacuum or steam distillation will yield a better quality product or if the process appears to be economically more favorable.

It should be noted at this point that not all crude oils yield the same distillation products because of the differences in composition (Charbonnier et al., 1969; Coleman et al., 1978; Speight, 2012). In fact, the nature of the crude oil dictates the processes that may be required for refining. Crude oil can be classified according to the nature of the distillation residue, which in turn depends on the relative content of hydrocarbon types: paraffins, naphthenes, and aromatics. For example, a paraffin-base crude oil produces distillation cuts with higher proportions of paraffins than an asphalt-base crude. The converse is also true; that is, an asphalt-base crude oil produces materials with higher proportions of cyclic compounds. A paraffin-base crude oil yields wax distillates rather than the lubricating distillates produced by the naphthenic-base crude oils. The residuum from paraffin-base crude oil is referred to as cylinder stock rather than asphaltic bottoms which is the name often given to the residuum from distillation of naphthenic crude oil. It is emphasized that, in these cases, it is not a matter of the use of archaic terminology but a reflection of the nature of the product and the crude oil from which it is derived.

In summary, the distillation section comprising the atmospheric unit and the vacuum unit (Figure 4.2) is required to have the greatest flexibility in terms of variable quality of feedstock and range of product yields (Parkash, 2003; Gary et al., 2007; Speight, 2014a, 2017; Hsu and Robinson, 2017). The maximum permissible temperature of the feedstock in the vaporizing furnace is the factor limiting the range of products in a single-stage (atmospheric) column. Thermal decomposition or cracking of the constituents begins as the temperature of the oil approaches 350°C (660°F) and the rate increases markedly above this temperature. This thermal decomposition is generally regarded as being undesirable because the coke-like material produced tends to be deposited on the tubes with consequent formation of hot spots and eventual failure of the affected tubes. In the processing of lubricating oil stocks, an equally important consideration in the avoidance of these high temperatures is the deleterious effect on the lubricating properties. However, there are occasions when cracking distillation might be regarded as beneficial and the still temperature will be adjusted accordingly. In such a case, the products will be named accordingly using the prefix cracked, e.g., cracked residuum, in which case the term pitch (Chapter 1) would be used.

4.3.1 Distillation at Atmospheric Pressure

The present-day crude oil distillation unit is a collection of distillation units but, in contrast to the early battery units, a tower is used in the typical modern refinery and brings about a fairly efficient degree of fractionation (separation).

Understanding the operation of the distillation process requires an understanding that the atmospheric distillation unit is the main fractionator of the feedstock. The important characteristics that distinguish this tower are (1) all of the heat available for the distillation enters the tower with the feedstock that is produced from a fired heater or a preheat train, (2) the tower has multiple heat removal zones using either pump-arounds or pump-downs, (3) multiple side-stream products are produced in the tower, and more particularly, (4) feedstock preparation to the downstream vacuum unit adds a third performance parameter.

After passing through the dewatering and desalting processes, the feedstock is heated in heat exchangers or reboilers (preheating of crude oil) (Figure 4.3).

Typically, a low level of preheating is not sufficient, since the feedstock oil must be partially evaporated to the extent that all products except the constituents of the atmospheric residuum must be in the vapor phase when the oil enters the atmospheric column. Thus, the furnace is required to raise the temperature between 330°C and 385°C (625°F and 725°F) depending on the constituents of the feedstock. The partially evaporated crude oil is transferred to the flash zone column located at a point below the distillation column and above what is called the stripping section. The size of the distillation column is determined by the number of plates and the amount of steam required for a successful operation. In addition, the amount of steam is determined by the volatile constituent of the feedstock and, as a result, the rising steam may require a large diameter column above the flash zone. At the bottom of the section, water vapor is injected into the column to remove any lingering volatile constituents from the atmospheric residuum and to reduce the partial pressure of hydrocarbon vapors in the flash zone and ascend into the column through the plates to the upper zone of the column.

The heating unit is known as a pipe still heater or pipe still furnace, and the heating unit and the fractional distillation tower make up the essential parts of a distillation unit or pipe still. The pipe still furnace heats the feed to a predetermined temperature – usually a temperature at which a predetermined portion of the feed will change into vapor. The vapor is held under pressure in the pipe in the furnace until it is discharged as a foaming stream into the fractional distillation tower. The unvaporized or liquid portion of the feed descends to the bottom of the tower to be pumped away as a bottom non-volatile product, while the vaporized material passes up the tower to be fractionated into gas oils, kerosene, and naphtha.

The walls and ceiling of a pipe still furnace are insulated with firebrick and the interior of the furnace is partially divided into two sections: a smaller convection section where the oil first enters the furnace and a larger section (fitted with heaters) where the oil reaches its highest temperature.

Another 20th century innovation in distillation is the use of heat exchangers which are also used to preheat the feed to the furnace. These exchangers are bundles of tubes arranged within a shell so that a feedstock passes through the tubes in the opposite direction from a heated feedstock passing through the shell. By this means, cold crude oil is passed through a series of heat exchangers where hot products from the distillation tower are cooled before entering the furnace and serving as a heated feedstock. This results in a saving of heater fuel and it is a major factor in the economical operation of modern distillation units.

All of the primary fractions from a distillation unit are equilibrium mixtures and contain some of the lower boiling constituents that are characteristics of a lower boiling fraction. The primary fractions are stripped of these constituents (stabilized) before storage or further processing.

Distillation columns are the most commonly used separation units in a refinery. Operation is based on the difference in boiling temperatures of the liquid mixture components and on recycling counter-current gas–liquid flow. The properly organized temperature distribution up the column results in different mixture compositions at different heights. While multi-component inter-phase mass transfer is a common phenomenon for all column types, the flow regimes are very different depending on the internal elements used. The two main types are a tray column and a packed column, the latter equipped with either random or structured packing. Different types of distillation columns are used for different processes, depending on the desired liquid holdup, capacity (flow rates), and pressure drop but each column is a complex unit combining many structural elements.

The present-day crude oil distillation unit is, in fact, a collection of distillation units that enable a fairly efficient degree of fractionation to be achieved. In contrast to the early units, which consisted of separate stills, a tower is used in the modern-day refinery. In fact, of all the units in a refinery, the distillation unit is required to have the greatest flexibility in terms of variable quality of feedstock and range of product yields. Thus, crude oil can be separated into gasoline, kerosene, diesel oil, gas oil, and other products by distillation at atmospheric pressure. Distillation is an operation in which vapors rising through fractionating decks in a tower come into close contact with liquid descending across the decks so that higher boiling components condense and concentrate at the bottom of the tower while the lighter ones are concentrated at the top or pass overhead.

It is common practice to use furnaces to heat the feedstock only when distillation temperatures above 205°C (400°F) are required. Lower temperature (such as that used in the redistillation of naphtha and similar low-boiling products) is provided by heat exchangers and/or steam reboilers. Thus, the desalted feedstock is generally pumped to the unit directly from a storage tank, and it is important that charge tanks be drained completely free from water before charging to the unit. The crude feedstock is heat exchanged against whatever other heat sources are available to recover maximum heat before crude is charged to the heater, which ultimately supplies all the heat required for the operation of the distillation unit.

The feed to a fractional distillation tower is heated by flow through pipe arranged within a large furnace. It is at this point that the incompatibility and/or the instability of blended crude oil may manifest itself (Chapters 1 and 5). The heating unit is known as a pipe still heater or pipe still furnace, and the heating unit and the fractional distillation tower make up the essential parts of a distillation unit or pipe still. The pipe still furnace heats the feed to a predetermined temperature, usually a temperature at which a calculated portion of the feed changes into vapor. The vapor is held under pressure in the pipe still furnace until it discharges as a foaming stream into the fractional distillation tower. Here, the vapors pass up the tower to be fractionated into gas oil, kerosene, and naphtha, while the non-volatile or liquid portion of the feed descends to the bottom of the tower to be pumped away as a bottom product.

Pipe still furnaces vary greatly in size, shape, and interior arrangement and can accommodate 25,000 bbl or more of crude oil per day. The walls and ceiling are insulated with firebrick, and gas or oil burners are inserted through one or more walls. The interior of the furnace is partially divided into two sections: a smaller convection section where the oil first enters the furnace and a larger section into which the burners discharge and where the oil reaches its highest temperature.

Heat exchangers are also used to preheat the feedstock before it enters the furnace. Thus, in order to save heat, cold crude oil may play a significant role in efficient operation of refineries by passing through a series of heat exchangers where hot products from the distillation tower are cooled before entering the furnace.

Steam reboilers may take the form of a steam coil in the bottom of the fractional distillation tower or in a separate vessel. In the latter case, the bottom product from the tower enters the reboiler where part is vaporized by heat from the steam coil. The hot vapor is directed back to the bottom of the tower and provides part of the heat needed to operate the tower. The non-volatile product leaves the reboiler and passes through a heat exchanger, where its heat is transferred to the feed to the tower. Steam may also be injected into a fractional distillation tower not only to provide heat but also to induce boiling to take place at lower temperatures. Reboilers generally increase the efficiency of fractionation, but a satisfactory degree of separation can usually be achieved more conveniently by the use of a stripping section. The stripping operation occurs in that part of the tower below the point at which the feed is introduced. The more volatile components are stripped from the descending liquid. Above the feed point (the rectifying section), the concentration of the less volatile component in the vapor is reduced.

If the feedstock has not been sufficiently dewatered, any remaining water can be entrained in the feedstock and it will vaporize in the exchangers and in the heater, leading to a high pressure drop through that equipment. The decks in the fractionating column could be damaged if a slug of water is to be charged to the unit as the quantity of steam generated by its vaporization is so much greater than the quantity of vapor obtained from the same volume of oil. Water expands in volume 1600 times upon vaporization at 100°C (212°F) at atmospheric pressure.

The feed to a fractional distillation tower is heated by flow through pipe arranged within a large furnace. The heating unit is known as a pipe still heater or pipe still furnace, and the heating unit and the fractional distillation tower make up the essential parts of a distillation unit or pipe still. The pipe still furnace heats the feed to a predetermined temperature, usually at a temperature where a calculated portion of the feed changes into vapor. The vapor is held under pressure in the pipe still furnace until it discharges as a foaming stream into the fractional distillation tower. Here, the vapors pass up the tower to be fractionated into gas oil, kerosene, and naphtha, while the non-volatile or liquid portion of the feed descends to the bottom of the tower to be pumped away as a bottom product.

Heat exchangers are used to preheat the crude oil feedstock before entry into the distillation unit. In order to reduce the cost of operating a crude unit, as much heat as possible is recovered from the hot streams by exchanging heat with the cold crude charge. The number of heat exchangers within the crude unit and cross heat exchange with other units will vary with unit design. A record should be kept of heat exchanger outlet temperatures so that fouling can be detected and possibly corrected before the capacity of the unit is affected.

Heat exchangers are bundles of tubes arranged within a shell so that a stream passes through them in the opposite direction from a stream passing through the shell. Thus, in order to save heat, cold crude oil may play a significant role in efficient operation of refineries by passing through a series of heat exchangers where hot products from the distillation tower are cooled before entering the furnace.

Crude entering the flash zone of the fractionating column flashes into the vapor that rises up the column and the liquid residue that drops downward. This flash is a very rough separation; the vapors contain appreciable quantities of heavy ends, which must be rejected downward into reduced crude, while the liquid contains lighter products, which must be stripped out. During the distillation of crude oil, light naphtha and gases are removed as vapor from the top of the tower; heavy naphtha, kerosene, and gas oil are removed as side-stream products; and reduce crude is taken from the bottom of the tower.

Having the heater transfer temperature reset flow of fuel to the burners controls the heat input. The heater transfer temperature is merely a convenient control, and the actual temperature, which has no great significance, will vary from 320°C (610°F) to as high as 430°C (805°F), depending on the type of crude and the pressure at the bottom of the fractionating tower. It is noteworthy that if the quantity of gasoline and kerosene in a crude is reduced, the transfer temperature required for the same operation will be increased, even though the lift is less. However, at such temperatures, the residence time of the crude oil and its fractions exerts considerable influence on the potential for racking reactions to occur. This is particularly important in determining the properties of the atmospheric residuum.

External reflux that is returned to the top of the fractionator passes downward against the rising vapors. Lighter components of the reflux are vaporized and return to the top of the column, while the heavier components in the rising vapors are condensed and return down the column. Thus, there is an internal reflux stream flowing from the top of the fractionator all the way back to the flash zone and becoming progressively heavier as it descends.

The products with higher boiling rates than the overhead material are obtained by withdrawing portions of the internal reflux stream. The endpoint of a side-stream fraction (side cut) will depend on the quantity withdrawn. If the side-stream fraction withdrawal rate is increased, the extra product is the material that was formerly flowing down the fractionator as internal reflux. Since the internal reflux below the draw-off is reduced, heavier vapors can now rise to that point and result in a heavier product. Changing the draw-off rate is the manner in which side-stream fractions are kept on endpoint specifications. The temperature of the draw-off decks indicates the endpoint of the product drawn at that point, and the draw-off rate can be controlled to hold a constant deck temperature, and as a result, a product that meets specifications.

The degree of fractionation is generally judged by measuring the number of degrees centigrade between the 95% point of the lighter product and the 5% point of the heavier product. Both the initial boiling point (IBP) and the final boiling point (FBP) can be used but the initial boiling point varies with the intensity and efficiency of the stripping operation. Fractionation can be improved by increasing the reflux in the fractionator, which is done by raising the transfer temperature. There may be occasions when the internal reflux necessary to achieve satisfactory fractionation between the heavier products is so great that if it were supplied from the top of the fractionator, the upper decks would flood. An intermediate circulating reflux (ICR) solves this problem. Some internal reflux is withdrawn, pumped through a cooler or exchanger, and returned colder a few decks higher in the column. This cold oil return condenses extra vapors to liquid and increases the internal reflux below that point. Improvement in the fractionation between the light and heavy gas oil can be achieved by increasing the heater transfer temperature, which would cause the top reflux to increase, then restore the top reflux to its former rate by increasing the circulating reflux rate. Even though the heater transfer temperature is increased, the extra heat is recovered by exchange with crude oil feedstock and, as a result, the heater duty will only increase slightly.

Sometimes, a fractionator will be pulled dry insofar as the rate at which a product is being withdrawn is greater than the quantity of internal reflux in the fractionator. All the internal reflux then flows to the stripper, the decks below the draw-off run dry, and therefore, no fractionation takes place, while at the same time there is insufficient material to maintain the level in the stripper, and the product pump will tend to lose suction. It is then necessary to either lower the product withdrawal rate or increase the internal reflux in the tower by raising the transfer temperature or by reducing the rate at which the next lightest product is being withdrawn.

Pipe still furnaces vary greatly in size, shape, and interior arrangement and can accommodate 25,000 bbl or more of crude oil per day. The walls and ceiling are insulated with firebrick, and gas or oil burners are inserted through one or more walls. The interior of the furnace is partially divided into two sections: a smaller convection section where the oil first enters the furnace and a larger section into which the burners discharge and where the oil reaches its highest temperature.

It is common practice to use furnaces to heat the feedstock only when distillation temperatures above 205°C (400°F) are required. Lower temperature (such as that used in the redistillation of naphtha and similar low-boiling products) is provided by heat exchangers and/or steam reboilers.

The stripping section is necessary because the flashed residue in the bottom of the fractionator and the side-stream products have been in contact with lighter boiling vapors. These vapors must be removed to meet flash point specifications and to drive the light ends into lighter and (usually) more valuable products.

Steam, usually superheated steam, is used to strip these light ends. Generally, sufficient steam is used to meet a flash point specification and, while a further increase in the quantity of steam may raise the IBP of the product slightly, the only way to substantially increase the IBP of a specific product is to increase the yield of the next lighter product. Provided, of course, the fractionator has enough internal reflux to accomplish an efficient separation of the feedstock constituents.

All of the stripping steam that is condensed in the overhead receiver must be drained off because refluxing water will upset the balance of activities in the fractionator. If the endpoint of the overhead product is very low, water may not pass the overhead and instead will accumulate on the upper decks and flood the tower thereby reducing efficiency and perhaps even forcing the tower to shut down. If the latter occurs, and if distillation is the first (other than desalting) process to which a crude oil is subjected in a refinery, the economic consequences for the refinery operation can be substantial.

In simple refineries, cut points can be changed slightly to vary yields and balance products, but the more common practice is to produce relatively narrow fractions and then process (or blend) them to meet product demand. Since all these primary fractions are equilibrium mixtures, they all contain some proportion of the lighter constituents characteristic of a lower boiling fraction and so are stripped of these constituents, or stabilized, before further processing or storage. Thus, gasoline is stabilized to a controlled butanes–pentanes content, and the overhead may be passed to super-fractionators, towers with a large number of plates that can produce nearly pure C_1–C_4 hydrocarbons (methane to butanes, CH_4 to C_4H_{10}) – these successive columns are known as de-ethanizers, de-propanizers, de-butanizers, and whatever separation columns are still necessary.

Kerosene and gas oil fractions are obtained as side-stream products from the atmospheric tower (primary tower), and these are treated in stripping columns (i.e. vessels of a few bubble trays) into which steam is injected and the volatile overhead from the stripper is returned to the primary tower. Steam is usually introduced by the stripping section of the primary column to lower the temperature at which fractionation of the heavier ends of the crude can occur.

The specifications for most crude oil products make it extremely difficult to obtain marketable material by distillation alone. In fact, the purpose of atmospheric distillation is to produce fractions that serve as feedstocks for intermediate refining operations and for blending. Generally, this process is carried out at atmospheric pressure, while light crude oils may be topped at an elevated pressure and the residue distilled at atmospheric pressure.

The topping operation differs from normal distillation procedures insofar as the majority of the heat is directed to the feed stream rather than reboiling the material in the base of the tower. In addition, products of volatility intermediate between that of the overhead fractions and bottoms (residua) are withdrawn as side-stream products. Furthermore, steam is injected into the base of the column and the side-stream strippers to adjust and control the fractions' initial boiling range (or point). Topped crude oil must always be stripped with steam to elevate the flash point or to recover the final portions of gas oil. The composition of the topped crude oil is a function of the temperature of the vaporizer (or flasher).

These characteristics necessitate that the main fractionators have closely related heat and material balances. Understanding their operation requires tracking how heat and material balances affect each other. However, the most important consequence for the analysis of atmospheric distillation in the atmospheric distillation unit is the limitation of heat input. The combination of (1) the heat limits of the feedstock and (2) the operating pressure sets the maximum vaporization in the flash zone (i.e. tower feed entry area). Most atmospheric distillation units operate at a specified feedstock limit, either maximum duty input or maximum temperature.

All products are cooled before being sent to storage. Low-boiling products should be restrained to temperatures below 60°C (140°F) in order to reduce vapor losses in storage, but the need to store higher boiling products below such a temperature is not as acute, unless facile oxidation of the product at higher temperatures is possible. If a product is being charged to another unit as feedstock, there may be an advantage in transmitting the hot product to the unit. However, caution is advised if there is a chance that a product leaving a unit at temperatures in excess of 100°C (212°F) would enter a tank with a water bottom. The hot oil could readily boil the water and cause the roof to detach from the tank, perhaps violently!

4.3.2 DISTILLATION AT REDUCED PRESSURE

Distillation under reduced pressure (more commonly referred to as vacuum distillation; Figure 4.4) as used in the crude oil refining industry, was developed to separate the less volatile products, such as lubricating oils, from the crude oil without subjecting these high-boiling products to cracking conditions.

The boiling point of the heaviest cut obtainable at atmospheric pressure is limited by the temperature (ca. 350°C; ca. 660°F) at which the residue starts to decompose (crack). When the feedstock is required for the manufacture of lubricating oils, further fractionation without cracking is desirable and this can be achieved by distillation under vacuum conditions.

From the atmospheric unit, the residuum (*bottoms*) fraction is sent to a vacuum distillation unit (Figure 4.4) where the operating conditions are usually 50–100 mm of mercury (atmospheric pressure: 760 mm of mercury). In order to minimize the potential for large fluctuations in pressure during vacuum distillation, the vacuum tower is necessarily of a larger diameter but generally lesser height than the atmospheric tower – some vacuum distillation units have diameters on the order of 45 feet. By this means, a high-boiling gas oil fraction (heavy gas oil, vacuum gas oil) is obtained as an overhead product at temperatures of about 150°C (300°F), and a higher boiling fraction (lubricating oil fraction) may be obtained at temperatures of 250°C–350°C (480°F–660°F). As for the atmospheric tower, caution is taken to prevent exposure of the feedstock to a temperature on the order of 370°C (700°F) with a controlled residence time to prevent undesirable cracking – unless some cracking is a desired option. As is the case with the atmospheric tower, steam is also injected into the vacuum tower. This helps reduce the partial pressure of the products and facilitates distillation in addition to stripping the residuum of any heavy gas oil (or lubricating oil) constituents.

The unit employed for distillation at reduced pressure (more commonly referred to as the vacuum distillation unit) consists of the vacuum furnace, vacuum tower, and the vacuum producing system. The topped crude (the non-volatile residue) for the atmospheric distillation unit is heated up in the vacuum furnace to a temperature that is controlled to be just below the temperature of thermal decomposition which is on the order of 385°C–455°C (730°F–850°F). The furnace outlet temperature is selected depending on the thermal reactivity (or coking propensity of the feedstock) and the desired level of separation in the column.

Steam ejectors or vacuum pumps are used to create a vacuum for evaporation of the light (low-boiling, relatively low-density) vacuum gas oil and heavy (high-boiling, relatively high-density) vacuum gas oil fractions. The temperature and pressure in the vacuum distillation unit also (like the atmospheric distillation unit) depend on whether steam is introduced, or whether the separation is carried out without the steam addition in the so-called dry towers, and operate at a reduced pressure on the order of 10–30 mm Hg (atmospheric pressure = 760 mm Hg) at the bottom of the tower. However, lower pressures and higher temperatures are used in dry towers. To minimize the pressure difference between the bottom and top of the column, some special packing materials are used instead of trays for providing contact between liquid and vapor streams to improve fractionation.

The fractions obtained by vacuum distillation of the *reduced crude* (*atmospheric residuum*) from an atmospheric distillation unit depend on whether or not the unit is designed to produce lubricating oil or VGO. In the former case, the overhead fractions include (1) heavy gas oil, which is used as feedstock for a catalytic cracking unit or, after suitable treatment, as lubricating oil blend stock, (2) lubricating oil, which is obtained as a side-stream product and can be separated further as three fractions: light, intermediate, and heavy lubricating oil, and (3) vacuum residuum (sometimes incorrectly called *asphalt*), which is used to produce asphalt or which may also be blended with gas oil to produce fuel oil. As with the atmospheric residuum, the vacuum residuum is not always stable and may suffer from instability and/or incompatibility issues (Speight, 2014a). Such issues will be evident if the vacuum residuum is used for asphalt preparation where the instability of the asphalt will be manifested through the tendency of the asphalt to separate from the aggregate because of poor binding characteristics.

Although a single fraction (cut) of VGO is allowed in some cases, producing drawing LVGO (low-boiling or low-density vacuum gas oil) and HVGO (high-boiling or high-density vacuum gas oil) separately is often more beneficial because the resultant HVGO temperature is produced at a temperature on the order of 90°C–120°C (160°F–125°F) higher than the corresponding draw temperature of a single VGO cut. Any more volatile constituents are removed from the residue by steam stripping. In addition, coke formation is reduced by circulating partially cooled bottoms to quench the liquid to a lower temperature. Because the heavy (high-boiling) crude fraction contains metal complexes (asphaltene constituents and porphyrins), which are catalyst poisons for downstream processes, sometimes a recirculation of wash oil in the bottom of the unit may be included to prevent these compounds from incorporating into the HVGO.

In some cases, the fractions obtained by vacuum distillation of the reduced crude (atmospheric residuum) from an atmospheric distillation unit depend on whether or not the unit is designed to produce lubricating oil or VGO. In the former case, the fractions include (1) heavy gas oil, which is an overhead product and is used as catalytic cracking stock or, after suitable treatment, a light lubricating oil; (2) lubricating oil (usually three fractions – light, intermediate, and heavy), which is obtained as a side-stream product; and (3) residuum, sometimes incorrectly referred to as asphalt which is the non-volatile product (Table 4.1) and may be used directly as, or to produce, asphalt and which may also be blended with gas oils to produce a heavy fuel oil.

In order to minimize large fluctuations in pressure in the vacuum tower, the units are necessarily of a larger diameter than the atmospheric units. Some vacuum distillation units have diameters on the order of 45 feet (14 m). By this means, a heavy gas oil may be obtained as an overhead product at temperatures of approximately 150°C (300°F), and lubricating oil cuts may be obtained at temperatures of 250°C–350°C (480°F–660°F), provided feed and residue temperatures are kept below the temperature of 350°C (660°F), the temperature above which cracking will occur. The partial

---

TABLE 4.1
Properties of various residua

Residuum	Gravity (°API)	Sulfur (% w/w)	Nitrogen (% w/w)	Nickel (ppm)	Vanadium (ppm)	Asphaltenes* (% w/w)	Carbon Residue** (% w/w)
Arabian Light. >650 F	17.7	3.0	0.2	10.0	26.0	1.8	7.5
Arabian Light, >1050 F	8.5	4.4	0.5	24.0	66.0	4.3	14.2
Arabian Heavy, >650 F	11.9	4.4	0.3	27.0	103.0	8.0	14.0
Arabian Heavy, >1050 F	7.3	5.1	0.3	40.0	174.0	10.0	19.0
Alaska, North Slope, >650 F	15.2	1.6	0.4	18.0	30.0	2.0	8.5
Alaska, North Slope, >1050 F	8.2	2.2	0.6	47.0	82.0	4.0	18.0
Kuwait, >650 F	13.9	4.4	0.3	14.0	50.0	2.4	12.2
Kuwait, >1050 F	5.5	5.5	0.4	32.0	102.0	7.1	23.1
Lloydminster (Canada), >650 F	10.3	4.1	0.3	65.0	141.0	14.0	12.1
Lloydminster (Canada), >1050 F	8.5	4.4	0.6	115.0	252.0	18.0	21.4
Taching, >650 F	27.3	0.2	0.2	5.0	1.0	4.4	3.8
Taching, >1050 F	21.5	0.3	0.4	9.0	2.0	7.6	7.9
Tia Juana, >650 F	17.3	1.8	0.3	25.0	185.0		9.3
Tia Juana, >1050 F	7.1	2.6	0.6	64.0	450.0		21.6

Heptane asphaltenes.
Conradson carbon residue; ASTM D189.
Please use tear sheet from:
Speight, J.G. 2017. Handbook of Petroleum Refining. CRC Press, Taylor & Francis Group, Boca Raton, FL. Table 7.7, p. 269.

pressure of the hydrocarbons is effectively reduced still further by the injection of steam. The steam added to the column, principally for the stripping of asphalt in the base of the column, is superheated in the convection section of the heater.

The boiling range of the highest boiling fraction that can be produced at atmospheric pressure is limited by the temperature at which the residue starts to decompose or crack. If the atmospheric residuum is required for the manufacture of lubricating oils, further fractionation without cracking may be desirable, and this may be achieved by distillation under vacuum. The residua produced by distillation under reduced pressure have properties markedly different from the residua produced by distillation at atmospheric pressure.

Vacuum distillation as applied to the crude oil refining industry is a technique that has seen wide use in crude oil refining. Vacuum distillation evolved because of the need to separate the less volatile products, such as lubricating oils, from the crude oil without subjecting these high boiling products to cracking conditions. The boiling point of the heaviest cut obtainable at atmospheric pressure is limited by the temperature (ca. 350°C; ca. 660°F) at which the residue starts to decompose or crack, unless cracking distillation is preferred. When the feedstock is required for the manufacture of lubricating oils, further fractionation without cracking is desirable and this can be achieved by distillation under vacuum conditions.

The distillation of high-boiling lubricating oil stocks may require pressures as low as 15–30 mm Hg (0.29–0.58 psi), but operating conditions are more usually 50–100 mm Hg (0.97–1.93 psi). Volumes of vapor at these pressures are large and pressure drops must be small to maintain control, so vacuum columns are necessarily of large diameter. Differences in vapor pressure of different fractions are relatively larger than for lower boiling fractions, and relatively few plates are required. Under these conditions, high-boiling gas oil may be obtained as an overhead product at temperatures of approximately 150°C (300°F). Lubricating oil fractions may be obtained as side-stream products at temperatures of 250°C–350°C (480°F–660°F). The feedstock and residue temperatures being kept below the temperature of 350°C (660°F), above which the rate of thermal decomposition increases and cracking occurs (Parkash, 2003; Gary et al., 2007; Speight, 2014a, 2017; Hsu and Robinson, 2017). The partial pressure of the hydrocarbon derivatives is effectively reduced yet further by the injection of steam. The steam added to the column, principally for the stripping of the higher molecular weight constituents (such as the residuum) in the base of the column, is superheated in the convection section of the heater.

At the point where the heated feedstock is introduced in the vacuum column (the flash zone), the temperature should be high and the pressure as low as possible to obtain maximum distillate yield. The flash temperature is restricted to approximately 420°C (790°F), however, in view of the cracking tendency of the feedstock constituents. Vacuum is maintained with vacuum ejectors and lately also with liquid ring pumps. In the older type high vacuum units, the required low hydrocarbon partial pressure in the flash zone could not be achieved without the use of lifting steam that acts in a similar manner as the stripping steam of atmospheric distillation units. This type of unit is often referred to as a wet unit.

One of the latest developments in vacuum distillation has been the deep vacuum flashers in which no steam is required. These dry units operate at very low flash zone pressures and low pressure drops over the column internals. For that reason, the conventional reflux sections with fractionation trays have been replaced by low pressure-drop spray sections. Cooled reflux is sprayed via a number of specially designed spray nozzles in the column counter-current to the up-flowing vapor. This spray of small droplets comes into close contact with the hot vapor, resulting in good heat and mass transfer between the liquid and the vapor phase.

When trays similar to those used in the atmospheric column are used in vacuum distillation, the column diameter may be extremely high, up to 45 feet. To maintain low pressure drops across the trays, the liquid seal must be minimal. The low holdup and the relatively high viscosity of the liquid limit the tray efficiency, which tends to be much lower than in the atmospheric column. The vacuum is maintained in the column by removing the non-condensable gas that enters the column by way of the feed to the column or by leakage of air.

The fractions obtained by vacuum distillation of reduced crude depend on whether the run is designed to produce lubricating oil or VGO. In the former case, the fractions include (1) high-boiling gas oil, an overhead product and is used as catalytic cracking stock or, after suitable treatment, a low-boiling lubricating oil; (2) gas oil, which is usually three fractions: low-boiling gas oil, intermediate-boiling gas oil, and high-boiling gas oil that are obtained as side-stream products; and (3) the residuum, which is the non-volatile product that may be used as asphalt directly or converted to asphalt by, for example, oxidation to asphalt that meets the necessary specifications.

The residuum may also be used as a feedstock for a coking unit or blended with gas oils to produce a high-boiling fuel oil (Parkash, 2003; Gary et al., 2007; Speight, 2014a, 2017; Hsu and Robinson, 2017). However, if the reduced crude is not required as a source of lubricating oils, the lubricating and high-boiling gas oil fractions are combined or, more likely, removed from the residuum as one fraction and used as a catalytic cracking feedstock.

The continued use of atmospheric and vacuum distillation has been a major part of refinery operations during this century and will undoubtedly continue to be employed, at least into the early decades of the 21st century, as the primary refining operation.

Three types of high-vacuum units for long residue upgrading have been developed for commercial application: (1) feedstock preparation units, (2) high-vacuum units, which are used to produce lubricating oil, and (3) high-vacuum units, which are used for the production of residuum for asphalt production.

The feedstock preparation units make a major contribution to deep conversion upgrading and produce distillate feedstocks for further upgrading in catalytic crackers, hydrocracking units, and coking units. To obtain an optimum waxy distillate quality, a wash oil section is installed between feed flash zone and waxy distillate draw-off. The wash oil produced is used as fuel component or recycled to feed. The flashed residue (short residue) is cooled by heat exchange against long residue feed. A slipstream of this cooled short residue is returned to the bottom of the high-vacuum column as quench to minimize cracking (maintain low bottom temperature).

High-vacuum units for the production of lubricating oil are specifically designed to produce high-quality distillate fractions and, therefore, precautions are taken to prevent thermal degradation of the distillates produced. The units are of the wet type. Normally, three sharply fractionated distillates are produced (spindle oil, low-density machine oil, and medium density machine oil). Cut points between those fractions are typically controlled on their viscosity quality. Spindle oil and low-density light machine oil are subsequently steam-stripped of any steam-volatile constituents in dedicated strippers. The distillates are further processed to produce lubricating base oil. The short residue is normally used as feedstock for the solvent deasphalting process to produce deasphalted oil, an intermediate for bright stock manufacturing. High-vacuum units for asphalt production are designed to produce straight-run asphalt and/or feedstocks for residuum blowing to produce blown asphalt that meets specifications. In principle, these units are designed on the same basis as feed preparation units, which may also be used to provide feedstocks for asphalt manufacturing.

Deep cut vacuum distillation that involves a revamp of the vacuum distillation unit to cut deeper into the residue is one of the first options available to the refiner. In addition to the limits of the major equipment, other constraints include (1) the VGO quality specification required by downstream conversion units, (2) the minimum flash zone pressure achievable, and (3) the maximum heater outlet temperature achievable without excessive cracking. These constraints typically limit the cut point (true boiling point) to 560°C–590°C (1,040°F–1,100°F) although units are designed for cut points (true boiling point) as high as 625°C (1,160°F).

In order to maximize the production of gas oil and lighter (lower boiling) components from the residuum of an atmospheric distillation unit (reduced crude), the residuum can be further distilled in a vacuum distillation unit. Residuum distillation is conducted at a low pressure in order to avoid thermal decomposition or cracking at high temperatures. A stock that boils at 400°C (750°F) at 0.1 psi (50 mm) would not boil until approximately 500°C (930°F) at atmospheric pressure and crude oil constituents commence thermal decomposition (cracking) at approximately 350°C (660°F)

(Speight, 2000, 2014a, 2017; Parkash, 2003; Gary et al., 2007; Hsu and Robinson, 2017). In the vacuum unit, almost no attempt is made to fractionate the products. It is only desired to vaporize the gas oil, remove the entrained residuum, and condense the liquid product as efficiently as possible. Vacuum distillation units that produce lubricating oil fractions are completely different in both design and operation.

In the vacuum tower, the reduced crude is charged through a heater into the vacuum column in the same manner as whole crude is charged to an atmospheric distillation unit. Although the flash zone of an atmospheric column may be at 14–18.5 psi (760–957 mm), the pressure in a vacuum column is very much lower, generally ranging from less than 0.8 psi (less than 40 mm Hg) in the flash zone to less than 0.2 psi (less than 10 mm Hg) at the top of the vacuum tower. The vacuum heater transfer temperature is generally used for control, even though the pressure drop along the transfer line makes the temperature at that point somewhat meaningless. The flash zone temperature has much greater significance.

The heater transfer and flash zone temperatures are generally varied to meet the vacuum bottoms specification, which is probably either a gravity (or viscosity) specification for fuel oil or a penetration specification for asphalt. The penetration of an asphalt is the depth in 1/100 cm to which a needle weighing 100 g sinks into a sample at 25°C (77°F) in 5 seconds (ASTM D5); the lower the penetration, the heavier the residuum or asphalt. If the flash zone temperature is too high, the crude can start to crack and produce gases that overload the ejectors and break the vacuum. It is then necessary to lower the temperature, and if a heavier residuum product is still required, an attempt should be made to obtain a better vacuum.

Slight cracking may occur without seriously affecting the vacuum and the occurrence of cracking can be established by a positive result from the Oliensis Spot Test (Speight, 2014a, 2015). This test is a convenient laboratory test that indicates the presence of cracked components as sediment by the separation of the sediment when a 20% solution of asphalt in naphtha is dropped on a filter paper. However, some crude oils yield a residuum that exhibits a positive test for the presence of sediment (solid phase) in the residuum. If a negative test result is required, operation at the highest vacuum and lowest temperature should be attempted. Since the degree of cracking depends on both the temperature and the time (residence time in the hot zone) during which the oil is exposed to that temperature, the level of the residuum in the bottom of the tower should be kept to a minimum and its temperature should be reduced by recirculating some of the residuum from the outlet of the residuum/crude oil heat exchanger to the bottom of the column. Quite often, when the level of the residuum rises, the column vacuum falls because of cracking due to increased residence time.

The flash zone temperature will vary widely and is dependent on the source of the crude oil, residuum specifications, the quantity of product taken overhead, and the flash zone pressure, and temperatures from below 315°C (600°F) to more than 425°C (800°F) have been used in commercial operations. Some vacuum distillation units are provided with facilities to strip the residuum with steam, which tends to lower the temperature required to fulfill an asphalt specification, but an excessive quantity of steam will overload the jets.

The distillation of high-boiling lubricating oil stocks may require pressures as low as 0.29–0.58 psi (15–30 mm Hg), but operating conditions are more usually 0.97–1.93 psi (50–100 mm Hg). Volumes of vapor at these pressures are large and pressure drops must be small to maintain control, so vacuum columns are necessarily of large diameter. Differences in vapor pressure of different fractions are relatively larger than for lower boiling fractions, and relatively few plates are required. Under these conditions, a heavy (high-boiling) gas oil may be obtained as an overhead product at temperatures of approximately 150°C (300°F). Fraction suitable for lubricating oil production may be obtained as side-stream products at temperatures of 250°C–350°C (480°F–660°F). The feedstock and residue temperatures are kept below the temperature of 350°C (660°F), above which the rate of thermal decomposition (cracking) increases (Speight, 2000, 2014a). The partial pressure of the hydrocarbons is effectively reduced yet further by the injection of steam. The steam added to the column, principally for the stripping of bitumen in the base of the column, is superheated in the convection section of the heater.

When trays similar to those used in the atmospheric column are used in vacuum distillation, the column diameter may be extremely high, up to 45 feet (14 m). To maintain low pressure drops across the trays, the liquid seal must be minimal. The low holdup and the relatively high viscosity of the liquid limit the tray efficiency, which tends to be much lower than in the atmospheric column. The vacuum is maintained in the column by removing the non-condensable gas that enters the column by way of the feed to the column or by leakage of air.

The fractions obtained by vacuum distillation of reduced crude depend on whether the run is designed to produce lubricating oil or VGO. In the former case, the fractions include (1) heavy gas oil, an overhead product and is used as catalytic cracking stock or, after suitable treatment, a light lubricating oil; (2) lubricating oil (usually three fractions: light, intermediate, and heavy), obtained as a side-stream product; and (3) residuum, the non-volatile product that may be used directly as asphalt or converted to asphalt. The residuum may also be used as a feedstock for a coking operation or blended with gas oils to produce a heavy fuel oil. However, if the reduced crude is not required as a source of lubricating oils, the lubricating and heavy gas oil fractions are combined or, more likely, removed from the residuum as one fraction and used as a catalytic cracking feedstock.

Three types of high-vacuum units for long residue upgrading have been developed for commercial application: (1) feedstock preparation units, (2) lube oil high-vacuum units, and (3) high-vacuum units for asphalt production.

The feedstock preparation units make a major contribution to deep conversion upgrading and produce distillate feedstocks for further upgrading in catalytic crackers, hydrocracking units, and coking units. To obtain an optimum waxy distillate quality, a wash oil section is installed between feed flash zone and waxy distillate draw-off. The wash oil produced is used as fuel component or recycled to feed. The flashed residue (short residue) is cooled by heat exchange against long residue feed. A slip-stream of this cooled short residue is returned to the bottom of the high-vacuum column as quench to minimize cracking (maintain low bottom temperature). Other units are also available and in use.

For example, lube oil high-vacuum units are specifically designed to produce high-quality distillate fractions for lube oil manufacturing. Special precautions are therefore taken to prevent thermal degradation of the distillates produced. The units are of the wet type. Normally, three sharply fractionated distillates are produced (spindle oil, light machine oil, and medium machine oil). Cut points between those fractions are typically controlled by their viscosity quality. Spindle oil and light machine oil are subsequently steam-stripped in dedicated strippers. The distillates are further processed to produce lubricating base oil. The short residue is normally used as feedstock for the solvent deasphalting process to produce deasphalted oil, an intermediate for bright stock manufacturing. High-vacuum units for asphalt production are designed to produce straight-run asphalt and/or feedstocks for residuum blowing to produce blown asphalt that meets specifications. In principle, these units are designed on the same basis as feed preparation units, which may also be used to provide feedstocks for asphalt manufacturing.

Deep cut vacuum distillation that involves a revamp of the vacuum distillation unit to cut deeper into the residue is one of the first options available to the refiner. In addition to the limits of the major equipment, other constraints include (1) the VGO quality specification required by downstream conversion units, (2) the minimum flash zone pressure achievable, and (3) the maximum heater outlet temperature achievable without excessive cracking. These constraints typically limit the cut point (true boiling point) to 560°C–590°C (1040°F–1100°F) although units are designed for cut points (true boiling point) as high as 627°C (1,160°F).

The fractions obtained by vacuum distillation of reduced crude depend on whether the run is designed to produce lubricating oil or VGO. In the former case, the fractions include (1) heavy gas oil, an overhead product and is used as catalytic cracking stock or, after suitable treatment, a light lubricating oil; (2) lubricating oil (usually three fractions: light, intermediate, and heavy), obtained as a side-stream product; and (3) residuum, the non-volatile product that may be used directly as asphalt or converted to asphalt. The residuum may also be used as a feedstock for a coking operation or blended with gas oils to produce a heavy fuel oil.

If the reduced crude is not required as a source of lubricating oils, the lubricating and heavy gas oil fractions are combined or, more likely, removed from the residuum as one fraction and used as a catalytic cracking feedstock. The continued use of atmospheric and vacuum distillation has been a major part of refinery operations during this century and will undoubtedly continue to be employed, at least into the early decades of the 21st century, as the primary refining operation.

The vacuum residuum (vacuum bottoms) must be handled more carefully than most refinery products since the pumps that handle hot heavy material have a tendency to lose suction. Recycling cooled residuum to the column bottom thereby reducing the tendency of vapor to form in the suction line can minimize this potential situation. It is also important that the residuum pump be sealed in such a manner so as to prevent the entry of air. In addition, because the majority of vacuum residua remain solid at ambient temperature, all vacuum residua handling equipment must either be kept active or flushed out with gas oil when it is shut down. In cases where steam tracing alone is insufficient to maintain the residuum fluid, high-pressure steam should be applied.

The vacuum residuum from a vacuum tower is sometimes cooled in open box units, as shell-and-tube units are not efficient in this service. It is often desirable to send residuum to storage at high temperature to facilitate blending. If it is desired to increase the temperature of the residuum, it is better to do so by lowering the level of water in the open box rather than by lowering the water temperature. If the water in the box is too cold, the residuum can solidify on the inside wall of the tube and insulate the hot residuum in the central core from the cooling water. Lowering the water temperature can actually result in a hotter product. When the residuum is sent to storage at over 100°C (212°F), care should be taken to ensure that the tank is absolutely free from water. Residuum coolers should always be flushed out with gas oil immediately once the residuum flow stops, since melting the contents of a cooler is a slow process.

The vapor rising above the flash zone will entrain some residuum that cannot be tolerated in cracking unit charge. The vapor is generally washed with gas oil product and sprayed into the slop wax section. The mixture of gas oil and entrained residuum is known as slop wax, and it is often circulated over the decks to improve contact, although the circulation rate may not be critical. The final stage of entrainment removal is obtained by passing the rising vapors through a metallic mesh demister blanket through which the fresh gas oil is sprayed.

Most of the gas oil spray is vaporized by the hot rising vapors and returned up the column. Some slop wax must be yielded in order to reject the captured entrainment. The amount of spray to the demister blanket is generally varied so that the yield of slop wax necessary to maintain the level in the slop wax pan is approximately 5% of the charge. If the carbon residue or the metals content of the HVGO is high, a greater percentage of slop wax must be withdrawn or circulated. Variation in the color of the gas oil product is a valuable indication of the effectiveness of entrainment control.

Slop wax is a mixture of gas oil and residuum, and it can be recirculated through the heater to the flash zone and reflashed. If, however, a crude contains volatile metal compounds, these will be recycled with the slop wax and can finally rise into the gas oil. Where there is an issue with volatile metals, it is necessary either to yield slop wax as a product or to make lighter asphalt, which will contain the metal compounds returned with the slop wax.

The scrubbed vapor rising above the demister blanket is the product, and no further fractionation is required. It is only desired to condense these vapors as efficiently as possible. This could be done in a shell-and-tube condenser, but as these are inefficient at low pressures, the high pressure drop through such a condenser would raise the flash zone pressure. The most efficient method is to contact the hot vapors with liquid product that has been cooled by pumping through heat exchangers.

Finally, confusion often arises because of different scales used to measure the vacuum. Positive pressures are commonly measured as kilograms per square centimeter gauge, which are kilograms per square centimeter above atmospheric pressure, which is 1.035 kg/cm^2 or 14.7 psi. Another means of measurement is to measure in millimeters of mercury in which atmospheric pressure (sea level) is 760 mm of mercury absolute and a perfect vacuum is 0 mm absolute.

4.4 DISTILLATION TOWERS

Distillation towers (distillation columns) are made up of several components, each of which is used either to transfer heat energy or enhance material transfer. A typical distillation column consists of several major parts: (1) a vertical shell where the separation of the components is carried out, (2) column internals such as trays, or plates, or packings that are used to enhance component separation, (3) a reboiler to provide the necessary vaporization for the distillation process, (4) a condenser to cool and condense the vapor leaving the top of the column, and (5) a reflux drum to hold the condensed vapor from the top of the column so that liquid (reflux) can be recycled back to the column. The vertical shell, which houses the column internals, together with the condenser and the reboiler constitutes a distillation column.

In a crude oil distillation unit, the feedstock liquid mixture is introduced usually near the middle of the column to a tray known as the feed tray. The feed tray divides the column into a top (enriching, rectification) section and a bottom (stripping) section. The feed flows down the column where it is collected at the bottom in the reboiler. Heat is supplied to the reboiler to generate vapor. The source of heat input can be any suitable fluid, although in most chemical plants this is normally steam. In refineries, the heating source may be the output streams of other columns. The vapor raised in the reboiler is reintroduced into the unit at the bottom of the column. The liquid removed from the reboiler is known as the bottoms.

The vapor moves up the column, and as it exits the top of the unit, it is cooled by a condenser. The condensed liquid is stored in a holding vessel known as the reflux drum. Some of this liquid is recycled back to the top of the column and this is called the reflux. The condensed liquid that is removed from the system is known as the distillate or top product. Thus, there are internal flows of vapor and liquid within the column as well as external flows of feeds and product streams, into and out of the column.

The tower is divided into a number of horizontal sections by metal trays or plates, and each is the equivalent of a still. The more trays, the more redistillation, and hence, the better the fractionation or separation of the mixture fed into the tower. A tower for fractionating crude oil may be 13 feet in diameter and 85 feet high according to a general formula:

$$c = 220d^2r$$

In this equation, c is the capacity in bbl/day, d is the diameter in feet, and r is the amount of residuum expressed as a fraction of the feedstock. Therefore, the more trays, the more redistillation, and hence, the better the fractionation or separation of the mixture fed into the tower.

A tower for fractionating crude oil may be 13 feet in diameter and 85 feet high, but a tower stripping unwanted volatile material from gas oil may be only 3 or 4 feet in diameter and 10 feet high. Towers concerned with the distillation of liquefied gases are only a few feet in diameter but may be up to 200 feet in height. A tower used in the fractionation of crude oil may have from 16 to 28 trays, but one used in the fractionation of liquefied gases may have 30–100 trays. The feed to a typical tower enters the vaporizing or flash zone, an area without trays. The majority of the trays are usually located above this area. The feed to a bubble tower, however, may be at any point from top to bottom with trays above and below the entry point, depending on the kind of feedstock and the characteristics desired in the products.

Liquid is collected on each tray to a depth of, say, several inches and the depth is controlled by a dam or weir. As the liquid level rises, excess liquid spills over the weir into a channel (downspout), which carries the liquid to the tray below.

The temperature of the trays is progressively cooler from bottom to top (Figure 4.5).

The bottom tray is heated by the incoming heated feedstock, although in some instances a steam coil (reboiler) is used to supply additional heat. As the hot vapors pass upward in the tower, condensation occurs onto the trays until refluxing (simultaneous boiling of a liquid and condensing of the

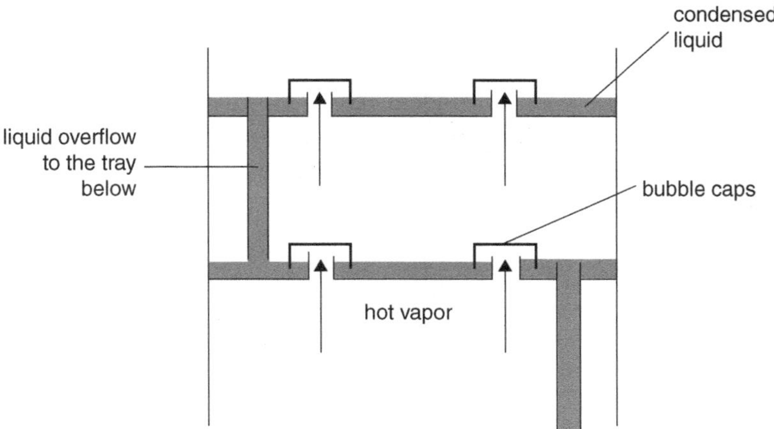

FIGURE 4.6 Representation of a bubble-cap tray.

vapor) occurs on the trays. Vapors continue to pass upward through the tower, while the liquid on any particular trays spills onto the tray below, and so on, until the heat at a particular point is too intense for the material to remain liquid. It then becomes vapor and joins the other vapors passing upward through the tower. The whole tower thus simulates a collection of several (or many) stills, with the composition of the liquid at any one point or on any one tray remaining fairly consistent. This allows part of the refluxing liquid to be tapped off at various points as side-stream products. Thus, during the distillation of crude oil, low-boiling naphtha and gases are removed as vapor from the top of the tower; high-boiling naphtha, kerosene, and gas oil are removed as side-stream products; and reduce crude is taken from the bottom of the tower.

The efficient operation of the distillation, or fractionating, tower requires the rising vapors to mix with the liquid on each tray. This is usually achieved by installing a short chimney on each hole in the plate and a cap with a serrated edge (bubble cap, hence bubble-cap tower) over each chimney (Figure 4.6). The cap forces the vapors to go below the surface of the liquid and to bubble up through it. Since the vapors may pass up the tower at substantial velocities, the caps are held in place by bolted steel bars.

In addition, a pump-around removes heat from the distillation column and lowers the vapor/liquid ratio in the section of the tower above the pump-around return. The pump-around rate is adjusted to control the temperature in a section of the tower. The energy efficiency of the distillation operation is also improved by using pump-around reflux. If sufficient reflux were produced in the overhead condenser to provide for all side-stream draw-offs as well as the required reflux, all of the heat energy would be exchanged at the bubble-point temperature of the overhead stream. By using pump-around reflux at lower points in the column, the heat transfer temperatures are higher and a higher fraction of the heat energy can be recovered by preheating the feedstock.

More specifically, pump-arounds take material from a lower tray, use it to preheat feed to a tower, and return the cooled liquid back to higher trays, providing internal reflux and rectification. Pump-arounds not only provide preheat/heat recovery capability but can also provide the same or greater flowrate through a column, reducing diameter and dechoking a process. On the other hand, pump-downs (sometimes referred to as pump-backs) are used to draw material from a higher tray down to a lower tray in order to aid internal rectification. This will also even out tower mass flowrates, reducing column diameter and oversizing a column increasing capital costs. Without this effect, the overhead condenser is relied upon to provide all of the reflux in the column, which produces a large amount of liquid throughout the column and necessitates a larger column diameter to achieve the necessary separation of the fractions.

These characteristics necessitate that the main fractionators have closely related heat and material balances. Understanding their operation requires tracking how heat and material balances affect each other. However, the most important consequence for the analysis of atmospheric distillation in the atmospheric distillation unit is the limitation of heat input. The combination of (1) the heat limits of the feedstock and (2) the operating pressure sets the maximum vaporization in the flash zone (i.e. tower feed entry area). Most atmospheric distillation units operate at a specified feedstock limit, either maximum duty input or maximum temperature.

Perforated trays are also used in fractionating towers. This tray is similar to the bubble-cap tray but has smaller holes (approximately 3 inches, 6 mm versus 2 inches, 50 mm). The liquid spills back to the tray below through weirs, but the speed of the rising vapors actually prevents the liquid from returning to the tray below through the holes. Needless to say, a minimum vapor velocity is required to prevent the return of the liquid through the perforations.

The topping operation differs from normal distillation procedures insofar as the majority of the heat is directed to the feed stream rather than reboiling the material in the base of the tower. In addition, products of volatility intermediate between that of the overhead fractions and bottoms (residua) are withdrawn as side-stream products. Furthermore, steam is injected into the base of the column and the side-stream strippers to adjust and control the fractions' initial boiling range (or point). Topped crude oil must always be stripped with steam to elevate the flash point or to recover the final portions of gas oil. The composition of the topped crude oil is a function of the temperature of the vaporizer (or flasher). In addition, the properties of the residuum are very dependent on the extent of volatiles removal by either atmospheric distillation or vacuum distillation.

A tower that is used to remove (strip) unwanted volatile material from gas oil may only be 3 or 4 feet in diameter and 10 feet high with less than 20 trays (Speight, 2014a). Towers concerned with the distillation of liquefied gases are only a few feet in diameter but can reach heights up to 200 feet. A tower used in the fractionation of crude oil may have 16–28 trays, but one used in the fractionation (superfractionation) of liquefied gases may have 30–100 trays. The feed to a typical tower enters the vaporizing or flash zone, which is an area without trays. The majority of the trays are usually located above this area. The feed to a bubble tower, however, may be at any point from top to bottom with trays above and below the entry point, depending on the kind of feedstock and the desired properties in the products.

4.4.1 Tray Tower

The tray tower (trayed tower, also called the staged-contact columns because of the manner in which vapor and liquid are contacted) typically combines the open flow channel with weirs, down comers, and heat exchangers. Free surface flow over the tray is disturbed by gas bubbles coming through the perforated tray and possible liquid leakage dropping through the upper tray.

Prior to the 1960s, most of the trays in a vacuum tower were conventionally designed to provide as low a pressure drop as possible. Many of these standard trays have been replaced by grid packing that provides very low pressure drops as well as a high tray efficiency. Up to this time, flash zone temperature reduction was enhanced by steam stripping of the residuum; but with the new grid packing, the use of steam to enhance flash temperature has been eliminated and most modern units are dry vacuum units.

Liquid collects on each tray to a depth of, say, several inches and the depth controlled by a dam or weir. As the liquid level rises, excess liquid spills over the weir into a channel (downspout), which carries the liquid to the tray below. The temperature of the trays is progressively cooler from bottom to top. The bottom tray is heated by the incoming heated feedstock, although in some instances a steam coil (reboiler) is used to supply additional heat. Condensation forms on the trays as the hot vapors pass upward in the tower until refluxing (simultaneous boiling of a liquid and condensing of the vapor) occurs on the trays. Vapors continue to pass upward through the tower, while the liquid on any particular tray spills onto the tray below, and so on, until the heat at a particular point is too intense for the material to remain liquid. It then becomes vapor and joins the other vapors passing upward through the tower. The whole tower thus simulates a collection of several (or many) stills, with the composition of the liquid at any one point or on any one tray remaining fairly consistent. This allows part of the refluxing liquid to be tapped off at various points as side-stream products.

The efficient operation of the distillation, or fractionating, tower requires the rising vapors to mix with the liquid on each tray. This is usually achieved by installing a short chimney on each hole in the plate and a cap with a serrated edge (bubble cap, hence bubble-cap tower) over each chimney (Figure 4.6). The cap forces the vapors to go below the surface of the liquid and to bubble up through it. Since the vapors may pass up the tower at substantial velocities, the caps are held in place by bolted steel bars.

As a result, flashed vapors rise up the fractionating column and through the trays, flowing counter-current to the internal reflux that is flowing down the column. The lightest product, which is generally gasoline, passes overhead and is condensed in the overhead receiver. If the crude oil contains any non-condensable gas, it will leave the receiver as a gas and can be recovered by other equipment, which should be operated to obtain the minimum flash zone pressure. The temperature at the top of the fractionator is a good measure of the endpoint of the gasoline and is controlled by returning some of the condensed gasoline (as reflux) to the top of the column. Increasing the reflux rate lowers the top temperature and results in the net overhead product having a lower endpoint. The loss in net overhead product must be removed on the next lower draw tray. This decreases the IBP of material from this tray. Increasing the heater transfer temperature increases the heat input and demands more reflux to maintain the same top temperature.

Usually, trays are horizontal, flat, especially prefabricated metal sheets, which are placed at a regular distance in a vertical cylindrical column. Trays have two main parts: (1) the part where vapor (gas) and liquid are being contacted – the contacting area and (2) the part where vapor and liquid are separated, after having been contacted – the downcomer area. Classification of trays is based on (1) the type of plate used in the contacting area, (2) the type and number of downcomers making up the downcomer area, (3) the direction and path of the liquid flowing across the contacting area of the tray, (4) the vapor (gas) flow direction through the (orifices in) the plate, and (5) the presence of baffles, packing, or other additions to the contacting area to improve the separation performance of the tray.

Common plate types for use in the contacting area are the following: (1) bubble-cap tray in which caps are mounted over risers fixed on the plate. The caps come in a wide variety of sizes and shapes, round, square, and rectangular (tunnel); (2) sieve trays come with different hole shapes (round, square, triangular, rectangular (slots), star), various hole sizes (from approximately 2 mm to approximately 25 mm), and several punch patterns (triangular, square, rectangular); and (3) the valve tray that is also available in a variety of valve shapes (round, square, rectangular, triangular), valve sizes, valve weights (light and heavy), orifice sizes, and either as fixed or floating valves.

Trays usually have one or more downcomers. The type and the number of downcomers used mainly depend on the amount of downcomer area required to handle the liquid flow. Single pass trays are trays with one downcomer delivering the liquid from the next higher tray, one bubbling area across which the liquid passes to contact the vapor, and one downcomer for the liquid to the next lower tray. Trays with multiple downcomers and hence multiple liquid passes can have a number of layout geometries. The downcomers may extend, in parallel, from wall to wall. The downcomers may be rotated 90° (or 180°) on successive trays. The downcomer layout pattern determines the liquid flow path arrangement and liquid flow direction in the contacting area of the trays. Giving a preferential direction to the vapor flowing through the orifices in the plate will induce the liquid to flow in the same direction. In this way, liquid flow rate and flow direction, as well as liquid height, can be manipulated. The presence of baffles, screen mesh, or demister mats and the addition of other devices in the contacting area can be beneficial for improving the contacting performance of the tray, viz. its separation efficiency.

The most important parameter of a tray is its separation performance and four parameters are of importance in the design and operation of a tray column: (1) the level of the tray efficiency in the normal operating range, (2) the vapor rate at the "upper limit", i.e. the maximum vapor load, (3) the vapor rate at the "lower limit", i.e. the minimum vapor load, and (4) the tray pressure drop.

The separation performance of a tray is the basis of the performance of the column as a whole. The primary function of, for instance, a distillation column is the separation of a feed stream in (at least) one top product stream and one bottom product stream. The quality of the separation performed by a column can be judged from the purity of the top and bottom product streams. The specification of the impurity levels in the top and bottom streams and the degree of recovery of pure products set the targets for a successful operation of a distillation column. It is evident that the tray efficiency is influenced by (1) the specific component under consideration (this holds especially for multi-component systems in which the efficiency can be different for each component), because of different diffusivities, diffusion interactions, and different stripping factors, and (2) the vapor flow rate: usually increasing the flow rate increases the effective mass transfer rate, while simultaneously decreasing the contact time. These counteracting effects lead to a roughly constant efficiency value for a tray in its normal operating range. Upon approaching the lower operating limit, a tray starts weeping and loses efficiency.

4.4.2 Packed Tower

In the packed tower (also called the continuous-contact), improved separation is possible by supplementing the use of trays by addition of packings which impart good vapor–liquid contact by increasing the surface area. Packings may be (1) dumped (random) packing, e.g., pall rings, or (2) structured packing, which is also known as ordered, arranged, or stacked packing.

In the packed tower reactor (packed column reactor), the liquid film flows down over the packing surface in contact with the upward gas flow. A small fragment of packing geometry can be accurately analyzed assuming the periodic boundary conditions, which allows calibration of the porous media model for a big packing segment. The packing in a distillation column creates a surface for the liquid to spread on thereby providing a high surface area for mass transfer between the liquid and the vapor.

Variations in both the atmospheric and vacuum distillation protocols, including the tower internals, are claimed to improve process efficiency. Also, the incorporation of a vacuum flasher into the distillation circuit is claimed to produce an increased yield of distillate materials as well as the usual vacuum residuum.

In summary, the continued use of atmospheric and vacuum distillation has been a major part of refinery operations during this century and will undoubtedly continue to be employed, at least into the middle and later decades of the 21st century, as the primary refining operation (Speight, 2011a).

The packed tower is considered preferable when (1) the temperature-sensitive mixtures are to be separated to avoid decomposition and/or polymerization, vacuum operation may then be necessary and the smaller liquid holdup and pressure drop theoretical stage of a packed column may be particularly desirable, (2) ceramic or plastic is a desirable material of construction from a non-corrosion and liquid-wettability standpoint, (3) refitting of a tray-type column is desired to increase loading, increase efficiency, and/or decrease pressure drop, (4) liquid rates are very low and/or vapor rates are high in which case structured packing may be particularly desirable, (5) the mixture to be separated is clear, non-fouling, and free of solids, cleaning of column internals will not be necessary, and (6) the mixture to be separated tends to form foam, which collapses more readily in a packed column.

4.5 PROCESS OPTIONS FOR HEAVY FEEDSTOCKS

In order to further distill heavy crude oil, extra heavy crude oil, and tar sand bitumen, reduced pressure is required to prevent thermal cracking and the process takes place in one or more vacuum distillation towers. Since the heavy feedstocks are expected to contribute a growing fraction of hydrocarbon fuels production, changes can be expected in terms of the actual unit internals, unit operation, and prevention of corrosion (Speight, 2014b). Innovations to the distillation units will most likely be more subtle than a complete restructuring of the distillation section of the refinery and will focus on (1) changes to the internal packing to prevent fouling within the distillation system and (2) the use of metal alloy systems to mitigate corrosion.

The principles of vacuum distillation resemble those of fractional distillation and, with the exception that larger diameter columns are used to maintain comparable vapor velocities at the reduced pressures, the equipment is also similar. The internal designs of some vacuum towers are different from atmospheric towers in that random packing and demister pads are used instead of trays. A typical first-phase vacuum tower may produce gas oil, lubricating oil base stock, and a high-boiling residuum for propane deasphalting. A second-phase tower operating at lower vacuum may distill surplus residuum from the atmospheric tower, which is not used for lube-stock processing, and surplus residuum from the first vacuum tower, which is not used for deasphalting.

The development of catalytic distillation or reactive distillation which employs catalyst and distillation devices in the same equipment is typically applied to application that involves reversible reactions in order to shift an unfavorable equilibrium by continuous reaction product withdrawal (DeCroocq, 1997). But catalytic distillation can also provide several advantages in selective hydrogenation of C_3, C_4, and C_5 cuts for the manufacture of petrochemicals (Speight, 2019). Inserting the catalyst in the fractionation column improves mercaptans removal, catalyst fouling resistance, and selective hydrogenation performances by modifying the reaction mixture composition along the column. Thus, there is the potential for applying a related concept to the deep distillation of heavy crude oil.

Vacuum towers are typically used to separate catalytic cracking feedstock from surplus residuum, and heavy oil and tar sand bitumen have fewer components distilling at atmospheric pressure and under vacuum than conventional crude oil. Nevertheless, some heavy oils still pass through the distillation stage of a refinery before further processing is undertaken. In addition, a vacuum tower has recently been installed at the Syncrude Canada plant to offer an additional process option for upgrading tar sand bitumen (Speight, 2005, 2013b, 2014a). The installation of such a tower as a means of refining heavy feedstocks (with the possible exception of the residua that are usually produced through a vacuum tower) is a question of economics and the ultimate goal of the refinery in terms of product slate. After distillation, the residuum from the heavy oil might pass to a cracking unit such as visbreaking or coking to produce saleable products. Catalytic cracking of the residuum or the whole heavy oil is also an option but is very dependent on the constituents of the feedstock and their interaction with the catalyst.

Thus, there is the potential for applying a related concept to the deep distillation of the heavy feedstocks. The continued and projected increased influx of heavy oil, extra heavy oil, and tar sand bitumen into refineries will require reassessment of the need for refinery distillation. Nevertheless, vacuum distillation is an option for tar sand bitumen processes in which the distillation unit is employed to collect as much valuable high-VGO as possible from the bitumen before the residuum is sent to a conversion unit. This option can assist in balancing the overall technical efficiency and economic efficiency of the bitumen refinery. Furthermore, if partial conversion (such as the use of visbreaking is an option for processing heavy feedstocks or partial upgrading during recovery of the feedstock) is possible, distillation will still find a use in refineries.

Moreover, as feedstocks change in composition, the distillation unit will be required to achieve higher degrees of efficiency to produce the precursors to hydrocarbon fuels as well as feedstocks for other units that will eventually produce hydrocarbon fuels through cracking. This will more likely be achieved by changes in the internals of the distillation units as well as changes to the overlay use of the units. The overall effects will be that refineries will be able to create the option to take deeper cuts into the crude oil feedstock leaving a harder residue to be used as feedstocks for the cracking units.

Catalytic distillation (reactive distillation) is a branch of reactive distillation which combines the processes of distillation and catalysis (Ng and Rempel, 2002; Harmsen, 2007). Catalytic distillation is a reactor technology that combines a heterogeneous catalytic reaction and the separation of reactants and products via distillation in a single reactor/distillation column. The heterogeneous catalyst provides the sites for catalytic reactions and also the interfacial surface for liquid/vapor separation. The distinct difference between the catalytic distillation column and the conventional distillation column lies in the placement of solid catalysts usually incorporated in a packing within the distillation column to provide a reaction section in addition to the traditional trays or random packings used for separations in the stripping the rectifying sections of the distillation column. Mass transfer characteristics of the catalytic distillation column packing in the reaction zone have significant influence on the product yield and selectivity. The benefits of catalytic distillation include energy and capital savings, enhanced conversion and product selectivity, longer catalyst lifetime, and reduction of waste streams.

Fouling and foaming are frequent occurrences in distillation towers. Chemical reactions and surface phenomena in fouling and foaming systems can further complicate their predictability. Several techniques for dealing with such unpredictable problems include monitoring of tower conditions, selection of tower internals, and pretreatment of recycle streams. These methods will be improved and developed to the point where they are operative in all distillation units. Furthermore, distillation efficiency is limited by the undesired coke deposition, resulting in a significant loss of distillation efficiency. When a residuum is heated to pyrolysis temperatures (>350°C, 650°F), there is typically an induction period before coke formation begins. To avoid fouling, refiners often stop heating well before coke forms, using arbitrary criteria, but cessation of the heating can result in less than maximum distillate yield.

Over the past three decades, a better understanding of the chemistry and physics of coking has evolved (Chapter 5) (Speight, 2014a), and improved designs based on primary internals have allowed an increase in the amount of gas oil produced with increases in cut point from approximately 520°C (970°F)–590°C (1095°F). As continuing inroads are made into the chemistry of coking, future distillation units will show improvements in the internal design leading to process equivalents of the laboratory spinning band distillation units. Thus, with the potential for an increase in the influx of heavy oil, tar sand bitumen, and biomass to refineries, there may be a resurgence of interest in the application of reactive distillation in refineries which is a process where the still is also a chemical reactor. Separation of the product from the reaction mixture does not need a separate distillation step, which saves energy (for heating) and materials. This technique is especially useful for equilibrium-limited reactions and conversion can be increased far beyond what is expected by the equilibrium due to the continuous removal of reaction products from the reactive zone. This helps reduce capital and investment costs and may be important for sustainable development due to a lower consumption of resources. The suitability of reactive distillation for a particular reaction depends on various factors such as volatility of the reactants and products along with the feasible reaction and distillation temperature. Hence, the use of reactive distillation for every reaction may not be feasible. Exploring the candidate reactions for reactive distillation is an area that needs considerable attention to expand the domain of reactive distillation processes.

However, the conditions in the reactive column are suboptimal both as a distillation column and a chemical reactor, since the reactive column combines these. In addition, the introduction of an in situ separation process in the reaction zone or vice versa leads to complex interactions between vapor–liquid equilibrium, mass transfer rates, diffusion and chemical kinetics, which poses a great challenge for design and synthesis of these systems. Side reactors, where a separate column feeds a reactor and vice versa, are better for some reactions if the optimal conditions of distillation and reaction differ too much.

4.6 POTENTIAL FOR CORROSION AND FOULING OF EQUIPMENT

This section deals with the consequences within the distillation section of the refinery of inefficient dewatering and desalting units.

After dewatering and desalting, the next step in the refining process is the separation of crude oil into various fractions or straight-run cuts by distillation in atmospheric and vacuum towers. The main fractions (often referred to as cuts) obtained have specific boiling point ranges and can be classified in order of decreasing volatility into gases, light distillates, middle distillates, gas oils, and residuum. The concentration of certain constituents by the distillation process can cause corrosion, sediment fouling, and affect flow rates. A properly designed distillation column can reduce the effects of these consequences, but corrosion and other effects are very prominent in reducing separation efficiency in the column.

Refinery distillation units run as efficiently as possible to reduce costs. One of the major issues that occurs in distillation units and decreases efficiency is corrosion of the metal components found throughout the process line of the hydrocarbon refining process. Corrosion causes the failure of parts in addition to dictating the shutdown schedule of the unit, which can cause shutdown of the refinery. Attempts to block such corrosive influences are a major issue of current refineries and will continue to be a major issue of future refineries (Speight, 2014b).

In both the atmospheric distillation column and the vacuum distillation column, corrosion by sulfur-containing derivatives and naphthenic acid derivatives occurs at similar temperatures. Corrosion by naphthenic acid derivatives occurs primarily in the high-velocity area of a crude distillation unit at temperatures ranging from 220°C to 400°C (430°F–750°F). However, as the temperature rises above 400°C (750°F), the corrosivity of the naphthenic acid derivatives decreases – possibly due to decarboxylation of the naphthenic acid derivatives (Francisco and Speight, 1990).

The presence of hydrogen chloride (HCl) is another common cause of corrosion in the atmospheric distillation unit. At temperatures above 120°C (250°F), hydrogen chloride can be formed by the decomposition of sodium chloride (NaCl), calcium chloride ($CaCl_2$), and magnesium chloride ($MgCl_2$). At this temperature, the water vapor in the overhead system condenses to water, which then absorbs the hydrogen chloride to produce hydrochloric acid. Water can also absorb ammonia, which can then combine with hydrogen chloride to form ammonium chloride (NH_4Cl). These salts are extremely acidic and can form a layer on the surface.

Thus, as a first step in the refining process, to reduce corrosion, plugging, and fouling of equipment and to prevent poisoning the catalysts in processing units, these contaminants must be removed by desalting (dehydration). However, the desalting operation does not always remove all of the corrosive elements and hydrogen chloride may be a product of the thermal treatment that occurs as part of the distillation process. Inadequate desalting can cause fouling of heater tubes and heat exchangers throughout the refinery. Fouling restricts product flow and heat transfer and leads to failures due to increased pressures and temperatures. Corrosion, which occurs due to the presence of hydrogen sulfide, hydrogen chloride, naphthenic (organic) acids, and other contaminants in the crude oil, also causes equipment failure. Neutralized salts (ammonium chlorides and sulfides), when moistened by condensed water, can cause corrosion.

Crude oils (prior to the dewatering and desalting operations) often contain inorganic salts such as sodium chloride, magnesium chloride, and calcium chloride in suspension or dissolved in entrained water (brine). These salts must be removed or neutralized before processing to prevent catalyst poisoning, equipment corrosion, and fouling. Salt corrosion is caused by the hydrolysis of some metal chlorides to hydrogen chloride (HCl) and the subsequent formation of hydrochloric acid when crude is heated. Hydrogen chloride may also combine with ammonia to form ammonium chloride (NH_4Cl), which causes fouling and corrosion.

Prior to entering the distillation unit, desalted crude feedstock is preheated and then flows to a direct-fired crude charge heater where it is introduced into the distillation column at pressures slightly above atmospheric and at temperatures ranging from 345°C to 370°C (650°F–700°F) where the lower boiling constituents flash into vapor. As the hot vapor rises in the tower, its temperature is reduced. Heavy fuel oil or residuum is taken from the bottom and, at successively higher points on the tower, various major products are drawn off (Parkash, 2003; Gary et al., 2007; Speight, 2014, 2017; Hsu and Robinson, 2017).

However, despite a seemingly efficient desalting operation, corrosion agents can still appear in various streams after desalting. For example, in the distillation unit, acid gases are formed of which hydrogen sulfide is notorious. Steam, which is injected into the crude tower to improve the fractionation, condenses in the upper part of the unit. The hydrogen sulfide dissolves in the condensate and forms a weak acid which is known to cause stress corrosion cracking in the top section of the tower and in the overhead condenser. This may lead to frequent renewal of tubing of the condenser and in severe cases to replacement of the entire crude tower top.

Corrosion and fouling (as well as foaming) do occur in distillation towers as problems that are not well understood. The difficulty is the unpredictable characteristics which are hard to repeat or correlate in lab experiments or industrial operations. Chemical reactions and surface phenomena in fouling and foaming systems can further complicate their predictability. Several techniques for dealing with such unpredictable problems include monitoring of tower conditions, selection of tower internals, and pretreatment of recycle streams. These methods will be improved and developed to the point where they are operative in all distillation units.

Distillation is one of the most energy-intensive operations in the petroleum refinery and is used throughout the refinery to separate process products. In order for distillation to commence in the refinery, the crude oil is heated in two stages. The preheat train heats the feedstock from ambient temperature to approximately 270°C (520°F – the coil inlet temperature) when it enters the furnace, where the crude oil is heated to the temperature (360°C–395°C, 680°F–740°F) required for distillation. Over time, fouling reduces the performance of the heat exchangers, increasing the amount of energy that has to be supplied. It is possible to bypass units to allow them to be cleaned, with an associated cost and temporary loss of performance.

If the crude oil blending operations have not been monitored carefully and suitable compatibility tests carried out prior to blending, phase separation and deposition of solids can occur prior to (or at) the point where the feedstock enters the tower. Deposition of solid at any point can bring about the necessity to shut down the tower and hence leading to the shutdown of a major part of the refinery – since many other units depend upon feedstock from the tower for operation.

New control technologies that provide a more accurate estimation of distillation column flooding will be required. A distillation column flooding predictor which is a pattern recognition system that identifies patterns of transient tower instabilities, which precede tray flooding, will be very advantageous. This would allow for more stable long-term operation, resulting in greater efficiency and throughput increase. If flooding occurs and there is inorganic sediment or organic solid in the feedstocks, there will also be blockages of the fluid flow on the column trays with the accompanying effects of bubble-cap inefficiencies. The same is true for the column with perforated trays – blockages by inorganic or organic sediment can cause shutdown of the distillation section.

In addition, the desalter wastewater (brine) and aqueous condensate from the overhead reflux drums of main fractionating column contain appreciable levels of hydrogen sulfide as does the hot-well water from the vacuum unit. The common practice is to strip hydrogen sulfide from the sour water (sour water stripping) before disposal. Fuel gas and the LPG stream which may contain hydrogen sulfide are (should be) subjected to an amine wash to remove hydrogen sulfide before the caustic wash. The sour water, fuel gas, and rich amine containing hydrogen sulfide are potential sources for corrosion.

Typically, corrosion inhibitors and neutralizers such as caustic soda or ammonia are injected with the aim of increasing the pH of the sour water. The presence of various acid gases and ammonia can result in solid salt depositing from which ammonium bisulfide is one of the main causes of alkaline sour water corrosion. Alkalinity (pH > 7.6) dramatically increases ammonium bisulfide corrosion and, as in desalting, the key to corrosion reduction is in accurate pH control. Proper neutralizer dosing will not only reduce corrosion but also reduce chemical consumption. Reductions in the use of corrosion inhibitors of more than 15% have been reported.

Failure of condenser tubes constitutes the largest cause of outages in the distillation section, so the choice of tube material is accordingly crucial (Chapters 2 and 6). The thermal conductivity of the tubes has to be reasonably high, they must have sufficient ductility to expand into the tube plate, and their corrosion performance should be well understood. Three types of materials present themselves as being adequate to the task: (1) copper-base alloys, (2) stainless steels, and (3) titanium – each of which possesses its own merits and limitations.

Corrosion in the atmospheric crude distillation unit overhead system stems primarily from the presence of hydrogen chloride – the most common source of hydrogen chloride (HCl, a low molecular weight volatile gas) is the decomposition of sodium chloride (NaCl), calcium chloride ($CaCl_2$), and magnesium chloride ($MgCl_2$) at temperatures exceeding 120°C (250°F) as well as from the decomposition of any organic chlorides. The hydrogen chloride moves into the crude unit overhead condensing systems where it is readily absorbed into condensing water.

Hydrogen chloride in the absence of water does not significantly corrode carbon steel (CS). The overhead system of the crude tower condenses water which absorbs hydrogen chloride and produces hydrochloric acid. The water also absorbs ammonia (NH_3), which when combined with ammonia forms ammonium chloride (NH_4Cl). In situations where the water returns to the vapor state, solid deposits of ammonium chloride form, thus creating the potential for under-deposit corrosion which occurs when corrosive salts form before a water phase is present. The strong acid hydrogen chloride reacts with ammonia and neutralizing amines to form salts that deposit on process surfaces. These salts are acidic and also readily absorb water from the vapor stream. The water acts as the electrolyte to enable these acid salts to corrode the surface and, typically, pitting typically occurs beneath these salts.

Various remedies are used to mitigate the acidic attack from condensed water containing hydrogen chloride, including neutralizing compounds like ammonia and organic amines, film-forming inhibitors, wash water systems, and careful control of temperature in the overhead circuit. To a lesser extent, chlorides can also enter the unit as entrained solids protected by an oil film.

Generally, the segments of the distillation section that are susceptible to corrosion include (but may not be limited to) preheat exchanger (due to the presence of hydrogen chloride and hydrogen sulfide), preheat furnace and bottoms exchanger (hydrogen sulfide and sulfur compounds), atmospheric tower and vacuum furnace (hydrogen sulfide, sulfur compounds, and organic acids), vacuum tower (hydrogen sulfide and organic acids), and overhead (hydrogen chloride, hydrogen sulfide, and water). Where high-sulfur (sour) crude oils are processed, severe corrosion can occur in furnace tubing and in both atmospheric and vacuum towers where metal temperatures exceed 235°C (450°F). Wet hydrogen sulfide will also cause cracks in steel. When processing high nitrogen crude oils, nitrogen oxides can form in the flue gases of furnaces and these gases are corrosive to steel when cooled to low temperatures in the presence of water.

There are three main ways to neutralize acidic aqueous solutions in the crude distillation unit: (1) injecting gaseous ammonia, (2) injecting ammonium hydroxide solution, and (3) injecting neutralizing amine solutions. Regardless of the neutralization technique applied, the pH is lower than the dew point of water. This adds more challenges in measuring pH when condensation occurs; this is the preferred region for the corrosion process to begin. Neutralization equations are the following:

$$HCl\,(aq) + NH_3\,(aq) \rightarrow NH_4Cl\,(aq)$$

$$HCl\,(aq) + RNH_2\,(aq) \rightarrow RNH_2 \cdot HCl\,(aq)\,\left(or\,RNH_3{}^+Cl^-\right)$$

One concern for neutralization technique is the difficulty of controlling the ammonia or amine flow rates, which depend on the varying levels of hydrogen chloride levels in the distillation unit. The neutralizer injection levels can be too low and the pH in the overhead can drop. Excess neutralizer levels, especially in the presence of hydrogen sulfide, contribute to precipitation of salts, such as ammonia or amine disulfides or chlorides. Once formed, these salts (molten or solid) deposit on pipe surfaces and they can also cause localized corrosion with a high rate of thickness loss. If salt formation occurs after condensation, then salt dissolution into water represents minimal corrosion.

In addition to acidic corrosion (by hydrogen chloride and hydrogen sulfide) at high temperature, another form of acidic corrosion in the atmospheric and vacuum distillation units is a major concern (Chapter 1). In high acid (high-TAN) crude processing, acidity increases significantly in the overhead of atmospheric and vacuum units, as these crudes generally contain higher concentrations of salt, sediments, and sometimes organic chlorides. Desalting these crudes is difficult and, therefore, salt carryover to the overhead is more likely. Naphthenic acid corrosion and high-temperature crude corrosivity, in general, are reliability issues in refinery distillation units. The presence of naphthenic acid and sulfur compounds considerably increases corrosion in the high temperature parts of the distillation units.

In the vacuum column, preferential vaporization and condensation of naphthenic acids increase the acid content of the condensates. The naphthenic acids are most active at their boiling point but the most severe corrosion generally occurs when the vapors condense to the liquid phase. In fact, the corrosion mechanism is mainly condensate corrosion and is directly related to content, molecular weight, and boiling point of the naphthenic acid. Corrosion is typically severe at the condensing point corresponding to high acid content and temperature.

Therefore, desalter management and corrosion control and monitoring are critical in the atmospheric distillation unit and the vacuum distillation unit (VDU). In fact, residence time of the crude oil in the desalter, the wash water injection rate, the mixing valve differential pressure, and the efficacy of the emulsion breaker all play vital roles in corrosion management.

The sections of the process susceptible to corrosion include (but may not be limited to) preheat exchanger (hydrogen chloride and hydrogen sulfide), preheat furnace and bottoms exchanger (hydrogen sulfide and sulfur compounds), atmospheric tower and vacuum furnace (hydrogen sulfide, sulfur compounds, and organic acids), vacuum tower (hydrogen sulfide and organic acids), and overhead (hydrogen sulfide, hydrogen chloride, and water). Where sour crudes are processed, severe corrosion can occur in furnace tubing and in both atmospheric and vacuum towers where metal temperatures exceed 230°C (450°F). Wet hydrogen sulfide will also cause cracks in steel. When processing high nitrogen crudes, nitrogen oxides can form in the flue gases of furnaces and these oxides are corrosive to steel when cooled to low temperatures (nitric and nitrous acids are formed) in the presence of water.

Naphthenic acid corrosion occurs in many refinery distillation units and the presence of naphthenic acids considerably increases corrosion in the high-temperature parts of the distillation units. The difference in process conditions, materials of construction, the crude oil blend processed in each refinery, and especially the frequent variation in crude diet increases the problem of correlating corrosion of a unit to a certain type of crude oil. In addition, crude oil composition from the same field can change with time.

In the vacuum column, preferential vaporization and condensation of naphthenic acids increase the total acid number (TAN) of the condensates and the corrosion is similar to corrosion in fractions with a high TAN and velocity has virtually no effect on the process. The naphthenic acids are most active at their boiling point but the most severe corrosion generally occurs on condensation. The corrosion mechanism is mainly a condensate corrosion and is directly related to content, molecular weight, and boiling point of the naphthenic acid. Corrosion is typically severe at the condensing point corresponding to high TAN and temperature.

Despite an effective desalting operation, corrosive materials can still appear during downstream processing, for example, the sour water corrosion that occurs in the crude oil distillation unit. During the process, acid gases are formed of which hydrogen sulfide is notorious. Steam, which is injected into the crude tower to improve the fractionation, condenses in the upper part of the unit and the hydrogen sulfide dissolves in the condensate and forms a weak acid which is known to cause stress corrosion cracking in the top section of the tower and in the overhead condenser.

Typically, corrosion inhibitors and neutralizers such as caustic soda or ammonia are injected with the aim of increasing the pH of the sour water. However, the presence of various acid gases and ammonia can result in solid salt depositing from which ammonium bisulfide (NH_4HS) is one of the main causes of alkaline sour water corrosion and a pH level that is higher than 7.6 can cause an increase in the corrosion by ammonium bisulfide corrosion. In addition to the atmospheric distillation, other downstream operations (including the VDU) involve sour water and can also suffer from corrosion issues.

High-temperature crude corrosivity of distillation units will continue to be a major concern to the refining industry. The presence of naphthenic acid and sulfur compounds considerably increases corrosion in the high-temperature parts of the distillation units, and equipment failures have become a critical safety and reliability issue. The difference in process conditions, materials of construction, blend processed in each refinery, and especially the frequent variation in crude diet increases the problem of correlating corrosion of a unit to a certain type of crude oil. In addition, a large number of interdependent parameters influence the high-temperature crude corrosion process.

Damage is typically in the form of unexpected high corrosion rates on alloys that would normally be expected to resist sulfidic corrosion. In many cases, even very highly alloyed materials have been found to exhibit sensitivity to corrosion under these conditions.

Naphthenic acid corrosion is differentiated from sulfidic corrosion by the nature of the corrosion (pitting and impingement) and by its severe attack at high velocities in crude distillation units. Crude feedstock heaters, furnaces, transfer lines, feed and reflux sections of columns, atmospheric and vacuum columns, heat exchangers, and condensers are among the types of equipment subject to this type of corrosion.

From a materials standpoint, carbon steel can be used for refinery components. Carbon steel is resistant to the most common forms of corrosion, particularly from hydrocarbon impurities at temperatures below 205°C (400°F), but other corrosive chemicals and high-temperature environments prevent its use everywhere. Common replacement materials are low alloy steel containing chromium and molybdenum, with stainless steel containing more chromium dealing with more corrosive environments. More expensive materials commonly used are nickel, titanium, and copper alloys. These are primarily saved for the most problematic areas where extremely high temperatures or very corrosive chemicals are present.

In fact, the increasing move in the crude oil slate (crude oil from tight-low-permeability formations notwithstanding), there is the need to design units that will be (as near as possible) not susceptible to corrosion. The sections most susceptible to corrosion from hydrogen chloride (HCl) and hydrogen sulfide (H_2S) include the preheat exchanger; preheat furnace and bottoms exchanger from hydrogen sulfide and sulfur compounds; atmospheric tower and vacuum furnace from hydrogen sulfide, sulfur compounds, and organic acids; vacuum tower from H_2S and organic acids; and overhead from hydrogen sulfide, hydrogen chloride, and water. Where sour crudes are processed, severe corrosion can occur in furnace tubing and in both atmospheric and vacuum towers when metal temperatures exceed 230°C (450°F). Wet hydrogen sulfide will also cause cracks in steel leading to spills, leaks, and fugitive emissions.

When processing high nitrogen crudes, nitrogen oxides can form in the flue gases of furnaces. Nitrogen oxides are corrosive to steel when cooled to low temperatures in the presence of water. The authors claim that fugitive emissions from these sources come from aging refineries and are not adequately accounted for. Chemicals are used to control corrosion by hydrogen chloride produced in distillation units. Ammonia (NH_3) may be injected into the overhead stream prior to initial condensation and/or an alkaline solution may be injected into the hot crude oil feed. If sufficient wash water is not injected, deposits of ammonium chloride can form causing serious corrosion. Crude feedstock may contain appreciable amounts of water in suspension. This water phase can separate during startup and, together with water remaining in the tower from steam purging, settle in the bottom of the tower. When the water is heated to its boiling point, an instantaneous vaporization explosion can occur upon contact with the oil in the unit. This can lead to devastating damage to the columns and catastrophic release of vapors and liquid.

Attempts to mitigate corrosion will continue to use a complex system of monitoring, preventative repairs, and careful use of materials. Monitoring methods include both off-line checks taken during maintenance and online monitoring. Off-line checks measure corrosion after it has occurred, allowing the engineer to determine when to replace the equipment based on the historical data he has gathered.

Online systems are a more modern development and are revolutionizing the way corrosion is, and will be, approached. There are several types of online corrosion monitoring technologies such as linear polarization resistance, electrochemical noise, and electrical resistance. Online monitoring has generally had slow reporting rates in the past (minutes or hours) which have been limited by process conditions and sources of error, but newer technologies can report rates up to twice per minute with much higher accuracy (referred to as real-time monitoring). This allows process engineers to treat corrosion as another process variable that can be optimized in the system. Immediate responses to process changes enable the control of corrosion mechanisms, allowing them to be minimized while also increasing production output.

Materials methods will include selecting the proper material for the application., Cheap materials are preferred in locations where corrosion is not a problem, but more expensive but longer lasting materials should be used where serious corrosion can occur. Other materials methods will be in the form of protective barriers between corrosive substances and the equipment metals. These can be either a lining of refractory material such as standard Portland cement or other special acid-resistant cements that are shot onto the inner surface of the vessel. Also available are thin overlays of more expensive metals that protect cheaper metal against corrosion without requiring lots of material.

Blending of crude feedstocks will continue to be used to mitigate the effects of corrosion as well as blending, inhibition, materials upgrading, and process control. Blending will be used to reduce the naphthenic acid content of the feed, thereby reducing corrosion to an acceptable level. However, while blending viscous crude oils with heavy and light crude oils can change shear stress parameters and might also help reduce corrosion, the potential for incompatibility between the constituents of viscous crude oil and light crude oil exists.

Chemicals are used to control corrosion by hydrochloric acid produced in distillation units. For example, caustic is added to desalting water to neutralize acids and reduce corrosion. They are also added to desalted crude in order to reduce the amount of corrosive chlorides in the tower overheads. They are used in some refinery treating processes to remove contaminants from hydrocarbon streams.

Ammonia may be injected into the overhead stream prior to initial condensation and/or an alkaline solution may be carefully injected into the hot crude oil feed. If sufficient wash water is not injected, deposits of ammonium chloride can form and cause serious corrosion. The crude feedstock may contain appreciable amounts of water in suspension which can separate during startup and, together with water remaining in the tower from steam purging, settle in the bottom of the tower. This water can be heated to the boiling point and create an instantaneous vaporization explosion upon contact with the oil in the unit.

More extreme treatment, in terms of the use of chemicals, would be the addition of inorganic materials such as calcium oxide or magnesium oxide to trap the sulfur compounds in the distillation unit. This would be especially helpful in the case of cracking distillation where additional volatile and corrosive sulfur species are generated. In either case, any excess oxides and the calcium sulfide or magnesium sulfide would collect in the residuum. There may be a benefit to have such inorganic materials in any asphalt produced from the residue.

Injection of corrosion inhibitors will provide some protection for specific fractions that are known to be particularly severe. However, monitoring will need to be adequate in this case to check on the effectiveness of the treatment. Process control changes may provide adequate corrosion control if there is the possibility of reducing charge rate and temperature.

Process changes will include any action to remove or at least reduce the amount of acid gas present and to prevent the accumulation of water on the tower trays. Material upgrading will include the lining of distillation tower tops with alloys resistant to hydrochloric acid. Design changes will be used to prevent the accumulation of water and will include coalescers and water draws. The application of chemicals will include the injection of a neutralizer to increase the pH and a corrosion inhibitor. The presence of many acids, such as naphthenic acids and carbon dioxide (in water), can buffer the environment and require greater use of neutralizers. However, the use of excessive amounts of neutralizers will cause plugging of trays and corrosion under the salt deposits.

Heavy feedstocks (of the types that are relevant to this chapter, i.e. heavy crude oil, extra heavy crude oil, and tar sand bitumen) contain inorganic salts such as sodium chloride, magnesium chloride, and calcium chloride in suspension or dissolved in entrained water (brine). These salts must be removed or neutralized before processing to prevent catalyst poisoning, equipment corrosion, and fouling. Salt corrosion is caused by the hydrolysis of some metal chlorides to hydrogen chloride (HCl) and the subsequent formation of hydrochloric acid when crude is heated. Hydrogen chloride may also combine with ammonia to form ammonium chloride (NH_4Cl), which causes fouling and corrosion.

The sections of the process susceptible to corrosion include (but may not be limited to) preheat exchanger (HCl and H_2S), preheat furnace and bottoms exchanger (H_2S and sulfur compounds), atmospheric tower and vacuum furnace (H_2S, sulfur compounds, and organic acids), vacuum tower (H_2S and organic acids), and overhead (H_2S, HCl, and water). Where sour crudes are processed, severe corrosion can occur in furnace tubing and in both atmospheric and vacuum towers when metal temperatures exceed 450°F. Wet hydrogen sulfide will also cause cracks in steel. When processing high nitrogen crudes, nitrogen oxides can form in the flue gases of furnaces and these oxides are corrosive to steel when cooled to low temperatures (nitric and nitrous acids are formed) in the presence of water.

Corrosion occurs in various forms in the distillation section of the refinery and is manifested by events such as pitting corrosion from water droplets, embrittlement from chemical attack if the dewatering and desalting unit has not operated efficiently, and stress corrosion cracking from sulfide attack.

High-temperature corrosion of distillation units will continue to be a major concern to the refining industry. The presence of naphthenic acid and sulfur compounds considerably increases corrosion in the high-temperature parts of the distillation units, and equipment failures have become a critical safety and reliability issue. The difference in process conditions, materials of construction, blend processed in each refinery, and especially the frequent variation in crude diet increases the problem of correlating corrosion of a unit to a certain type of crude oil. In addition, a large number of interdependent parameters influence the high-temperature crude corrosion process.

Naphthenic acid corrosion is differentiated from sulfidic corrosion by the nature of the corrosion (pitting and impingement) and by its severe attack at high velocities in crude distillation units. Crude feedstock heaters, furnaces, transfer lines, feed and reflux sections of columns, atmospheric and vacuum columns, heat exchangers, and condensers are among the types of equipment subject to this type of corrosion.

Attempts to mitigate corrosion will continue to use a complex system of monitoring, preventative repairs, and careful use of materials. Monitoring methods include both off-line checks taken during maintenance and online monitoring. Off-line checks measure corrosion after it has occurred, allowing the engineer to determine when to replace the equipment based on the historical data he has gathered.

Blending of refinery feedstocks (with the inherent danger of phase separation and incompatibility) will continue to be used to mitigate the effects of corrosion as well as blending, inhibition, materials upgrading, and process control. Blending will be used to reduce the naphthenic acid content of the feed, thereby reducing corrosion to an acceptable level. However, while blending of heavy and light crude oils can change shear stress parameters and might also help reduce corrosion, potential for incompatibility between the constituents of heavy and light crude oils exists.

In summary, refinery distillation may appear to be waning in terms of processing such heavy feedstocks, yet it is definitely not out.

4.7 THE FUTURE

The distillation unit (along with the catalytic cracking unit) is the core unit of the modern crude oil refinery (Parkash, 2003; Gary et al., 2007; Speight, 2014a, 2017; Hsu and Robinson, 2017). In general, refineries in the future will optimize energy use through more efficient heat exchange and heat integration, better controls, and adopting energy-saving approaches to very energy-intensive process units (e.g., furnaces, distillation towers).

Heavy crude oil and/or sour crude oil as well as extra heavy crude oil and tar sand bitumen, which require more energy-intensive processing than conventional sweet (low-density, low-sulfur) crude oils, are expected to contribute to a growing fraction of hydrocarbon fuels production. As existing reserves of oil are depleted and there is greater worldwide competition for the light, low-sulfur crude oils, refiners will increasingly utilize heavy crude oil and/or sour crude oil to meet demand.

The influx of such crude oils is causing noticeable changes to the operation of distillation units in terms of the actual unit internals, unit operation, and prevention of corrosion (Speight, 2014b). Innovations to the distillation units will most likely be more subtle than a complete restructuring of the distillation section of the refinery and will focus on (1) energy savings, (2) changes to the internal packing to prevent fouling within the distillation system, and (3) the use of metal alloy systems to mitigate corrosion.

Some of the process options and improvements mentioned below may already be under consideration; but where they have not been considered, the potential for adoption will be high.

In the near future, retrofitting will be used to increase the efficiency of the process by maximizing the use of existing equipment. Examples of retrofit objectives are to (1) increase the throughput, (2) change the feedstock, (3) increase the production or the quality of the products, (4) reduce the energy demand, and (5) reduce the atmospheric emissions. The retrofit methodology for crude oil distillation systems will necessitate the use of rigorous modeling and optimization procedures to optimize the process conditions and to explore structural modifications to increase the capacity and the energy efficiency of the system.

Another issue that has received, and will continue to receive, attention is the formation of a carbonaceous deposit (coke) which tends to occur at the inner side of the crude oil tube in the furnace during the preheating process. The coking and coating of the inner tube cause the coke to accumulate, the product quality to deteriorate, the pressure drop across the furnace to increase, the energy consumption to increase, and the tube wall carbonization to corrode.

Generally, decoking is executed on demand based on the measured heat loss or performed every fixed period of time. However, due to different rates of coke formation of various crude oils, as well as the difference in cost of product, process operation, and maintenance operation, both on-demand and fixed maintenance practices are not optimal methods for decoking. Modeling of the coke formation process and the rate of coke accumulation at the inner tube of the furnace will be needed for the optimization, prediction, and scheduling of the decoking operation.

This will require a novel approach that allows the distillation and heat integration arrangements to be designed and optimized simultaneously. In a retrofit operation, the need to improve the design for increased energy efficiency or for increased throughput brings special challenges. In addition to unit design, changes to operating conditions throughout the system, column internals, and heat exchange arrangements must all be optimized simultaneously (Table 4.2).

Also, there are numerous heaters and exchangers in the atmospheric and VDUs that are sources of ignition, spills, leaks, and fugitive emissions. Variations in pressure, temperature, or liquid levels may occur if automatic controls fail. Control of temperature, pressure, and reflux within operating parameters is needed to prevent thermal cracking within the towers. Relief systems must be used for overpressure and operations monitored to prevent the crude from entering the reformer charge.

Energy efficiency opportunities exist in the heating side and by optimizing the distillation column – such opportunities will no doubt be put into practice. For example, optimization of the reflux

TABLE 4.2

Future Actions for the Distillation Unit

Action	Outcome
Optimize the operating conditions	Reduction in the energy consumption
Modify the design	Reduction in the energy consumption
Different distillation configurations	Increase unit efficiency
Optimize the configuration of new units	Increase unit efficiency

ratio of the distillation column can produce significant energy savings. The efficiency of a distillation column is determined by the characteristics of the feed. If the characteristics of the feed have changed over time or compared to the design conditions, operational efficiency can be improved.

Damaged or worn internals result in increased operation costs. As the internals become damaged, efficiency decreases and pressure drops rise. This causes the column to run at a higher reflux rate over time. With an increased reflux rate, energy costs will increase accordingly. Replacing the trays with new ones or adding a high-performance packing can have the column operating like the day it was brought online. If operating conditions have seriously deviated from designed operating conditions, the investment may have a relatively short payback. When replacing the trays, it will be worthwhile to consider new efficient tray designs. New tray designs can result in enhanced separation efficiency and decreased pressure drop, which will result in reduced energy consumption. When considering new tray designs, the number of trays should be optimized.

An energy-efficient design for a crude distillation unit has recently been developed by Elf in which the crude preheater and the distillation column were redesigned. The crude preheat train was separated in several steps to recover fractions at different temperatures. The distillation tower was redesigned to work at low pressure and the outputs were changed to link to the other processes in the refinery and product mix of the refinery. The design resulted in reduced fuel consumption and better heat integration. Such options will be more obvious in future refineries.

Effective integration of controls and practices to increase energy efficiency (e.g., pipe insulation) would result in higher levels of energy optimization. Refineries will maximize their ability to produce energy on-site by increasing the use of cogeneration to generate both heat and power, and in some cases, would be producing electricity for sale back to the local grid. In many cases, high-efficiency turbines and steam generators would be used to achieve a high thermal efficiency in cogeneration and power generation systems.

As feedstocks change in composition, the distillation unit will be required to achieve higher degrees of efficiency to produce the precursors to hydrocarbon fuels as well as feedstocks for other units that will eventually produce hydrocarbon fuels through cracking. This will more likely be achieved by changes in the internals of the distillation units as well as changes to the overlay use of the units. The overall effects will be that refineries will be able to create the option to take deeper cuts into the crude oil feedstock leaving a harder residue to be used as feedstocks for the cracking units.

Reboilers consume a large part of total refinery energy use as part of the distillation process. By using chilled water, the reboiler duty can in principle be lowered by reducing the overhead condenser temperature.

Fouling and foaming are frequently at the top of a list of well-known but little understood problems in distillation towers. The difficulty is the unpredictable characteristics which are hard to repeat or correlate in lab experiments or industrial operations. Chemical reactions and surface phenomena in fouling and foaming systems can further complicate their predictability. Several techniques for dealing with such unpredictable problems include monitoring of tower conditions, selection of tower internals, and pretreatment of recycle streams. These methods will be improved and developed to the point where they are operative in all distillation units.

New control technologies that provide a more accurate estimation of distillation column flooding will be required. A distillation column flooding predictor which is a pattern recognition system that identifies patterns of transient tower instabilities, which precede tray flooding, will be very advantageous. This would allow for more stable long-term operation, resulting in greater efficiency and throughput increase.

There is a tendency to focus on the debottlenecking of the column internals, which is basically employing the traditional approach to improve the hydraulic performance of the column internals, i.e. to get more hydraulic capacity through an existing distillation column.

One method to avoid column debottlenecking is to retrofit the column with the internals that allow more vapor or liquid traffic to pass through without the loss of efficiency. In this kind of project, major changes to the existing column are required to replace the existing internals with the new one that debottleneck the column. For example, advanced trays (Nye trays and multiple downcomer trays) and structured packings require the removal of the existing tray support rings and the existing internals. The use of better internals obviously promotes debottlenecking but this is normally achieved at the expense of a huge cost. Of concern to us are the costs of new types of internals, costs of removing the existing internals, and the opportunity cost of extensive plant downtime.

In the near future, the useful capacity of column internals, either with trays, structured packing, or random packing, will be maximized and optimized with customized feed distributors, downcomers, and other fabricated components.

Refinery distillation efficiency is limited by the undesired coke deposition, resulting in a significant loss of distillation efficiency. When a residuum is heated to pyrolysis temperatures (>350°C, 650°F), there is typically an induction period before coke formation begins. To avoid fouling, refiners often stop heating well before coke forms, using arbitrary criteria, but cessation of the heating can result in less than maximum distillate yield.

Over the past three decades, a better understanding of the chemistry and physics of coking has evolved (Chapter 3) and improved designs based on primary internals have allowed an increase in the amount of gas oil produced with increases in cut point from approximately 520°C (970°F) to 590°C (1095°F). As continuing inroads are made into the chemistry of coking, future distillation units will show improvements in the design of the internals leading to process equivalents of the laboratory spinning band distillation units.

The concept of a non-conventional distillation column is not new. This type of column has found its application in the crude oil refining, such as the use of side strippers, and the cryogenic air separation, such as the use of side rectifiers. Other forms of these columns such as the pre-fractionator arrangement, the fully thermally coupled, column and the dividing wall column, however, have started to attract the interests of both process designers and operating companies. The justifications for their use have been prompted by less energy consumption and greater efficiency when compared with the conventional fractionation system.

With the potential for an increase in the influx of heavy crude oil, extra heavy oil, and tar sand bitumen to refineries, there may be a resurgence of interest in the application of reactive distillation in refineries. Although invented in 1921, the industrial application of reactive distillation did not take place before the 1980s.

Reactive distillation is a process where the still is also a chemical reactor. Separation of the product from the reaction mixture does not need a separate distillation step, which saves energy (for heating) and materials. This technique is especially useful for equilibrium-limited reactions and conversion can be increased far beyond what is expected by the equilibrium due to the continuous removal of reaction products from the reactive zone. This helps reduce capital and investment costs and may be important for sustainable development due to a lower consumption of resources.

The suitability of reactive distillation for a particular reaction depends on various factors such as volatilities of reactants and products along with the feasible reaction and distillation temperature. Hence, the use of reactive distillation for every reaction may not be feasible. Exploring the candidate reactions for reactive distillation is an area that needs considerable attention to expand the domain of reactive distillation processes.

The benefits of applying reactive distillation include (1) increased speed and improved efficiency, (2) lower costs – reduced equipment use, energy use, and handling, (3) less waste and fewer byproducts, and (4) improved product quality – reducing opportunity for degradation because of less heat.

In a similar vein, there is also a greater potential for the application of extractive distillation.

Extractive distillation is the distillation in the presence of a miscible, high boiling, relatively non-volatile component, and the solvent, which forms no azeotrope with the other components in the mixture. The method is used for mixtures having a low value of relative volatility, nearing unity. Such mixtures cannot be separated by simple distillation, because the volatility of the two components in the mixture is nearly the same, causing them to evaporate at nearly the same temperature at a similar rate, making normal distillation impractical.

The method of extractive distillation uses a separation solvent, which is generally non-volatile, has a high boiling point, and is miscible with the mixture, but doesn't form an azeotropic mixture. The solvent interacts differently with the components of the mixture thereby causing their relative volatilities to change. This enables the new three-part mixture to be separated by normal distillation. The original component with the greatest volatility separates out as the top product. The bottom product consists of a mixture of the solvent and the other component, which can again be separated easily because the solvent doesn't form an azeotrope with it. The bottom product can be separated by any of the methods available.

It is important to select a suitable separation solvent for this type of distillation. The solvent must alter the relative volatility by a wide enough margin for a successful result. The quantity, cost, and availability of the solvent should be considered. The solvent should be easily separable from the bottom product, and it should not react chemically with the components or the mixture or cause corrosion in the equipment (Speight, 2014b).

Membranes may offer future alternatives to distillation. Current membrane systems will probably be most effective in hybrid distillation processes to perform a first, crude, low-energy, low-cost separation, leaving the polishing operation for distillation. If high selectivity could be achieved with membranes, there is the potential to replace distillation in many separation processes. For example, the separation of olefins (ethylene, propylene) from paraffins (ethane, propane) on a commercial scale is accomplished almost exclusively via cryogenic distillation. This separation process is very energy-intensive, and hence, a less energy-intensive method for separating olefins from paraffins would therefore be extremely beneficial. The separation of olefins from paraffins could be accomplished with less energy if selective facilitated transport membranes were to be used. A complete olefin/paraffin separation with membrane processes alone would require membranes with high selectivity.

Improved isomer separations could significantly reduce the energy consumption required for the manufacture of certain chemical products. The isomer separations that could be achieved using membranes include the separation of p-xylene from other xylenes. Molecular sieve membranes could also be used in the membrane reactor configuration for the manufacture of p-xylene and linear olefins and paraffins if their corresponding isomerization reactions were enhanced. In this case, a chemical synthesis would be performed in conjunction with a closely coupled, but separate, membrane separation device. For example, para-xylene could be selectively produced by an equilibrium redistribution of mixed isomeric xylenes coupled with a selective transport of the product through a membrane (Parkash, 2003; Gary et al., 2007; Speight, 2014a, 2017; Hsu and Robinson, 2017).

REFERENCES

ASTM. 2021. Annual Book of Standards. ASTM International, West Conshohocken, PA.
ASTM D6. 2021. Standard Test Method for Loss on Heating of Oil and Asphaltic Compounds. Annual Book of Standards. ASTM International, West Conshohocken, PA.

ASTM D20. 2021. Standard Test Method for Distillation of Road Tars. Annual Book of Standards. ASTM International, West Conshohocken, PA.

ASTM D86. 2021. Standard Test Method for Distillation of Petroleum Products at Atmospheric Pressure. Annual Book of Standards. ASTM International, West Conshohocken, PA.

ASTM D189. 2021. Standard Test Method for Conradson Carbon Residue of Petroleum Products. Annual Book of Standards. ASTM International, West Conshohocken, PA.

ASTM D1160. 2021. Standard Test Method for Distillation of Petroleum Products at Reduced Pressure. Annual Book of Standards. ASTM International, West Conshohocken, PA.

ASTM D2715. 2021. Standard Test Method for Volatilization Rates of Lubricants in Vacuum. Annual Book of Standards. ASTM International, West Conshohocken, PA.

ASTM D2887. 2021. Standard Test Method for Boiling Range Distribution of Petroleum Fractions by Gas Chromatography. Annual Book of Standards. ASTM International, West Conshohocken, PA.

ASTM D2892. 2021. Standard Test Method for Distillation of Crude Petroleum (15-Theoretical Plate Column). Annual Book of Standards. ASTM International, West Conshohocken, PA.

ASTM D3710. 2021. Standard Test Method for Boiling Range Distribution of Gasoline and Gasoline Fractions by Gas Chromatography. Annual Book of Standards. ASTM International, West Conshohocken, PA.

ASTM D5307. 2021. Standard Test Method for Determination of Boiling Range Distribution of Crude Petroleum by Gas Chromatography. Annual Book of Standards. ASTM International, West Conshohocken, PA.

Carbognani, L., Díaz-Gómez, L., Oldenburg, T.B.P., and Pereira-Almao, P. 2012. Determination of molecular masses for petroleum distillates by simulated distillation. *CT&F - Ciencia, Tecnología y Futuro*, 4(5): 43–55.

Charbonnier, R.P., Draper, R.G., Harper, W.H., and Yates, Y. 1969. Analyses and Characteristics of Oil Samples from Alberta. Information Circular No. IC 232. Department of Energy Mines and Resources, Mines Branch, Ottawa, Ontario, Canada.

Coleman, H.J., Shelton, E.M., Nicholls, D.T., and Thompson, C.J. 1978. Analysis of 800 Crude Oils from United States Oilfields. Report No. BETC/RI-78/14. Technical Information Center, Department of Energy, Washington, DC.

DeCroocq, D. 1997. Major scientific and technical challenges about development of new processes in refining and petrochemistry. *Revue Institut Français de Pétrole*, 52(5): 469–489.

Diwekar, U.M. 1995. *Batch Distillation: Simulation, Optimal Design, and Control*. Taylor & Francis Inc., Philadelphia, PA.

Eser, S., and Riazi, M.R. 2013. Crude oil refining processes. In: *Petroleum Refining and Natural Gas Processing*, M.R. Riazi, S. Eser, and J.L. Peña (Editors). ASTM International, West Conshohocken, PA, pp. 103–104.

Fajobi, M.A., Loto, R.T., and Oluwole, O.O. 2019. Corrosion in crude distillation overhead system: A review. *Journal Bio- and Tribo-Corrosion*, 5: 1–9. https://link.springer.com/article/10.1007/s40735-019-0262-4.

Francisco, M.A., and Speight, J.G. 1990. Studies in petroleum composition IV: Changes in the nature of chemical constituents during crude oil distillation. *Revue de l'Institut Francais du Petrole*, 45: 733–740.

Gary, J.G., Handwerk, G.E., and Kaiser, M.J. 2007. *Petroleum Refining: Technology and Economics*. CRC Press, Taylor & Francis Group, Boca Raton, FL.

Harmsen, G.J. 2007. Reactive distillation: The front-runner of industrial process intensification. *A Full Review of Commercial Applications, Research, Scale-Up, Design and Operation. Chemical Engineering and Processing*, 46: 774–780.

Hsu, C.S., and Robinson, P.R. 2017. *Practical Advances in Petroleum Processing, Volumes 1 and 2*. Springer, New York.

Jones, D.S.J. 1995. *Elements of Petroleum Processing*. John Wiley & Sons Inc., Chichester, West Sussex; Hoboken, NJ.

Long, R.B. 1995. *Separation Processes in Waste Minimization*. CRC Press, Taylor & Francis Group, Boca Raton, FL. Chapter 12.

MacAllister, D.J., and DeRuiter, R.A. 1985. Further Development and Application of Simulated Distillation for Enhanced Oil Recovery. Paper No. SPE 14335. 60th Annual Technical Conference. Society of Petroleum Engineers. Las Vegas. September 22–25.

Neer, L.A., and Deo, M.D. 1995. Simulated distillation of oils with a wide carbon number distribution. *Journal of Chromatographic Science*, 33: 133–138.

Ng, F.T.T., and Rempel, G.L. 2002. *Catalytic Distillation. Encyclopedia of Catalysis*, John Wiley & Sons Inc., Hoboken, NJ.

Parkash, S. 2003. *Refining Processes Handbook*. Gulf Professional Publishing, Elsevier, Amsterdam.

Romanowski, L.J., and Thomas, K.P. 1985. Steamflooding of Preheated Tar Sand. Report No. DOE/FE/60177-2326. United States Department of Energy, Washington, DC.

Schwartz, H.E., Brownlee, R.G., Boduszynski, M.M., and Su, F. 1987. Simulated distillation of high-boiling petroleum fractions by capillary supercritical chromatography and vacuum thermal gravimetric analysis. *Analytical Chemistry*, 59: 1393–1401.

Speight, J.G. 2000. *The Desulfurization of Heavy Oils and Residua*, 2nd Edition. Marcel Dekker Inc., New York.

Speight, J.G. 2005. Natural Bitumen (Tar Sands) and Heavy Oil. In Coal, Oil Shale, Natural Bitumen, Heavy Oil and Peat, from *Encyclopedia of Life Support Systems (EOLSS)*, +Developed under the Auspices of the UNESCO, EOLSS Publishers, Oxford, UK. http://www.eolss.net.

Speight, J.G. 2011a. *The Refinery of the Future*. Gulf Professional Publishing, Elsevier, Oxford.

Speight, J.G. (Editor). 2011b. *The Biofuels Handbook*. Royal Society of Chemistry, London.

Speight, J.G. 2012. *Crude Oil Assay Database*. Knovel; Elsevier, New York. Online version available at: http://www.knovel.com/web/portal/browse/display?_EXT_KNOVEL_DISPLAY_bookid=5485&VerticalID=0.

Speight, J.G. 2013a. *Heavy and Extra Heavy Oil Upgrading Technologies*. Gulf Professional Publishing Company, Elsevier, Oxford.

Speight, J.G. 2013b. *The Chemistry and Technology of Coal*, 3rd Edition. CRC Press, Taylor & Francis Group, Boca Raton, FL.

Speight, J.G. 2014a. *The Chemistry and Technology of Petroleum*, 5th Edition. CRC Press, Taylor & Francis Group, Boca Raton, FL.

Speight, J.G. 2014b. *Oil and Gas Corrosion Prevention*. Gulf Professional Publishing, Elsevier, Oxford.

Speight, J.G. 2015. *Handbook of Petroleum Product Analysis*, 2nd Edition. John Wiley & Sons Inc., Hoboken, NJ.

Speight, J.G. 2017. *Handbook of Petroleum Refining*. CRC Press, Taylor & Francis Group, Boca Raton, FL; Marcel Dekker Inc., New York.

Speight, J.G. 2019. *Handbook of Petrochemical Processes*. CRC Press, Taylor & Francis Group, Boca Raton, FL.

Speight, J.G. 2020. *Synthetic Fuels Handbook: Properties, Processes, and Performance*, 2nd Edition. McGraw-Hill, New York.

Speight, J.G. 2021. *Chemistry and Technology of Alternate Fuels*. World Scientific Publishing Co. Pte. Ltd., Singapore and Hackensack, NJ.

5 Ancillary Distillation Processes

5.1 INTRODUCTION

The crude oil refining industry employs a wide variety of processes and none is more certain (after the dewatering and desalting processes) for crude oil and heavy crude oil (extra heavy crude oil and tar sand bitumen are exceptions) than the distillation units. However, the separation into bulk fraction (Table 5.1) by atmospheric distillation or by vacuum distillation into a variety of fractions (also referred to as "cuts" based on different temperatures at which they evaporate and condense – cut points) is not always the end of the distillation trail. The processing flow scheme for any refinery is largely determined by the composition of the refinery feedstock blend (and the properties of the individual crude oil in the blend) and the chosen slate of petroleum products, and the representation of a crude oil refinery by a general flow scheme (Figure 5.1) is used only to give a general indication of various processing options. The arrangement of these processes will vary among refineries, and few, if any, employ all of these processes.

The processes used to refine a crude oil (or a blend of crude oils) must be selected and products manufactured to give a balanced operation in which crude oil is converted into a variety of products in amounts that are in accord with the demand for each. For example, the manufacture of products from the lower boiling portion of crude oil automatically produces a certain amount of higher boiling components. If the latter cannot be sold as, say, heavy fuel oil, these products will accumulate until refinery storage facilities are full. To prevent the occurrence of such a situation, the refinery must be flexible and be able to change operations as needed. This usually means more processes:

TABLE 5.1

Examples of the Boiling Fractions of Crude Oil

Fraction	Boiling Range[a]	
	°C	°F
Gas[b]	−160 to −1	−260 to +30
Light naphtha	−1 to 150	30–300
Heavy naphtha	150–205	300–400
Kerosene	205–260	400–500
Light gas oil	260–315	400–600
Heavy gas oil	315–370	600–700
Atmospheric residuum	>370	>700
Light vacuum gas oil	370–425	700–800
Heavy vacuum gas oil	425–510	800–950
Vacuum residuum	>510	>950[c]

[a] For convenience, boiling ranges are converted to the nearest 5°. In addition, the boiling ranges may change from refinery to refinery by 5°–10°.

[b] Methane, ethane, propane, and butane.

[c] In some refineries, the residuum is >565°C (>1050°F) which adds a further increment (510°F–565°F, 950°F–1050°F) to the heavy vacuum gas oil.

Source: Please use tear sheet from: Speight, J.G. 2014. *The Chemistry and Technology of Petroleum*, 5th Edition. CRC Press, Taylor & Francis Publishers, Boca Raton, FL. Table 15.1, p. 392.

DOI: 10.1201/9781003184959-5

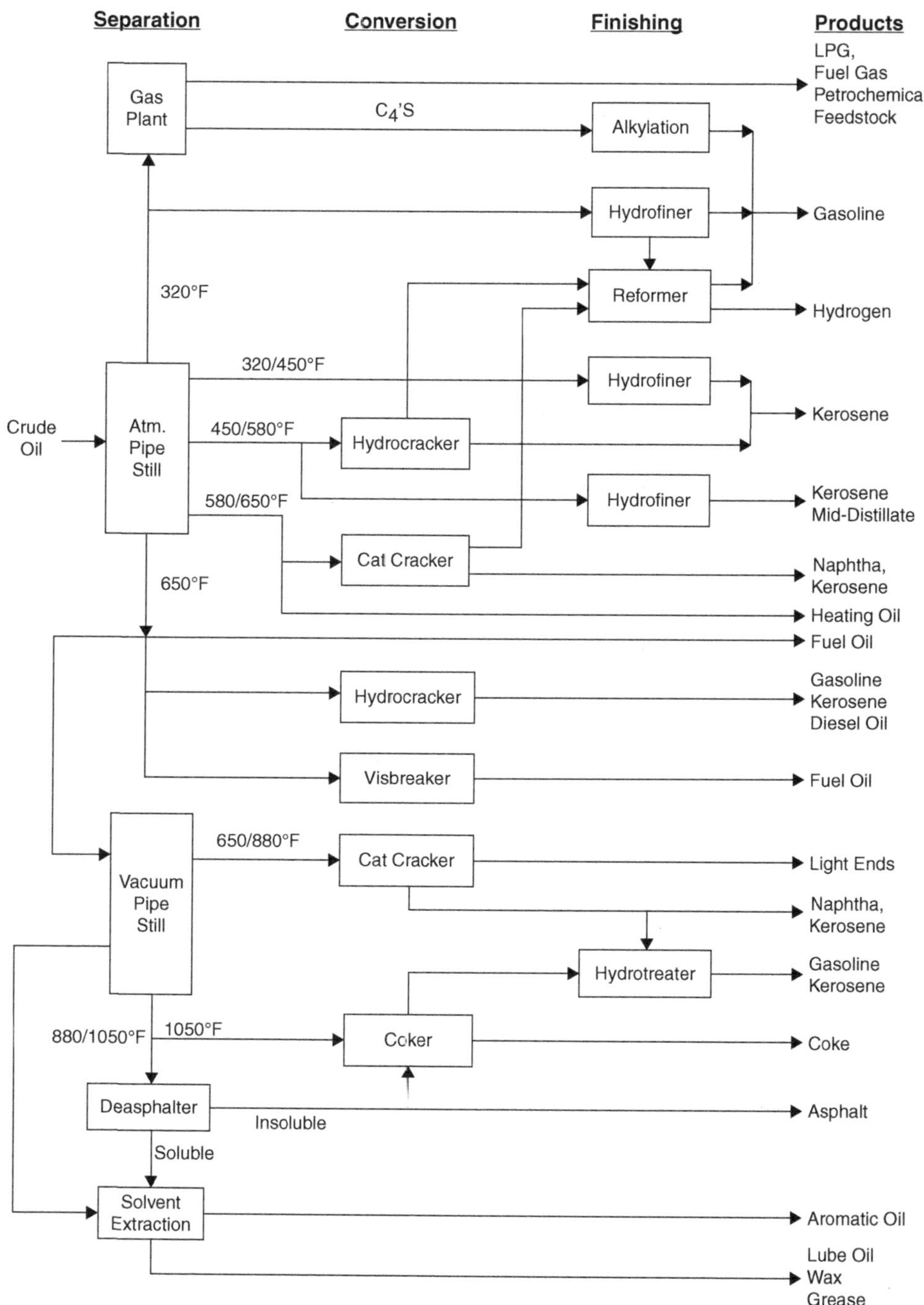

FIGURE 5.1 Schematic overview of a refinery. (Speight, J.G. 2017. *Handbook of Petroleum Refining*. CRC Press, Taylor & Francis Group, Boca Raton, FL, Figure 7.1, p. 251.)

thermal processes to change an excess of heavy fuel oil into more gasoline with coke as the residual product, or a vacuum distillation process to separate the heavy oil into lubricating oil blend stocks and asphalt.

Briefly, crude oil refining is the separation of crude oil into fractions and the subsequent treating of these fractions to yield marketable products (Parkash, 2003; Gary et al., 2007; Speight, 2014, 2017).

In fact, a refinery is essentially a group of manufacturing plants which vary in number with the variety of products produced (Figure 5.1).

By the way of introduction and as a continuation from an earlier chapter, the crude distillation unit is at the front end of the crude oil refinery (Chapters 1 and 4). The dewatering and desalting processes (Chapter 2) occur prior to crude oil distillation section. The desalting process of crude oil is imperative to ensure the good quality of crude oil, that is, to remove impurities before its transfer to refining. This procedure minimizes or eliminates contaminants such as sulfur, water, salts, and even mechanical impurities, which ensures the efficient operation of refinery equipment pipelines. The distillation section offers a more physical than a chemical process and is characterized by mass-thermal transfer of materials, which leads to the production of various bulk fractions (Table 5.1).

The distillation section which comprises the atmospheric distillation unit and the vacuum distillation unit operates consecutively. Also, regardless of the profile of the refinery, great importance is given to the crude distillation unit. Thus, the distillation unit section has the capability to produce a variety of bulk fractions (i.e. products), if well modeled and organized, makes it possible to obtain already more light products, the refinement of which requires more processes; however, before being subject to conversion processes (such as catalytic cracking and hydrocracking), further purification may be needed and this is the point at which ancillary distillation processes are employed (Parkash, 2003; Gary et al., 2007; Speight, 2014, 2017; Hsu and Robinson, 2017).

Atmospheric distillation and vacuum distillation provide the primary fractions from crude oil that are used as feedstocks for other refinery processes to convert crude oil into products. Many of these subsequent processes involve fractional distillation and some of the procedures are so specialized and used with such frequency that they are identified by name.

A multitude of separations are accomplished by distillation, but its most important and primary function in the refinery is its use of the distillation tower and the temperature gradients therein (Chapter 3) for the separation of crude oil into fractions that consists of varying amounts of different components and, thence, different yields of the bulk distillation fractions (Coleman et al., 1978; Speight, 2012; Adiko and Mingasov, 2021). Thus, it is possible to obtain products ranging from gaseous materials taken off the top of the distillation column to a non-volatile atmospheric residuum (atmospheric bottoms, reduced crude) with correspondingly lower boiling materials (gas, gasoline, naphtha, kerosene, and gas oil) taken off at intermediate points with each crude oil providing different amounts of various fractions (Chapter 3) (Speight, 2000, 2014, 2017; Parkash, 2003; Gary et al., 2007; Hsu and Robinson, 2006).

The reduced crude may then be processed by vacuum distillation or by steam distillation to separate the high-boiling lubricating oil fractions without the danger of decomposition, which occurs at high (>350°C, 660°F) temperatures (Speight, 2000, 2014, 2017; Parkash, 2003; Gary et al., 2007; Hsu and Robinson, 2006). Indeed, atmospheric distillation may be terminated with a lower boiling fraction (boiling cut) if it is thought that vacuum or steam distillation will yield a better quality product or if the process appears to be economically more favorable.

It should be noted at this point that not all crude oils yield the same distillation products because of the differences in composition (Speight, 2000, 2014, 2017; Parkash, 2003; Gary et al., 2007; Hsu and Robinson, 2006). In fact, the nature of the crude oil dictates the processes that may be required for refining. Crude oil can be classified according to the nature of the distillation residue, which in turn depends on the relative content of hydrocarbon types: paraffins, naphthenes, and aromatics. For example, a paraffin-base crude oil produces distillation cuts with higher proportions of paraffins than asphalt base crude. The converse is also true; that is, an asphalt base crude oil produces materials with higher proportions of cyclic compounds. A paraffin-base crude oil yields wax distillates rather than the lubricating distillates produced by the naphthenic-base crude oils. The residuum from paraffin-base crude oil is referred to as cylinder stock rather than asphaltic bottoms which is the name often given to the residuum from distillation of naphthenic crude oil. It is emphasized that, in these cases, it is not a matter of the use of archaic terminology, but a reflection of the nature of the product and the crude oil from which it is derived.

TABLE 5.2

Production of Starting Materials for Petrochemical Processing

Feedstock	Process	Products
Petroleum	Distillation	Light ends
		Methane
		Ethane
		Propane
		Butane
	Catalytic cracking	Ethylene
		Propylene
		Butylenes
		Higher olefins
	Coking	Ethylene
		Propylene
		Butylenes
		Higher olefins
Natural gas	Refining/gas processing	Methane
		Ethane
		Propane
		Butane

TABLE 5.3

Sources of Naphtha

Process	Primary Product	Secondary Process	Secondary Product
Atmospheric distillation	Light naphtha	Cracking	Petrochemical feedstocks
	Heavy naphtha	Catalytic cracking	Light naphtha
	Gas oil	Catalytic cracking	Light naphtha
	Gas oil	Hydrocracking	Light naphtha
Vacuum distillation	Gas oil	Catalytic cracking	Light naphtha
		Hydrocracking	Light naphtha
	Residuum	Hydrocracking	Light naphtha
		Coking	Light naphtha

As the basic elements of crude oil, hydrogen and carbon form the main input into a refinery, which combine to form thousands of individual constituents, and the economic recovery of these constituents varies with the individual crude oil according to its particular individual qualities and the processing facilities of a particular refinery. In general, crude oil, once refined, yields three basic groupings of products that are produced when it is broken down into cuts or fractions (Table 5.1). The gas and gasoline cuts form the lower boiling products and are usually more valuable than the higher boiling fractions and provide gas (liquefied petroleum gas), naphtha, aviation fuel, motor fuel, and feedstocks for the petrochemical industry (Table 5.2).

Naphtha, a precursor to gasoline and solvents, is produced from the light and middle range of distillate cuts (sometimes referred to collectively as light gas oil) and is also used as a feedstock for the petrochemical industry (Table 5.3). The middle distillates refer to products from the middle boiling range of crude oil and include kerosene, diesel fuel, distillate fuel oil, and light gas oil.

Waxy distillate and lower boiling lubricating oils are sometimes included in the middle distillates. The remainder of the crude oil includes the higher boiling lubricating oils, gas oil, and residuum (the non-volatile fraction of the crude oil). The residuum can also produce heavy lubricating oils and waxes but is more often sued for asphalt production. The complexity of crude oil is emphasized insofar as the actual proportions of light, medium, and heavy fractions vary significantly from one crude oil to another.

However, in commercial atmospheric distillation units and vacuum distillation units, the separation of the feedstock to discrete boiling fraction in distillation is not perfect. For example, a fraction with a true boiling point cut of 150°C–205°C (approximately 300°F–400°F) will have material (referred to as tails) that boils below 150°C (300°F) and other material that boils above 205°C (400°F). Because of these tails, the yield of the required product must be reduced to stay within the desired product quality limits. Furthermore, the size and shape of the tails of each product depends on the characteristics of the unit from which it was produced. The factors affecting the fractionation are the number of trays between the product draw trays, tray efficiency, reflux ratio, operating pressure, and boiling ranges of the products.

Because of the presence of the tails within each side-stream distillation fraction, there is the need to reprocess the side-stream fractions using any one (or more of a variety of processes) to ensure that the requirements of the distillation fractions are met. These processes (formerly known as ancillary distillation processes) are presented in the following section.

5.2 ANCILLARY PROCESSES

In the crude oil refinery, crude oil is altered into usable, consumable products such as gasoline, diesel, jet fuel, fuel oil, and other petroleum products. However, these desired products are not produced during the distillation process but are the products of other processes, which typically starts with a variety of conversion processes as well as with the processes known as ancillary distillation processes.

In the atmospheric distillation process, most towers have 25–35 trays between the flash zone and the tower top. The number of trays in various sections of the tower depends on the properties of fractions desired from the crude column. The allowable pressure drop for trays is approximately 0.1–0.2 psi per tray. Generally, a pressure drop of 5 psi is allowed between the flash zone and the tower top. Flash zone pressure is set as the sum of reflux drum pressure and combined pressure drop across condenser and trays above the flash zone. A pressure drop of 5 psi between the flash zone and the furnace outlet is generally allowed. In spite of the varied parameters, the products isolated as side-stream producers may (or typically) require further distillation to remove higher and lower boiling constituents that are not a desired part of a particular side-stream. In the case of side-stream strippers (using, for example, steam as the stripping agent), the small side-stream stripper columns (which are a very simple side-stream ancillary process), typically contain four to six plates, where lower boiling hydrocarbons are stripped out and the flash point of the product adjusted to the requirements.

As a simple example, distillate products such as naphtha (a blend stock for gasoline production) and kerosene (a blend stock for diesel fuel) are withdrawn from the column as side-stream and usually contain material from adjacent fractions. As examples, the naphtha fraction may contain some soluble gases and kerosene constituents, while the kerosene fraction may contain some naphtha and the light diesel cut may contain some light gas oil constituents. These side-stream fractions are then submitted to an ancillary process to remove the unwanted lower boiling constituents and the unwanted higher boiling constituents.

A wide assortment of processes and equipment not directly involved in the refining of crude oil is used in functions vital to the operation of the refinery. Examples are boilers, waste water treatment facilities, hydrogen plants, cooling towers, and sulfur recovery units. Products from auxiliary facilities (clean water, steam, and process heat) are required by most process units throughout the refinery. However, in the current context, the focus is on the assorted processes that complement the atmospheric distillation unit and the vacuum distillation unit, often referred to collectively as ancillary distillation process. While there are several such ancillary units, each one has a specific purpose and is only applied to the distillation product on an as-needed basis.

In simple distillation, a multi-component liquid mixture is slowly boiled in a heated zone and the vapors are continuously removed as they form, and at any instant in time, the vapor is in equilibrium with the liquid remaining on the still. Because the vapor is always richer in the more volatile components than the liquid, the liquid composition changes continuously with time, becoming more and more concentrated in the least volatile species. A simple distillation residue curve is a means by which the changes in the composition of the liquid residue curves on the pot change over time (Speight, 2000, 2014, 2015, 2017; Parkash, 2003; Gary et al., 2007; Hsu and Robinson, 2017). A residue curve map is a collection of the liquid residue curves originating from different initial compositions. Residue curve maps contain the same information as phase diagrams, but residue curve maps represent this information in a way that is more useful for understanding how to synthesize a distillation sequence to separate a mixture.

The composition of any fraction from the distillation of crude oil sample is assessed (or determined) by approximation of a true boiling point (TBP) curve. The method used is basically a batch distillation operation, using a large number of stages, usually greater than 60, and high reflux to distillate ratio (greater than 5). The temperature at any point on the temperature-volumetric yield curve represents the true boiling point of the hydrocarbon material present at the given volume percent point distilled. TBP distillation curves are generally run only on the crude as well as on various fractions from the distillation section of the refinery.

All of the residue curves originate at the light (lowest boiling) pure component in a region, move toward the intermediate boiling component, and end at the heavy (highest boiling) pure component in the same region. The lowest temperature nodes are termed as unstable nodes, as all trajectories leave from them; while the highest temperature points in the region are termed stable nodes, as all trajectories ultimately reach them. The point that the trajectories approach from one direction and end in a different direction (as always is the point of intermediate boiling component) is termed saddle point. Residue curve that divides the composition space into different distillation regions is called distillation boundaries.

Many different residue curve maps are possible when azeotropes are present. Ternary mixtures containing only one azeotrope may exhibit six possible residue curve maps that differ depending on the binary pair that formed the azeotrope and whether it is minimum or maximum boiling. By identifying the limiting separation achievable by distillation, residue curve maps are also useful in synthesizing separation sequences by combining distillation with other methods.

Intermediate products (also called side-streams product) are withdrawn at several points from the column. The side-streams are known as intermediate products because they have properties between those of the top or overhead product and those of the lower boiling constituents and purification of the side-stream is required. This can be accomplished by any one of several ancillary distillation units as described in the following sub-sections and these ancillary process (alphabetically) are (1) azeotropic distillation, (2) batch distillation, (3) extractive distillation, (4) pressure-swing distillation, (5) reactive distillation, (6) rerunning or redistillation, (7) steam distillation, (8) stripping, (9) stabilization and light end removal, and (10) superfractionation. The choice of any one (or more) of these ancillary processes is dependent on the properties of the original side-stream fraction as well as on the requirements (or specifications) of the finished product.

5.2.1 Azeotropic Distillation

In the early days of application of azeotropic distillation to be successful in the refinery, it was typically required of an azeotrope-forming substance that (1) it has a boiling point close to or not more than 30°C–40°C (54°F–72°F) away from the boiling range of the hydrocarbons to be separated,

(2) it is completely soluble in water and preferentially less soluble in the hydrocarbon at room temperature, so that the removal of the azeotrope-forming substance from the hydrocarbon with which it is associated in the azeotropic distillate may be accomplished easily by extraction with water, (3) it can be completely soluble in the hydrocarbon at the distillation temperature and for some degrees below this temperature. This is to ensure that two phases shall not form in the condenser and reflux regulator which would make the latter difficult to operate, and, when packed columns are used instead of bubble-cap columns, the liquid phase in the rectifying section shall be entirely homogeneous, thus avoiding the possibility of irregular segregation of two liquid phases in the packed section, (4) be readily obtainable in a sufficiently pure state at reasonable cost, and (5) be nonreactive with hydrocarbons or with the material of the still (Mair et al., 1941).

Thus, the separation of components with similar volatility may become economic if an entrainer can be found that effectively changes the relative volatility. It is also desirable that the entrainer (i.e. the separating agent used to enhance the separation of closely boiling compounds in azeotropic distillation or in extractive distillation) be reasonably stable, nontoxic, and readily recoverable from the components. In practice, it is probably this last-named criterion that limits severely the application of extractive and azeotropic distillation. The majority of successful processes are those in which the entrainer and one of the components separate into two liquid phases on cooling if direct recovery by distillation is not feasible.

For example, all compounds have definite boiling temperatures, but a mixture of chemically dissimilar compounds sometimes causes one or both of the components to boil at a temperature other than that expected. Thus, benzene boils at 80°C (176°F), but if it is mixed with hexane, it distills at 69°C (156°F). Thus, an azeotropic mixture boils at a temperature lower than the boiling point of either of the components.

In fact, although there are exceptions, nearly all polar organic molecules of the proper volatility form, with hydrocarbons of the paraffinic, naphthenic, aromatic, and olefinic classes, mixtures that have a total vapor pressure greater than that of the more volatile pure component, yielding minimum-boiling azeotropic mixtures. The organic compounds that form azeotropic mixtures with these hydrocarbons include those containing hydroxyl, carboxyl, cyanide, amino, nitro, and other structural groups that tend to produce polarity in organic molecules.

Depending on the boiling point of entrainer, three types of entrainer are observed which are (1) a heavy entrainer, which has a higher boiling point as compared to the components of an azeotropic mixture, (2) an intermediate entrainer, which has a boiling point in between boiling point of both the components of the mixture, and (3) a light entrainer, which has a lower boiling point with respect to both components of mixture. In many cases, the use of a light entrainer may give a more favorable outcome in a batch stripper (Deorukhkar et al., 2016). In fact, an entrainer can be used in any one of several ancillary distillation processes (Table 5.4) (Deorukhkar et al., 2016):

Two main types of azeotropes exist, i.e. the homogeneous azeotrope, where a single liquid phase is in the equilibrium with a vapor phase, and the heterogeneous azeotropes, where the overall liquid composition which forms two liquid phases is identical to the vapor composition. Most methods of distilling azeotropes and low relative volatility mixtures rely on the addition of specially chosen chemicals to facilitate the separation.

The separation of components with similar volatility may become economical if an entrainer can be found that effectively changes the relative volatility. It is also desirable that the entrainer be reasonably cheap, stable, nontoxic, and readily recoverable from the components. In practice, it is probably this last criterion that severely limits the application of extractive and azeotropic distillation. The majority of successful processes, in fact, are those in which the entrainer and one of the components separate into two liquid phases on cooling if direct recovery by distillation is not feasible. A further restriction in the selection of an azeotropic entrainer is that the boiling point of the entrainer be in the range 10°C–40°C (18°F–72°F) below that of the components. Thus, although the entrainer is more volatile than the components and distills off in the overhead product, it is present in a sufficiently high concentration in the rectification section of the column.

TABLE 5.4

Types of Distillation Processes in Which an Entrainer Can Be Employed

Homogeneous Azeotropic Distillation

Entrainer is completely miscible with the component in the original mixture

Forms homogeneous azeotrope with component in the original mixture

Carried out in single feed column

Heterogeneous Azeotropic Distillation

Entrainer forms heterogeneous azeotrope with one or more components in the original mixture which is carried out in a
distillation column with a decanter

Extractive Distillation

Carried out in two feed column in which entrainer is introduced above the original mixture feed point and largely
removed as bottom product

Entrainer is used to enhance the relative volatility of low volatility component to precede the separation of mixture

Entrainer has high boiling point (heavy entrainer) as compared to the original mixture component

Does not form any type of azeotrope

Reactive Distillation

Entrainer reacts with one or more components from the original mixture

Nonreacting component is produced as a distillate

Entrainer is recovered from reverse reaction

Salt Distillation

A type of extractive distillation

The relative volatility is altered by the addition of salt as an entrainer in the top reflux

---

Azeotropic distillation (Figure 5.2) is the use of a third component to separate two close-boiling
components by means of the formation of an azeotropic mixture between one of the original components and the third component to increase the difference in the boiling points and facilitate separation by distillation.

The five methods for separating azeotropic mixtures are the following: (1) extractive distillation
and homogeneous azeotropic distillation in which the liquid separating agent is completely miscible,
(2) heterogeneous azeotropic distillation, or more commonly, azeotropic distillation, where the liquid
separating agent – referred to as the entrainer – forms one or more azeotropes with the other components in the mixture and causes two liquid phases to exist over a wide range of compositions which is
the key to enable the distillation sequence work, (3) distillation using ionic salts, where the salts disso-
ciate in the liquid mixture and alter the relative volatilities sufficiently so that the separation becomes
possible, (4) pressure-swing distillation where a series of columns operating at different pressures are
used to separate binary azeotropes which change appreciably in composition over a moderate pressure
range or where a separating agent which forms a pressure-sensitive azeotrope is added to separate a
pressure-insensitive azeotrope, and (5) reactive distillation where the separating agent reacts preferentially and reversibly with one of the azeotropic constituents after which the reaction product is distilled
from the nonreacting components and the reaction is reversed to recover the initial component.

In fact, there is an increasing interest in the development of reactive separation processes which
combine reaction and separation mechanisms into a single integrated unit. Such processes bring
several important advantages among which are increase of reaction yield and selectivity, overcom-
ing thermodynamic restrictions, e.g., azeotropes, and considerable reduction in energy, water and
solvent consumption. Important examples of reactive separations are reactive distillation and reactive absorption. Due to strong interactions of chemical reaction and heat and mass transfer, the

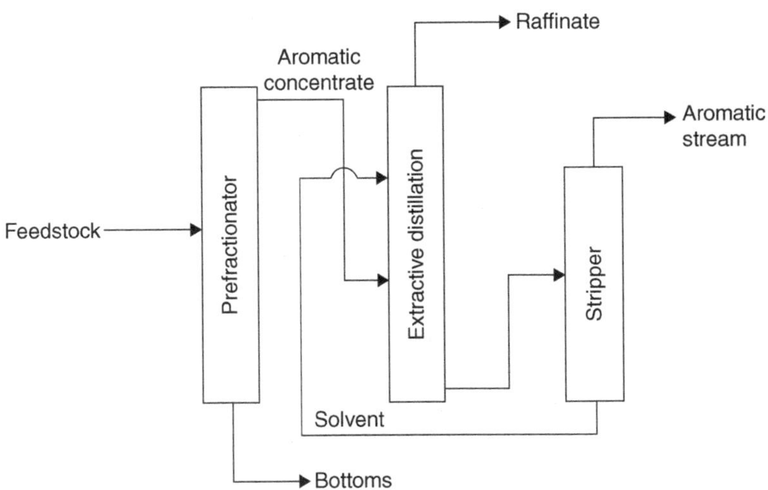

FIGURE 5.2 Extractive distillation for aromatics recovery. (Speight, J.G. 2017. *Handbook of Petroleum Refining*. CRC Press, Taylor & Francis Publishers, Boca Raton, FL, Figure 7.18, p. 283.)

process behavior of reactive separation processes may be adaptable not only to crude oil refineries (especially in the petrochemical section although other options within the refinery are also possibilities) but also for the separation of alternate fuels (Noeres et al., 2003; Sundmacher and Kienle, 2003; Harmsen, 2007; Kiss, 2009, 2011; Cardenas-Guerra et al., 2010; Kiss et al., 2018).

The separation of a desired compound calls for azeotropic distillation. All compounds have definite boiling temperatures, but a mixture of chemically dissimilar compounds sometimes causes one or both of the components to boil at a different temperature than expected.

5.2.2 Batch Distillation

Batch distillation is the distillation process in which the feedstock is distilled in batches insofar as a mixture (in this context, a crude oil or a crude oil product) is distilled to separate into its component fractions before the distillation still is charged with more mixture and the process is repeated. Thus, when small amounts of material or varying product compositions are required, batch distillation has several advantages over continuous distillation. In batch distillation, the feedstock is loaded into the reboiler, the steam is turned on, and after a short startup period, product can be withdrawn from the top of the column. When the distillation is finished, the heat is shut off and the material left in the reboiler is removed. Then, a new batch can be started. Usually, the distillate is the desired product.

The process is widely used for the separation of specialty and fine chemicals as well as for the recovery of small quantities of solvent during the production of high-purity products with added value. The main advantage of the batch distillation process is that it has a much simpler process structure (one column) than for a continuous operation and/or flexible in the range of product changes and also product amount variations.

In this process, the vapor passes upward through the column and is condensed into liquid at the top of the column. Part of the liquid is returned to the column as reflux and the remainder withdrawn as distillate. The main steps are operated discontinuously. In contrast with a continuous process, a batch process does not deliver its product continuously but in discrete manner. In practice, most batch processes are made up of a series of batch and semi-continuous steps. A semi-continuous step runs continuously with periodic startups and shutdowns.

The two basic modes of batch distillation are (1) constant reflux and variable product composition, (2) variable reflux and constant product composition of the key component, and (3) optimal reflux and optimal product composition.

Similar operating modes are also observed in the emerging batch distillation columns. For example, a stripper can also have three operating modes: (1) constant reboil ratio, (2) variable reboil ratio, and (3) optimal reboil ratio. The combination of the three reflux and three reboil modes results in at least nine possible operating policies. The operating modes of a multivessel column can be derived based on the middle vessel column, but this column configuration requires additional considerations with respect to operating variables such as the holdup in each vessel. The total reflux mode can also be considered especially in the middle vessel and multivessel columns.

In the continuous distillation processes, the distillation column is constantly being fed with the stream to be separated. No interruptions occur unless there is a problem with the column or surrounding process units. The process is the most commonly used and is capable of handling high throughputs.

More specifically, for continuous distillation operation, two columns are in operation for the continuous pressure-swing distillation system at two different pressures. Feedstock streams with different concentrations have to be put into the suitable column, depending on the concentration under or above the azeotropic point. For concentrations under the azeotropic point, the feedstock is introduced into the low pressure column; however for concentrations above the azeotropic point, the feedstock has to be introduced into the high pressure column. In both columns, pure product is withdrawn from the bottom. At the top of the columns, there are azeotropic mixtures with concentrations depending on the pressure in the column. The distillate stream is recycled into the other column, so there is a mass integration between the columns.

Semi-continuous operation involves only a single distillation column, which has lower investment costs and shorter downtimes when the mixture to be separated changes. Liquid levels are maintained, where reboiler and condenser are in charged condition. An optimal-control algorithm is employed to determine desirable campaigns and to schedule pressure switch-over policies.

Batch distillation is often preferable to continuous distillation where relatively small quantities of material are to be handled at regularly scheduled periods, as may occur in the petrochemical section of a refinery. The basic difference between batch distillation and continuous distillation is that in continuous distillation the feed is continuously entering the column, while in batch distillation the feedstock is charged into the reboiler at the beginning of the operation. Also, the batch distillation process is versatile insofar as a run may last from a few hours to several days. The process is used when the plant does not run continuously and is often used when the same equipment distills several different products at different times. If distillation is required only occasionally, batch distillation is the usual choice.

An alternative is called the inverted batch distillation (Wankat, 2016) because bottoms are withdrawn continuously while distillate is withdrawn only at the end of the distillation. In this case, the charge is placed in the accumulator and a reboiler with a small holdup is used. Inverted batch distillation is seldom used, but it is useful when quite pure bottoms product is required. Additional configurations include middle-vessel, multivessel, extractive, and hybrid (Wankat, 2016).

5.2.3 EXTRACTIVE DISTILLATION

The extractive distillation process, like the azeotropic distillation process, involves the addition of an entrainer. Thus, extractive distillation (Figure 5.3) is the use of a third component to separate two close-boiling components in which one of the original components in the mixture is extracted by the third component and retained in the liquid phase to facilitate separation by distillation. In this process, the difference in volatility of the components to be separated is enhanced by the addition of a solvent or an entrainer (Gerbaud et al., 2019).

The process uses a separation solvent, which is generally non-volatile, has a high boiling point, and is miscible with the mixture, but doesn't form an azeotropic mixture. The solvent interacts differently with the components of the mixture thereby causing their relative volatilities to change which enables the new three-part mixture to be separated by normal distillation. The original component with the greatest volatility separates out as the top product. The bottom product consists of

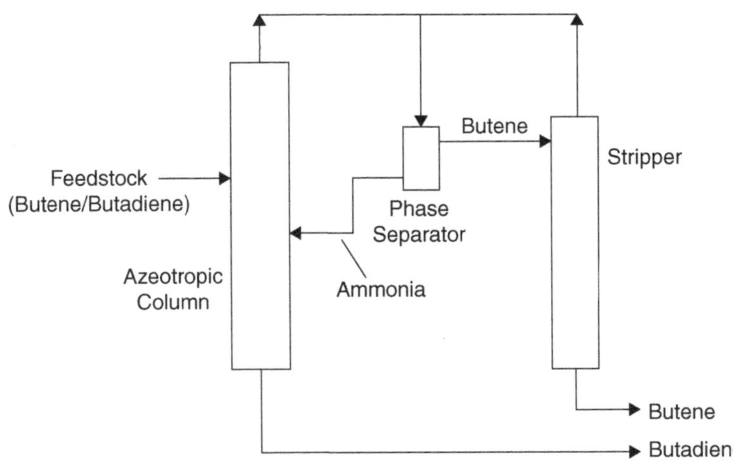

FIGURE 5.3 Separation of butene and butadiene by azeotropic distillation. (Speight, J.G. 2017. *Handbook of Petroleum Refining.* CRC Press, Taylor & Francis Publishers, Boca Raton, FL, Figure 7.19, p. 284.)

a mixture of the solvent and the other component, which can again be separated easily because the solvent does not form an azeotrope with it.

Extractive distillation processes enable the separation of non-ideal mixtures, including minimum or maximum boiling azeotropes and low relative volatility mixtures. Unlike azeotropic distillation, the entrainer fed at another location than the main mixture induces an extractive section within the column. A general feasibility criterion shows that intermediate and light entrainers as well as heterogeneous entrainers are suitable along common heavy entrainers. Entrainer selection rules rely upon selectivity ratios and residue curve map topology including volatility curves. For each type of entrainer, we define extractive separation classes that summarize feasibility regions, achievable products, and entrainer – feed flowrate ratio limits.

Depending on the separation class, a direct or an indirect split column configuration will allow to obtain a distillate product or a bottom product. Batch and continuous process operations differ mainly by the feasible ranges for the entrainer – feed flow rate ratio and reflux ratio. The batch process is feasible under total reflux and can orient the still path by changing the reflux policy. Optimization of the extractive process must give consideration to the extractive column along with the entrainer regeneration column that requires energy and may limit the product purity in the extractive column through recycle (Gerbaud et al., 2019).

It is important to select a suitable separation solvent for this type of distillation and the solvent must alter the relative volatility by a wide enough margin for a successful result. Also, the solvent should be easily separable from the bottom product and should not react chemically with the components or the mixture or cause corrosion of the equipment.

Using acetone-water as an extractive solvent for butanes and butenes, butane is removed as overhead from the extractive distillation column with acetone-water charged at a point close to the top of the column. The bottom product of butenes and the extractive solvent are fed to a second column where the butenes are removed as overhead. The acetone-water solvent from the base of this column is recycled to the first column.

Extractive distillation may also be used for the continuous recovery of individual aromatics, such as benzene, toluene, or xylene(s), from the appropriate crude oil fractions (Figure 5.4).

Prefractionation concentrates a single aromatic cut into a close-boiling cut after which the aromatic concentrate is subjected to extractive distillation with a solvent (usually phenol) for benzene or toluene recovery. Mixed cresylic acids (cresol derivatives and methyl phenol derivatives) are used as the solvent for xylene recovery.

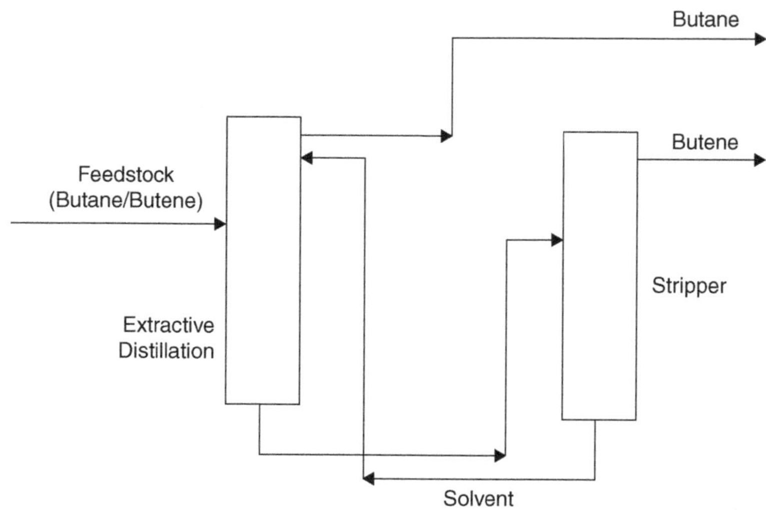

FIGURE 5.4 Separation of butane and butene by extractive distillation. (Speight, J.G. 2017. *Handbook of Petroleum Refining.* CRC Press, Taylor & Francis Publishers, Boca Raton, FL, Figure 7.19, p. 284.)

In general, none of the fractions or combinations of fractions separated from crude oil are suitable for immediate use as crude oil products. Each must be separately refined by treatments and processes that vary with the impurities in the fraction and the properties required in the finished product. The simplest treatment is the washing of a fraction with a lye solution to remove sulfur compounds. The most complex is the series of treatments – solvent treating, dewaxing, clay treating, or hydrorefining, and blending – required to produce lubricating oils. On rare occasions, no treatment of any kind is required. Some crude oils yield a light gas oil fraction that is suitable as furnace fuel oil or as a diesel fuel.

Extractive distillation is successful because the solvent is specially chosen to interact differently with the components of the original mixture, thereby altering their relative volatilities. Because these interactions occur predominantly in the liquid phase, the solvent is continuously added near the top of the extractive distillation column so that an appreciable amount is present in the liquid phase on all of the trays below. The mixture to be separated is added through second feed point further down the column. In the extractive column, the component having the greater volatility, not necessarily the one having the lowest boiling point, is taken overhead as a relatively pure distillate. The other component leaves with the solvent via the column bottoms. The solvent is separated from the remaining components in a second distillation column and then recycled back to the first column.

One of the most important steps in developing a successful (economical) extractive distillation sequence is selecting a good solvent. In general, selection criteria for the solvent include the following, the process (1) should significantly enhance the natural relative volatility of the key component, (2) should not require an excessive ratio of solvent to non-solvent because of cost of handling in the column and auxiliary equipment, (3) should not lead to the formation of two phases, and (4) allow the desired product to be easily separable from the bottom product.

No single solvent or solvent mixture satisfies all of the criteria for use in extractive distillation. In general, none of the fractions or combinations of fractions separated from crude oil are suitable for immediate use as crude oil products. Each fraction must be separately refined by processes that vary with the impurities in the fraction and the properties required in the finished product (Parkash, 2003; Gary et al., 2007; Speight, 2014, 2017; Hsu and Robinson, 2017). The simplest treatment is the washing of a fraction with a lye solution to remove sulfur compounds. The most complex is the series of treatments – solvent treating, dewaxing, clay treating or hydrorefining, and blending – required to produce lubricating oils. On rare occasions, no treatment of any kind is required. Some crude oils yield a light gas oil fraction that is suitable as furnace fuel oil or as a diesel fuel.

Two processes illustrate the similarities and differences between azeotropic distillation and extractive distillation. Both have been used for the separation of C$_4$ hydrocarbons (Figures 5.3 and 5.4). Thus, butadiene and butene may be separated by the use of liquid ammonia, which forms an azeotrope with butene. The ammonia–butene azeotrope overhead from the azeotropic distillation is condensed, cooled, and allowed to separate into a butene layer and a heavier ammonia layer. The butene layer is fed to a second column, where the ammonia is removed as a butene–ammonia azeotrope, and the remaining butene is recovered as bottom product. The ammonia layer is returned to the lower section of the first azeotropic distillation column. Butadiene is recovered as bottom product from this column.

5.2.4 FLASH VAPORIZATION

Flash vaporization (sometimes referred to as flash evaporation) is the partial or total evaporation that occurs when a saturated liquid stream is subjected to a reduction in pressure by passing the stream through a throttling valve or other throttling device.

If the throttling valve or device is located at the entry into a pressure vessel so that the flash evaporation occurs within the vessel, the vessel is often referred to as a flash drum. Thus:

$$\text{Feedstock} \rightarrow \text{heater} \rightarrow \text{flash drum} \rightarrow \text{vapor plus liquid}$$

If the saturated liquid is a single-component liquid (such as, for example, liquid propane), a part of the liquid immediately "flashes" into vapor (i.e. evaporation occurs). Both the vapor and the residual liquid are cooled to the saturation temperature (the saturation point) of the liquid at the reduced pressure (which is often referred to as "auto-refrigeration"). If the saturated liquid is a multi-component liquid (for example, a mixture of propane and butane isomers), a part of the liquid will also immediately flash into a vapor and the flashed vapor will be richer in the more volatile components than in the remaining liquid.

In this process, a heated liquid mixture is passed through a throttling valve in order to vaporize the liquid mixture. In this process, the feedstock, which can be a mixture of constituents in liquid phase or in partially vapor–liquid phase, is heated in order to increase its temperature. The heated feedstock is passed through a throttling valve which causes partial vaporization of the feedstock. If the feed is a liquid mixture, then vapors are produced, and if partially vaporized feed is passed through the valve, then more vapors are produced.

The partially vaporized feedstock is then passed to a flash drum and the pressure of the drum is adjusted accordingly so that separation also occurs inside the flash drum. During separation, some quantity of liquid may get carried away along with vapor (by entrainment) and the entrained droplets are separated by the use of de-entrainment mesh pads inside the drum. The vapor is obtained from the top of the drum (the top product) and the liquid is collected from the base of the drum (the bottom product).

Equilibrium flash vaporization involves heating a flowing feed and the separation of the liquid and vapor in a flash drum. The equilibrium flash vaporization of a multi-component liquid approximates a simple distillation process using a single equilibrium stage and is more complex than the flash evaporation of single-component liquid. In this process, the feedstock is heated as it flows continuously through a heating coil (Nelson, 1958; Edmister, 1961; Maxwell, 1968). Vapor formed travels along in the tube with the remaining liquid until separation is permitted in a vapor separator or vaporizer. By conducting the operation at various outlet temperatures, a curve of percent vaporized vs. temperature may be plotted. Also, this distillation can be run at a pressure above atmospheric pressure as well as under vacuum. Equilibrium flash vaporization (EFV) curves are run chiefly on crude oil or reduced crude samples being evaluated for vacuum column feed.

While flash vaporization as a condensate stabilization method is older technology that is not used in a modern gas processing plant, variations of the process are used as part of crude oil production facilities for stabilizing crude oil prior to transportation of the crude oil.

5.2.5 PRESSURE-SWING DISTILLATION

The pressure-swing distillation process is a method for separating a pressure-sensitive azeotrope by utilizing two columns operated in sequence at two different pressures. Thus, if a binary azeotrope mixture loses azeotropic behavior when the pressure in the system is changed, separation can be achieved without using an additional entrainer. Pressure-swing distillation is a method for separating a pressure-sensitive azeotrope that utilizes two columns operated in sequence at two different pressures. Pressure-swing distillation can be applied to both minimum-boiling and maximum boiling homogeneous azeotropic mixtures.

Whenever a liquid is boiled, it is often at a constant pressure, and when conventional distillation columns are employed to treat such systems, then the separation won't take place even if an infinite number of stages is used. Usually, when pressure changes, then the location of the azeotrope point also changes, and for the same liquid mixture, the composition at which the mixture forms an azeotrope is different for different pressures. Since a change in pressure can alter relative volatility of the liquid mixture, for many azeotropes, it is observed that if the pressure is changed to a certain level, then the azeotrope point disappears from the system. The system becomes completely zeotropic and it can be separated in a distillation column.

The process is one of the methods which are employed in process and chemical industries to separate azeotropes (liquid mixtures which when boiled gives no separation) and the composition of components in the gas phase is equal to the composition of components in the liquid phase. In fact, whenever a liquid is boiled, it is at a constant pressure, and when ordinary distillation columns are used to treat such systems, the separation will not occur even if an infinite number of stages is used.

The principle of the pressure-swing distillation process is based on the principle that at a constant pressure, the mixtures that form azeotrope may have more than one point of azeotropy but it is quite common for azeotropes to have only one point of azeotropy. Usually, when pressure changes, then the location of the point of azeotropy also changes. It means for the same liquid mixture, the composition at which the mixture forms an azeotrope is different for different pressures. Since a change in pressure can alter relative volatility of the liquid mixture, for many azeotropes, it is observed that if the pressure is changed to a certain level, then the azeotropy disappears from the system. The system becomes completely zeotropic and it can be separated in a distillation column.

The pressure-swing azeotropic distillation process uses two columns operating at two different pressures to separate azeotropic mixtures by taking high-purity product streams from one end of the columns and recycling the streams from the other end with compositions near the two azeotropes. This configuration mostly used due to economic reasons. In this process, changes in pressure significantly in the distillation column shift the composition of the azeotrope. The larger the shift, the smaller the required recycle flow rates, so the smaller the energy requirements in the two reboilers.

Pressure-swing distillation does not utilize any additives and no extra component is added into the feed or the column in order to make changes in the liquid mixture. In the continuous operation, the separation is performed using two columns which are maintained at two different pressures. The first column is maintained at a low pressure and the second column is maintained at a high pressure. The distillate from second column is recycled and fed back to first column.

There are three types of operation modes in the pressure-swing distillation process which are (1) batch operation, (2) semi-continuous operation, and (3) continuous operation.

The batch process is one of the best known distillation processes, and in the current context of the crude oil industry, it might be used in petrochemical operations. A main advantage is that it has a much simpler process structure (one column) than for a continuous operation and/or flexible in the range of product changes and also product amount variations. The semi-continuous operation involves only a single distillation column. Liquid levels are maintained, where reboiler and condenser are in charged condition. In the continuous process, two columns are in operation for the continuous pressure-swing distillation system at two different pressures. Feed streams with different concentrations have to be introduced into the suitable column, depending on whether the concentration is below or above the azeotropic point. For concentrations under the azeotropic point, the feed is introduced into the low pressure column; however, for concentrations above the azeotropic point, the feed has to be introduced into the high pressure column. In both columns, pure product is withdrawn from the bottom. At the top of the columns, there are azeotropic mixtures with concentrations depending on the pressure in the column. The distillate stream is recycled into the other column, so there is a mass integration between the columns.

The advantages of the process are the following: (1) no additional material is needed hence any challenges to deal with the availability or usage of a substance are also eliminated and (2) there are high energy savings in continuous operations because heat transfer can be optimized by utilizing heat integration (Kanse et al., 2019).

5.2.6 REACTIVE DISTILLATION

Reactive distillation is a process in which chemical reaction and distillation occur simultaneously in one single reactor. The process is a combination of reaction and distillation in a single functional unit that can also be effectively used to obtain desired selectivity in multi-reaction schemes involving highly non-ideal mixtures. Reactive distillation is a process that has many advantages, such as (1) the requirement of fewer process units, (2) less heat addition, (3) improved product removal, and (4) degradation of azeotropes (Lopes et al., 2012).

Since the product of the reaction is continuously removed from the reaction equilibrium, a full conversion of otherwise equilibrium-limited reactions can be achieved. In addition to the simultaneous removal and purification of the product, the process does not require a separate reagent recovery and reagent recycle step. Also, exothermic reactions also give rise to the possibility of using in situ heat for evaporation of the liquid, thus reducing the external heat demand in the reboiler.

Thus, reactive distillation (also called catalytic distillation) is a process where the chemical reactor is also the still. Separation of the product from the reaction mixture does not need a separate distillation step which saves energy and materials. The process combines the processes of distillation and catalysis (Ng and Rempel, 2002; Harmsen, 2007; Kiss et al., 2018). Removing one or more of the products is one of the principles behind reactive distillation. The likelihood of the process being employed as a petrochemical-oriented process is high.

Catalytic reactive distillation (sometimes referred simply as catalytic distillation) is a form of reactive distillation which combines the processes of distillation and catalysis to selectively separate mixtures within solutions. The main function of the process is to maximize the yield of catalytic organic reactions.

More specifically, the reactive distillation process relies on the synergy generated when combining catalyzed reactions and distillation into a single unit. However, since both operations take place in the same unit at the same time, there must be a good match between the operating parameters required for reaction and distillation. This overlap is usually limited by the properties of the components (such as boiling points), catalytic activity and catalyst selectivity, as well as the equipment design, among others.

The process is a reactor technology that combines a heterogeneous catalytic reaction and the separation of reactants and products via distillation in a single reactor/distillation column. The heterogeneous catalyst provides the sites for catalytic reactions and also the interfacial surface for liquid/vapor separation. The distinct difference between the catalytic distillation column and the conventional distillation column lies in the placement of solid catalysts usually incorporated in a packing within the distillation column to provide a reaction section in addition to the traditional trays or random packings used for separations in the stripping and rectifying sections of the distillation column. Mass transfer characteristics of the catalytic distillation column packing in the reaction zone have significant influence on the product yield and selectivity.

The benefits of catalytic distillation include energy and capital savings, enhanced conversion and product selectivity, longer catalyst lifetime, and reduction of waste streams. The first commercial application of catalytic distillation was for the production of methyl tertiary butyl ether (MTBE). There are many other possible applications of catalytic distillation such as the hydration of olefins, alkylation reactions, esterification reactions, hydrolysis, aldol condensation, hydrogenation, desulfurization, and oligomerization of olefins, all of which can be used in the petrochemical section of a refinery.

For example, catalytic distillation finds application for reversible reactions, such as MTBE and ethyl tributyl ether (ETBE) synthesis, so as to shift an unfavorable equilibrium by continuous reaction product withdrawal (Diwekar, 1995; DeCroocq, 1997). Catalytic distillation can also provide several advantages in selective hydrogenation of C_3, C_4, and C_5 cuts for petrochemical use. Inserting the catalyst in the fractionation column improves mercaptan derivatives removal, catalyst fouling resistance, and selective hydrogenation performances by modifying the reaction mixture composition along the column.

The separation of the product from the reaction mixture does not need a separate distillation and the method is useful for equilibrium-limited reactions such as esterification and ester hydrolysis. Conversion can be increased beyond what is expected by the equilibrium due to the continuous removal of reaction products from the reactive zone.

The conditions in the reactive column are sub-optimal both as a chemical reactor and as a distillation column, since the reactive column combines these. The introduction of an in situ separation process in the reaction zone or vice versa leads to complex interactions between (1) vapor–liquid equilibrium, (2) mass transfer rates, and (3) diffusion and chemical kinetics, which poses a great challenge for design and synthesis of these systems. Side reactors, where a separate column feeds a reactor and vice versa, are better for some reactions, if the optimal conditions of distillation and reaction differ too much.

Looking to the future in what might be termed as a biorefinery that is an adjunct to (or an integral part of) a crude oil refinery, enzymatic reactive distillation (ERD) is a bio-reactive process, in which enzymes, immobilized on the internal surface of a column, help to overcome chemical reaction and phase equilibrium limitations. A biocatalytic hydrophobic silica coating for commercial structured packing can be used to enable enzymatic reactions in reactive distillation columns (Wierschem et al., 2017).

5.2.7   Rerunning

Rerunning is a general term that covers, in the current context, the redistillation of any fraction of crude oil or a product from crude oil, and it usually indicates that a large part of the material is distilled overhead. The process is used for the purification of a liquid by means of a second, subsequent, or multiple distillations of the product. In contrast, the stripping process removes only a relatively small amount of material as an overhead product. A rerun tower may be associated with a crude distillation unit that produces wide boiling range naphtha as an overhead product. By separating the wide-cut fraction into a light (lower boiling) naphtha fraction and heavy (high-boiling) naphtha fractions, the rerun tower acts in effect as an extension of the crude distillation tower.

The product from chemical treating process of various fractions may be rerun to remove the treating chemical or its reaction products. If the volume of material being processed is small, a shell still may be used instead of a continuous fractional distillation unit. The same applies to gas oils and other fractions from which the front end or tail must be removed for special purposes.

5.2.8   Steam Distillation

Steam distillation (sometimes incorrectly referred to as steam stripping) is a process in which live steam (i.e. the steam is in direct contact with the distilling system) is used in either a batch operation or a continuous operation. The process has been optimized for use in the refinery for recovering or purifying useful compounds from chemical solutions, purifying process streams, and in the treatment of wastewater.

The process is employed (1) to separate relatively small amounts of volatile impurity from a large amount of material, (2) to separate appreciable quantities of higher boiling materials, (3) to recover high-boiling materials from small amounts of impurities, which have a higher boiling point, (4) in cases where the material to be distilled is thermally unstable or reacts with other components associated with it at the boiling temperature, (5) in cases where the material cannot be distilled by indirect heating even under low pressure because of the high boiling temperature, and (6) in cases where direct-fired heaters cannot be used because of fire hazards.

The basis of the steam distillation process rests on the fact that water forms immiscible mixtures with most organic substances, and these mixtures will boil at a temperature below that of either water or the other materials. This is a desirable feature when the organic compound has a high boiling point at which it may be unstable or decompose. Thus, steam distillation can be used when the boiling of the substance to be extracted is higher than that of water, and the starting material cannot be heated to that temperature because of decomposition or other unwanted reactions.

5.2.9 STRIPPING

Steam stripping (in contract to stream distillation) is a process that is used in the refinery and in petrochemical plants to remove volatile contaminants, such as hydrocarbon derivatives and other volatile organic compounds from a product, even from wastewater. It typically consists of passing a stream of superheated steam through the product (or wastewater). A side-stream stripper is a device used to perform further distillation on purification of a liquid stream (side-stream) from any one of the trays (or plates) of a bubble tower, usually with the use of steam.

In contrast and as the term is used in this text, steam distillation is used to isolate a product from a mixture, whereas steam stripping is used to remove volatile constituents from a product.

Stripping is a fractional distillation operation carried out on each side-stream product immediately after it leaves the main distillation tower. Since perfect separation is not accomplished in the main tower, unwanted components are mixed with those of the side-stream product. The purpose of stripping is to remove the more volatile components and thus reduce the flash point of the side-stream product. Thus, a side-stream product enters the top tray of a stripper, and as it spills down the four to six trays, steam injected into the bottom of the stripper removes the volatile components. The steam and volatile components leave the top of the stripper to return to the main tower. The stripped side-stream product leaves at the bottom, and after being cooled in a heat exchanger, goes to storage. Since strippers are short, they are arranged one above another in a single tower; each stripper, however, operates as a separate unit.

In the stripping process, a distillation tower has a plurality of liquid side-stream lines and a multi-stage side-stream stripper which includes a respective stripping section for each side-stream line housed in a common, upright, cylindrical shell that allows vapor to pass freely from each stage to the one above. Partial vaporization of each side-stream is achieved by applying a vacuum to the top of the stripper shell and/or introducing strip gas at the bottom. Because the vapor passes serially through the stripping sections from the bottom of the stripper to the top, the need to supply strip gas separately to the stripping sections and/or apply vacuum individually is avoided. The separation between the side-stream products is improved by including, in each stage, a rectification zone positioned above the stripping section.

A tower stripping unwanted volatile material from gas oil may be only 3 or 4 feet in diameter and 10 feet high with less than 20 trays (Parkash, 2003; Gary et al., 2007; Speight, 2014, 2017; Hsu and Robinson, 2017). Towers concerned with the distillation of liquefied gases are only a few feet in diameter but may be up to 200 feet in height. A tower used in the fractionation of crude oil may have 16–28 trays, but one used in the fractionation (superfractionation) of liquefied gases may have 30–100 trays. The feed to a typical tower enters the vaporizing or flash zone, which is an area without trays. The majority of the trays are usually located above this area. The feed to a bubble tower, however, may be at any point from top to bottom with trays above and below the entry point, depending on the kind of feedstock and the characteristics desired in the products.

In summary, in the steam stripping unit, the bottom stripping steam is used to recover the light components from the bottom liquid. In the flash zone of an atmospheric distillation column, approximately 50%–60% v/v of crude is vaporized. The unvaporized crude travels down the stripping section of the column containing four to six plates and stripped off any low-boiling point distillates still contained in the atmospheric residuum (the reduced crude) by superheated steam. The steam rate used is usually on the order of 5–10 lbs/bbl of stripped product and the flash point of the stripped stream can be adjusted by varying the stripping steam rates.

5.2.10 Stabilization and Light Ends Removal

Stabilization is a term used in the crude oil refinery industry that is achieved by subjecting "live" crude oil (or a live crude oil product) to temperature and pressure conditions in a fractionation vessel which drives off low-boiling hydrocarbon components to form a "dead" crude oil (or "dead" crude oil product) with a lower vapor pressure. The term "light ends" refers to refinery products that condense at the top of the distillation tower during the distillation (Chapter 4). Stabilization by flash vaporization (flash evaporation) is a relatively simple operation that employs two or three flash tanks. This process is similar to stage separation utilizing the equilibrium principles between vapor and condensate phases. Equilibrium vaporization occurs when the vapor and condensate phases are in equilibrium at the temperature and pressure of separation.

The stabilizer column is a fractionation tower that uses trays or packing (structured packing or random packing) in the column to achieve a close contact between the vapor and liquid phases thereby permitting the transfer of mass and heat from one phase to the other. Typically, the trays are orifice-type devices designed to disperse the gas uniformly on the tray and through the liquid on the tray, and the trays are commonly spaced 24 inches apart. The three most widely used trays are the valve, bubble cap, and perforated. Standard random packing, available in numerous sizes, geometric shapes, and materials, is composed of solids randomly packed in the tower. Structured packing is made of folded perforated plates welded together.

Thus, the stabilizer column is a distillation column that is used to remove what is normally a relatively small amount of light ends from a refinery product after which the product is designated as "stabilized" (Parkash, 2003; Gary et al., 2007; Speight, 2014, 2017; Hsu and Robinson, 2017). For example, light (low-boiling) naphtha might be run (passed) through a naphtha stabilizer to remove any hydrocarbons that are gaseous at room temperature (often referred to as light ends) but are dissolved in the naphtha (Brahim et al., 2017). An alternative is to use an absorption process (Chen et al., 2018).

If an extremely close separation is required, the column may be provided with an upper section reflux system which makes the column similar to a distillation column. As the reflux liquid flows down through the column, lower boiling components are removed leaving a richer mixture in higher boiling component. The overhead gas from the stabilizer passes through a back pressure control valve which maintains the pressure in the stabilizer. The stabilized crude oil, which is composed of pentane and higher boiling (higher molecular weight) hydrocarbon derivatives, is drawn from the base of the stabilizer and is cooled; this stabilized crude flows to tanks for storage or to a pipeline for transport to (for example) a crude oil refinery. The stabilization tower may typically operate at approximately 50–200 psi. If the crude oil contains high levels of hydrogen sulfide, the stabilizer may be operated at the lower end of the pressure range, whereas stabilization of a crude oil with low levels of hydrogen sulfide can take place at a higher pressure.

The gaseous and more volatile liquid hydrocarbons produced in a refinery are collectively known as the light ends (light hydrocarbons or low-boiling hydrocarbon derivatives such as methane, ethane, propane, and butane) (Parkash, 2003; Gary et al., 2007; Speight, 2014, 2017; Hsu and Robinson, 2017). Light ends are produced from crude oil in relatively small quantities and in large quantities when gasoline is manufactured by cracking and reforming. When a naphtha or gasoline component at the time of its manufacture is passed through a condenser, most of the light ends do not condense and are withdrawn and handled as a gas. A considerable part of the light ends, however, can remain dissolved in the condensate, thus forming a liquid with a high vapor pressure, which may be categorized as unstable and stabilization is required (Abdel-Aal et al., 2016).

Liquids with high vapor pressures may be stored in refrigerated tanks or in tanks capable of withstanding the pressures developed by the gases dissolved in the liquid. The more usual procedure, however, is to separate the light ends from the liquid by a distillation process generally known as stabilization. Enough of the light ends are removed to make a stabilized liquid, that is, a liquid with a low enough vapor pressure to allow for its storage in ordinary tanks without loss of vapor. The simplest stabilization process is a stripping process. Light naphtha from a crude tower, for example, may be pumped into the top of a tall, small-diameter fractional distillation tower operated under a pressure of 50–80 psi. Heat is introduced at the bottom of the tower by a steam reboiler. As the naphtha cascades down the tower, the light ends separate and pass up the tower to leave as an overhead product. Since reflux is not used, considerable amounts of liquid hydrocarbons pass overhead with the light ends.

Stabilization is usually a more precise operation than that just described. An example of more precise stabilization can be seen in the handling of the mixture of hydrocarbons produced by cracking. The overhead from the atmospheric distillation tower that fractionates the cracked mixture consists of light ends and cracked gasoline with light ends dissolved in it. If the latter is pumped to the usual type of tank storage, the dissolved gases cause the gasoline to boil, with consequent loss of the gases and some of the liquid components. To prevent this, the gasoline and the gases dissolved in it are pumped to a stabilizer maintained under a pressure of approximately 100 psi and operated with reflux. This fractionating tower makes a cut between the highest boiling gaseous component (butane) and the lowest boiling liquid component (pentane). The bottom product is thus a liquid free of all gaseous components, including butane, and hence, the fractionating tower is known as a debutanizer (Figure 5.5).

The debutanizer bottoms (gasoline constituents) can be safely stored, whereas the overhead from the debutanizer contains the butane, propane, ethane, and methane fractions. The butane fraction,

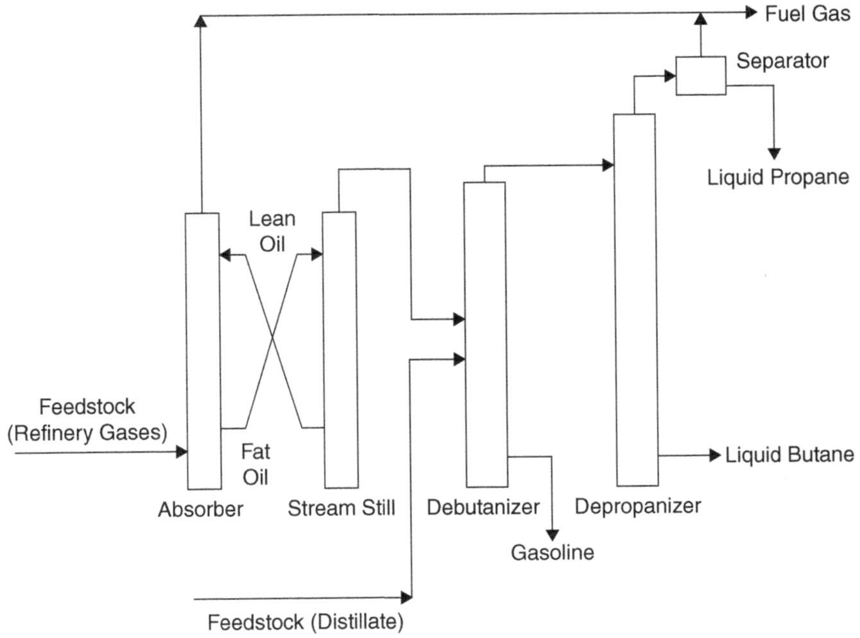

FIGURE 5.5 A light-ends plant. (Speight, J.G. 2017. *Handbook of Petroleum Refining.* CRC Press, Taylor & Francis Publishers, Boca Raton, FL, Figure 7.17, p. 280.)

which consists of all the hydrocarbons containing four carbon atoms, is particularly needed to give easy starting characteristics to gasoline. It must be separated from the other gases and blended with gasoline in amounts that vary with the season: more in the winter and less in the summer. Separation of the butane fraction is effected by another distillation in a fractional distillation tower called a depropanizer, since its purpose is to separate propane and the lighter gases from the butane fraction.

The depropanizer is very similar to the debutanizer, except that it is smaller in diameter because of the smaller volume being distilled and is taller because of the larger number of trays required to make a sharp cut between the butane and propane fractions. Since the normally gaseous propane must exist as a liquid in the tower, a pressure of 200 psi is maintained. The bottom product, known as the butane fraction, stabilizer bottoms, or refinery casinghead, is a high vapor pressure material that must be stored in refrigerated tanks or pressure tanks. The depropanizer overhead, consisting of propane and lighter gases, is used as a petrochemical feedstock or as a refinery fuel gas, depending on the composition.

A depentanizer is a fractional distillation tower that removes the pentane fraction from a debutanized (butane-free) fraction. Depentanizers are similar to debutanizers and have been introduced recently to segregate the pentane fractions from cracked gasoline and reformate. The pentane fraction when added to a premium gasoline makes this gasoline extraordinarily responsive to the demands of an engine accelerator.

The gases produced as overhead products from crude distillation, stabilization, and depropanization units may be delivered to a gas absorption plant for the recovery of small amounts of butane and higher boiling hydrocarbons. The gas absorption plant essentially consists of two towers in which one tower is the absorber where the butane and higher boiling hydrocarbons are removed from the lighter gases.

The gas mixture enters at the bottom of the tower and rises to the top. As it does this, it contacts the lean oil, which absorbs the butane and higher boiling hydrocarbons but not the lower boiling hydrocarbons. The latter leave the top of the absorber as dry gas. The lean oil that has become enriched with butane and higher boiling hydrocarbons is now termed fat oil. This is pumped from the bottom of the absorber into the second tower, where fractional distillation separates the butane and higher boiling hydrocarbons as an overhead fraction and the oil, once again lean oil, as the bottom product.

The condensed butane and higher boiling hydrocarbons are included with the refinery casinghead bottoms or stabilizer bottoms. The dry gas is frequently used as fuel gas for refinery furnaces. It contains propane and propylene, however, which may be required for liquefied petroleum gas for the manufacture of polymer gasoline or petrochemicals. Separation of the propane fraction (propane and propylene) from the lighter gases is accomplished by further distillation in a fractional distillation tower similar to those previously described and particularly designed to handle liquefied gases. Further separation of hydrocarbon gases is required for petrochemical manufacture.

Finally, if the objective of the process is to stabilize condensate, the stabilization is carried out using a fractionation tower in which nonlinear level control is maintained for the upstream flash or feed tank to provide a steady feed rate. Typically, a bottom temperature, preferably pressure compensated, is used to control the input of heat to the reboiler. An inferential property predictor can be added to drive the temperature set point in between laboratory updates. The reflux temperature, when used, should be controlled by a sensitive tray above the feed tray. Tower pressure should be driven as low as possible to enhance separation subject to constraints on an overhead compressor.

5.2.11 SUPERFRACTIONATION

The term superfractionation is sometimes applied to a highly efficient fractionating tower used to separate ordinary crude oil products. For example, in order to increase the yield of furnace fuel oil, heavy naphtha may be redistilled in a tower that is capable of making a better separation of the naphtha and the fuel oil components. The latter, obtained as a bottom product, is diverted to furnace fuel oil.

Fractional distillation as normally carried out in a refinery does not completely separate one crude oil fraction from another. One product overlaps another, depending on the efficiency of the fractionation, which in turn depends on the number of trays in the tower, the amount of reflux used, and the rate of distillation. Kerosene, for example, normally contains a small percentage of hydrocarbons that (according to their boiling points) belong to the naphtha fraction and a small percentage that should be in the gas oil fraction. Complete separation is not required for the ordinary uses of these materials, but certain materials, such as solvents for particular purposes (hexane, heptane, and aromatics), are required as essentially pure compounds. Since they occur in mixtures of hydrocarbons, they must be separated by distillation and with no overlap of one hydrocarbon with another. This requires highly efficient fractional distillation towers specially designed for the purpose and referred to as superfractionators. Several towers with 50–100 trays operated with high reflux ratios may be required to separate a single compound with the necessary purity.

5.3　POTENTIAL FOR CORROSION AND FOULING OF EQUIPMENT

Although by the time the distillation products reach the ancillary units, it is hoped that corrosion will be minimal, if at all. However, corrosion in the ancillary distillation units does occur.

In fact, corrosion in the crude oil distillation overhead systems is a common problem in refineries (Schempp et al., 2017). The primary mechanism of concern in the overhead system is the acidic attack on unit metallurgy due to the presence of high concentrations of hydrogen chloride in condensing water. Other corrosion and operational concerns include corrosion and fouling usually resulting from unintended consequences of acid neutralization and/or inefficient desalter operations.

The crude distillation unit (CDU; also called the atmospheric unit) may also be exposed to increased levels of chlorides and sulfur species, leading to substantial corrosion issues related to the species that are present in the atmospheric distillation unit. These corrosion issues include: (1) corrosion due to sublimating species such as ammonium chloride; (2) aqueous corrosion due to hydrochloric acid, HCl, or acidic sulfur species; and (3) fouling issues related to the buildup of sublimating species.

Often, the principal cause for these problems may be correlated to an excessively high chloride content in the crude oil after the desalter operation or inadequate controls to ensure conditions for fractionation operations above aqueous dew point. The difficulty in predicting and assessing the contribution and severity of these corrosion/fouling problems stems from the significant complexity of the chemistry involved and the inadequate documented experience correlating speciation of contaminants from the crude oil feedstock to the corrosion problem.

REFERENCES

Abdel-Aal, H.K., Aggour, M.A., and Fahim, M.A. 2016. *Petroleum and Gas Filed Processing.* CRC Press, Taylor & Francis Publishers, Boca Raton, FL.

Adiko, S.-B., and Mingasov, R.R. 2021. Crude Distillation Unit (CDU). In: *Analytical Chemistry - Advancement, Perspectives and Applications*, A.N. Srivastava (Editor). IntechOpen. https://www.intechopen.com/chapters/72948.

Brahim, A.O., Abderafi, S., and Bounahmidi, T. 2017. Modeling the stabilization column in the petroleum refinery. *Energy Procedia*, 139: 61–66.

Cardenas-Guerra, J.C., Lopez-Arenas, T., Lobo-Oehmichen, R., and Perez-Cisneros, E.S. 2010. A reactive distillation process for deep hydrodesulfurization of diesel: Multiplicity and operation aspects. *Computers and Chemical Engineering*, 34: 196–209.

Chen, J.Y., Pan, M., He, C., Zhang, B.J., and Chen, Q.L. 2018. New gasoline absorption–stabilization process for separation intensification and flowsheet simplification in refineries. *Industrial & Engineering Chemistry Research*, 57(43): 14707–14717.

Coleman, H.J., Shelton, E.M., Nicholls, D.T., and Thompson, C.J. 1978. Analysis of 800 Crude Oils from United States Oilfields. Report No. BETC/RI-78/14. Technical Information Center, Department of Energy, Washington, DC.

DeCroocq, D. 1997. Major scientific and technical challenges about development of new processes in refining and petrochemistry. *Revue Institut Français de Pétrole*, 52(5): 469–489.

Deorukhkar, O.A., Rahangdale, T.B., and Mahajan, Y.S. 2016. Entrainer selection approach for distillation column. *International Journal of Chemical Engineering Research*, 8(1): 29–38.

Diwekar, U.M. 1995. *Batch Distillation: Simulation, Optimal Design, and Control.* Taylor & Francis Inc., Philadelphia, PA.

Edmister, W.C. 1961. *Applied Hydrocarbons Thermodynamics.* Gulf Publishing Company, Houston, TX.

Gary, J.G., Handwerk, G.E., and Kaiser, M.J. 2007. *Petroleum Refining: Technology and Economics*, 5th Edition. CRC Press, Taylor & Francis Group, Boca Raton, FL.

Gerbaud, V., Rodriguez, I., Hegely, L., Lang, P., Dénes, F., and You, X. 2019. Review of extractive distillation. Process design, operation, optimization and control. *Chemical Engineering Research and Design*, 141: 229–271.

Harmsen, G.J. 2007. Reactive distillation: The front-runner of industrial process intensification. A full review of commercial applications, research, scale-up, design and operation. *Chemical Engineering and Processing*, 46: 774–780.

Hsu, C.S., and Robinson, P.R. 2006. *Practical Advances in Petroleum Processing, Volumes 1 and 2.* Springer, New York.

Kanse, N.G., Matondkar, D., Bane, S., and Matondkar, P. 2019. Overview of pressure-swing distillation process for separation of azeotropic mixture. *IJRAR-International Journal of Research and Analytical Reviews*, 6(1): 674–678.

Kiss, A.A. 2009. Novel process for biodiesel by reactive absorption. *Separation & Purification Technology*, 69: 280–287.

Kiss, A.A. 2011. Heat-integrated reactive distillation process for synthesis of fatty esters. *Fuel Processing Technology*, 92: 1288–1296.

Kiss, A.A., Jobson, M., and Gao, X. 2018. Reactive distillation: Stepping up to the next level of process intensification. *Industrial & Engineering Chemistry Research*, 58: 1–14.

Lopes, M.S., Savioli Lopes, M., Maciel Filho, R., Wolf Maciel, M.R., and Medina, L.C. 2012. Cracking of petroleum residues by reactive molecular distillation. *Procedia Engineering*, 42: 329–334. https://www.research-gate.net/publication/257725441_Cracking_of_Petroleum_Residues_by_Reactive_Molecular_Distillation.

Mair, B.J., Glasgow jr., A.R., and Rossini, F.D. 1941. Separation of hydrocarbons by azeotropic distillation. Research Paper RP1402. *Journal of Research of the National Bureau of Standards*, United States. Department of Commerce, Washington DC, Volume 27, pp. 39–63. https://nvlpubs.nist.gov/nistpubs/jres/27/jresv27n1p39_A1b.pdf.

Maxwell, J.B. 1968. *Data Book on Hydrocarbons.* Van Nostrand Scientific Publishers, Princeton, NJ.

Nelson, W.L. 1958. *Petroleum Refinery Engineering.* McGraw-Hill Publishing Company, New York.

Ng, F.T.T., and Rempel, G.L. 2002. *Catalytic Distillation. Encyclopedia of Catalysis*, John Wiley & Sons Inc., Hoboken, NJ.

Noeres, C., Kenig, E.Y., and Gorak, A. 2003. Modelling of reactive separation processes: reactive absorption and reactive distillation. *Chemical Engineering and Processing*, 42: 157–178.

Parkash, S. 2003. *Refining Processes Handbook.* Gulf Professional Publishing, Elsevier, Amsterdam, pp. 1–15.

Schempp, P., Köhler, S., Menzebach, M., Preuss, K., and Tröger, M. 2017. Corrosion in the crude distillation unit overhead line: Contributors and solutions. *Proceedings of European Corrosion Congress, EUROCORR 2017.* September 3–7, 2017, Prague, Czech Republic. Paper No. 88826. https://www.researchgate.net/publication/319449375_Corrosion_in_the_crude_distillation_unit_overhead_line_Contributors_and_solutions.

Speight, J.G. 2000. *The Desulfurization of Heavy Oils and Residua*, 2nd Edition. Marcel Dekker Inc., New York.

Speight, J.G. 2012. *Crude Oil Assay Database.* Knovel, Elsevier, New York. Online version available at: http://www.knovel.com/web/portal/browse/display?_EXT_KNOVEL_DISPLAY_bookid=5485&VerticalID=0.

Speight, J.G. 2014. *The Chemistry and Technology of Petroleum*, 5th Edition. CRC Press, Taylor & Francis Group, Boca Raton, FL.

Speight, J.G. 2015. *Handbook of Petroleum Product Analysis*, 2nd Edition. John Wiley, Hoboken, NJ.

Speight, J.G. 2017. *Handbook of Petroleum Refining.* CRC Press, Taylor & Francis Group, Boca Raton, FL.

Sundmacher, K., and Kienle, A., 2003. *Reactive Distillation – Status and Future Directions.* Wiley-VCH Verlag GmbH & Co. KGaA, Weinheim.

Wankat, P.C. 2016. *Separation Process Engineering: Includes Mass Transfer Analysis*, 4th Edition. Pearson Education, Hoboken, NJ.

Wierschem, M., Schlimper, S., Heils, R., Smirnova, I., Kiss, A. A., Skiborowski, M., and Lutze, P. 2017. Pilot-scale validation of enzymatic reactive distillation for butyl butyrate production. *Chemical Engineering Journal*, 312: 106–117.

6 Importance in the Refinery

6.1 INTRODUCTION

Crude oil refining is the separation of crude oil into fractions and the subsequent treating of these fractions to yield marketable products through the use of a series of unit processes where each unit process carries out a separate function (Parkash, 2003; Gary et al., 2007; Speight, 2014a, 2017; Hsu and Robinson, 2017). In addition, the corrosion and fouling issues that arise for a particular process (say, for example, the distillation process) will (if allowed to continue unabated) tend to be cumulative throughout the refinery system.

Refinery processes must be selected and products manufactured to give a balanced operation in which the crude oil feedstock is converted into a variety of products in amounts that are in accord with the demand for each (Parkash, 2003; Gary et al., 2007; Speight, 2014a, 2017; Hsu and Robinson, 2017). Thus, a refinery is assembled as a group of integrated manufacturing plants which vary in number with the variety of products produced (Chapter 1). Corrosion causes the failure of equipment items as well as dictates the maintenance schedule of the refinery, during which all or part of the refinery must be shut down. Although significant progress in understanding corrosion has been made, it is also clear that the problem continues to exist and will become progressively worse through the introduction of more heavy crude oils, opportunity crudes, and high-acid crudes into refineries (Pruneda et al., 2005; Bhatia and Sharma, 2006; Collins and Barletta, 2012).

The end of the widespread production of liquid fuels and other products from fossil fuel source within the current refinery infrastructure is considered by some observers to occur during the present century, even as early as within the next five decades but, as with all projections, this is very dependent not only on the remaining reserves but also on petro-politics which is related to the amount of crude oil recovered from various reservoirs (Speight, 2011a; Speight and Islam, 2016). During this time, fossil fuels (including natural gas and crude oil from tight formations) will be the mainstay of the energy scenarios of many countries.

The desalting process, which is actually the first refining process that is applied to the crude oil, removes salt, water, and solid particles that would otherwise lead to operational problems during refining such as corrosion, fouling of equipment, or poisoning of catalysts. In fact, refining actually commences with the production of fluids from the well or reservoir and is followed by pretreatment operations that are applied to the crude oil either at the wellhead to make it suitable for transportation to the refinery or at the refinery itself where further pretreatment of the crude oil occurs before submitting the feedstock for distillation. Crude oil pretreatment programs were originally developed to address desalting challenges presented by heavy crude oils in the 1980s and have, of necessity, continued (Kremer, 2000).

Thus, upon arrival at the refinery, the crude oil is typically placed into storage tanks where the crude oil and any accompanying water can segregate by gravity settlement of the water. Many observers believe that the distillation unit is the primary mode of refinery operations although the dewatering and desalting unit (Chapter 2) is the first refining operation to which the incoming feedstock is subjected (Parkash, 2003; Gary et al., 2007; Speight, 2014a, 2017; Hsu and Robinson, 2017).

Because of the disadvantages of not having a robust and efficient dewatering–desalting system in place, it is necessary to point out the types of corrosion and fouling that can occur and the consequences that follow when corrosion occurs. Thus, the focus of this chapter is to present a more detailed presentation of the corrosion and fouling that can occur in crude oil refineries to emphasize the need for feedstock cleanup as soon as the feedstocks (that are used to produce the feedstock blend) enter the refinery and the consequences of inefficient dewatering and desalting processes.

DOI: 10.1201/9781003184959-6

Removal of salts is important for reducing corrosion in the distillation tower and downstream processing units, and this chapter addresses the issue of the non-removal or partial removal of these salts.

6.2   DEWATERING AND DESALTING

When crude oil is recovered from a reservoir, it is not the pristine liquid that appears in a multitude of photographs; instead, it is mixed with a variety of substances, including gases, water, and dirt (minerals), which must be removed prior to transportation of the crude oil to the refinery; otherwise, corrosion and fouling will occur(Barnett, 1988). Often, some level of pretreatment is necessary at the wellhead because crude oil to be shipped by pipeline or, for that matter, by any other form of transportation must meet rigid specifications with regard to water and salt (Minerals) content. In some instances, sulfur content, nitrogen content, viscosity, and solids content (usually referring to the level of inorganic contaminants) may also need to be specified before the crude oil is accepted for transportation by the pipeline company.

Thus, field separation, which occurs at a field site near the recovery operation, is the first attempt to remove the gases, water, and dirt that accompany crude oil coming from the ground. The separator may be no more than a large vessel that gives a quieting zone for gravity separation into three phases: gases, crude oil, and water containing entrained dirt. If the crude oil from the separators contains water and dirt, water washing can remove much of the water-soluble minerals and entrained solids. If these crude oil contaminants are not removed, they can cause operating problems during refinery processing, such as equipment plugging and corrosion as well as catalyst deactivation. To meet the water content specified by pipeline companies, dehydrators are used to remove much of the remaining formation water and a portion of emulsified water. While salt content of produced crude depends primarily on salt content of formation water, salt content of dehydrated crude depends on its bottom sediment and water content (commonly referred to as the BS&W). Light crude leaving the dehydrators typically contains 0.05–0.2 vol% bottom sediment and water, medium crude 0.1%–0.4% v/v, and heavy crude 0.3%–3% v/v. These values are sufficiently low to meet the usual shipping specifications of pipeline companies, but dehydrators are usually followed by field desalters before the crude can be shipped to a refinery. However, the amount of bottom sediment and water in opportunity crude oils and in high-acid crude oils can show greater variations.

Thus, the objective of desalting process is to remove chloride salts and other minerals from the crude oil by water washing. Depending on the salt content of the crude oil, a one- or two-step process could be applied to reduce the salt content to the minimal level. Since the filterable solids, resin and asphaltene constituents, and salts are concentrated in the bottom fractions, the more difficult feedstocks tend to be (1) heavy crude oils and (2) tar sand bitumen diluted with lighter hydrocarbons or synthetic crudes to meet pipeline gravity and viscosity specifications (Andersen and Pederson, 1999; Derakhshesh et al., 2013). However, there are also lighter crudes (e.g., opportunity crudes) that can be problematic owing to their high filterable solids contents. The large and growing number of different feedstocks produced, and the variability in quality observed, requires increased vigilance by refiners seeking to understand and to overcome the impact of these feedstocks on desalter operations. Furthermore, the specific processing challenges depend on the physical and chemical characteristics of the feeds, their behavior when blended with other crudes, and the configurations and capabilities of the refinery process units. For example, feedstocks that have been blended with tar sand bitumen content will have a high content of resin and asphaltene constituents may have high asphaltene as well as high inorganic solids content, or difficult to remove salts. These contaminants can have a negative impact on desalter performance and can lead to fouling in downstream process units.

At the refinery, desalting is a water-washing operation performed at the production field and at the refinery site for additional crude oil cleanup (Chapter 2). For this reason, the dewatering and desalting unit may be referred to as pre-distillation cleaning or pre-distillation refining. If the crude oil from the separators contains water and dirt, water washing can remove much of the water-soluble minerals and entrained solids. If these crude oil contaminants are not removed, they can cause operating problems during refinery processing, such as equipment plugging and corrosion as well as catalyst deactivation.

The usual practice is to blend crude oils of similar characteristics, although fluctuations in the properties of the individual crude oils may cause significant variations in the properties of the blend over a period of time (Chapter 3). Although blending several crude oils prior to refining can eliminate the frequent need to change the processing conditions that may be required to process each of the crude oils individually, the incompatibility and instability of the blends, which can occur if, for example, a paraffinic crude oil is blended with heavy (viscous) crude oil, sediment formation (typically the precipitation of asphaltene constituents) can occur in the unrefined feedstock or in the products, thereby complicating the refinery process (Chapter 3) (Mushrush and Speight, 1995, 1998; Wiehe et al., 2001; Speight, 2014a).

In fact, simplification of the refining procedure is not always the end result. Blending two or more crude oils can cause unstable mixtures that precipitate species such as resin constituents or, more likely, asphaltene constituents and result in rapid fouling (Wilson and Polley, 2001). The crude oil incompatibility and the precipitation of asphaltene constituents on crude oil blending can cause significant fouling and coking in crude preheat train. For this reason, it is necessary to develop and use crude oil compatibility tests that will assist in predicting the proportions and order of blending of oils and avoid incompatibility under the relevant operating conditions (Crittenden et al., 1992; Polley et al., 2000, 2002, 2009; Wiehe et al., 2001; Saleh et al., 2004, 2005a, b; Yeap et al., 2004; Rodriguez and Smith, 2007; Coletti and Macchietto, 2009; Macchietto et al., 2009; Coletti et al., 2010; Costa et al., 2011a, b).

Besides reducing overhead corrosion, desalting reduces salt buildup and under-deposit corrosion in preheat exchangers, flash zone, upper sections and side-stream piping of the atmospheric column, and upper sections of any pre-flash and vacuum columns on the unit (Georgiadis et al., 2000; Barletta and White, 2007; Chambers et al., 2011). Desalting also decreases the amount of bottom sediment and water (BS&W) in the crude charge and the amount of suspended metal compounds that are sent to downstream units via reduced crude (atmospheric resid) and vacuum resid (Gutzeit, 1977, 2000, 2008). The evolution of refinery equipment during the past five decades has provided refiners with a variety of options, including options for dewatering and desalting refinery feedstocks (Kronenberger, 1984; Barnett, 1988; De Croocq, 1997; Lindemuth et al., 2001; Parkash, 2003; Gary et al., 2007; Speight, 2014a, 2017; Hsu and Robinson, 2017).

Crude unit cold preheat exchangers are usually shell and tube exchangers, with the crude in the tube side and crude unit products on the shell side (McDaniels and Olowu, 2016). These products (such as naphtha, kerosene, and light gas oil) are typically at a temperature on the order of 90°C–175°C (200°F–350°F). The more efficient the heat exchange is between the shell and tubes, the more the crude will be heated before entering the desalter. Heat helps the demulsifying process in the desalter. Heat decreases the viscosity of crude oil and weakens bonds formed by surfactants. This is especially important when running highly paraffinic crudes or highly asphaltenic crudes. Asphaltenes act as surfactants, and paraffinic waxes and asphaltenes can encapsulate water, salts, and solids, carrying them over into the crude unit and creating fouling and corrosion problems.

Inadequate desalting can cause fouling of heater tubes and heat exchangers throughout the refinery, which restricts product flow and heat transfer and leads to failures due to increased pressures and temperatures (Dickakian and Seay, 1988; Murphy and Campbell, 1992; Dos Santos Liporace and de Oliveira, 2005). Corrosion and the ensuing fouling, which occur due to the presence of hydrogen sulfide, hydrogen chloride, naphthenic (organic) acids, and other contaminants in the crude oil, cause equipment failure (Derungs, 1956; Blanco and Hopkinson, 1983; Chambers et al., 2011). Neutralized salts (ammonium chlorides and sulfides), when moistened by condensed water, can cause corrosion. Over-pressurizing the unit is another potential hazard that causes failures. In addition, where elevated operating temperatures are used when desalting sour crudes, hydrogen sulfide will be present. There is the possibility of exposure to ammonia, dry chemical demulsifiers, caustics, and/or acids during this operation.

Corrosive components in feedstocks, as well as in complete refinery processes, are a major cause of heat-exchanger tube failure which are usually related to a complex deposition mechanism formation and growth. Deposit layer thickening hinders heat transfer and obstructs the internal diameter of refinery and external pipelines.

The problems associated with heat-exchanger fouling have been known since the beginning of heat-exchanger use in refineries, and despite the best efforts to reduce or eliminate heat-exchanger fouling, deposit growth still occurs in some cases. Periodic heat exchanger cleaning is necessary to restore the heat exchanger to efficient operation. Scheduled and unscheduled shutdowns for cleaning can be very expensive because the startup may be very time-consuming. Thus, anything that can be done to reduce these shutdowns along with the cleaning procedure is of great benefit. When corrosion combines with fouling, the problem is more serious and complicated.

The drum water drainage experiment showed the amount of chlorides to be in the range of 120–440 ppm, supporting the mechanism discussed previously. The chloride content is composed of sodium, calcium, and magnesium chloride. Magnesium and calcium chloride start to hydrolyze at 120°C and 220°C (250°F and 430°F), respectively. Sodium chloride does not hydrolyze at such normal temperatures in the reboiler.

$$CaCl_2 + 2H_2O \rightarrow Ca(OH)_2 + 2HCl$$

$$MgCl_2 + 2H_2O \rightarrow Mg(OH)_2 + 2HCl$$

The reboiler tower outlet temperature is approximately 230°C (445°F), and therefore, the evolved hydrochloric acid is distilled up to the overhead system. The initial condensate forms after the vapor – containing a high percentage of hydrogen chloride – leaves the column. Due to the high-acid concentration dissolved in the water, the pH of the first condensate is quite low. Therefore, at dew point, water condensate corrodes the low carbon steel air cooler. Corrosion by acidic chloride condensates is driven by the hydrogen ion concentration (pH) via the reaction:

$$Fe + 2HCl \rightarrow FeCl_2 + H_2$$

Although the corrosive attack source is hydrogen chloride, the corrosion product is iron sulfide, not iron chloride. Iron sulfide is precipitated by the reaction between hydrogen sulfide and soluble iron chlorides from the corrosion reaction between hydrogen chloride and iron:

$$FeCl_2 + H_2 \rightarrow FeS_2 + 2HCl$$

The occurrence of hydrogen chloride in any stream will increase the corrosion rate, although the relation between the hydrogen chloride concentration in gas streams and corrosion rate is not always linear and the presence of water plays a role. In addition, the effects of hydrogen chloride are promoted by an increase in metal temperature.

Several chemical treatment programs have been developed to improve the overall desalting system performance while processing many types of opportunity crudes (Kremer, 2006a, b; Speight, 2014a). Most programs have been adapted for use with heavy WCSB feedstocks, beginning with the earliest production of conventional heavy crudes and continuing on with the latest variations of feedstock blended with tar sand bitumen. The most common chemical treatment regime includes the familiar application of an oil-soluble emulsion-breaking chemical injected into the crude oil upstream of the desalter. For treating heavy feedstocks, this first mode of treatment is often supplemented by the application of water-based wetting agents fed to the fresh wash-water stream, in order to improve the solids handling capabilities of the desalter.

Desalting chemicals improve overall desalting efficiency, reduce water and solids carryover with desalted crude, and reduce oil carry-under with brine effluent. Most desalting chemicals are demulsifiers that help break up the tight emulsion formed by the mix valve and produce relatively clean phases of desalted crude and brine effluent. Demulsifiers are usually purchased from the same process additives suppliers that supply anti-foulants, filming amine corrosion inhibitors, liquid organic neutralizers, and similar products for controlling overhead corrosion and fouling problems on crude units (or elsewhere in the refinery) (Sloley, 2013a, b). If necessary, demulsifiers can be custom formulated for high water removal rates from crudes, but at the cost of poor solids wetting and likely oil carry-under with the brine discharge. They can also be formulated for high oil removal rates from brine, but at the cost of water carryover with desalted crude. When formulated for high solids wetting rates, brine quality often decreases and water carryover with desalted crude increases.

Typical commercial demulsifiers are blends of several types of chemical compounds. Primary products normally contain 20%–25% v/v active ingredients in a solvent carrier, but some products are also available as concentrates with up to 80% v/v active ingredients to reduce shipping and handling costs. A secondary product may be needed for some applications. Oil-soluble or oil-based products are usually injected directly into the crude charge, while water-soluble or water-based products are injected into incoming desalter water. In either case, a simple pipe nipple connection with shutoff and ball check valves can be used; no injection quills or spray nozzles are needed. With heavy crudes, special demulsifiers have also been added at the crude holding tanks approximately 2–3 hours before the crude is sent to the desalter. Many demulsifiers are application specific; bench testing of competitive products may be required to find a product that produces the best results. Therefore, most demulsifier suppliers use small, portable bench test desalters for evaluating various formulations and dosages in the refinery laboratory. In addition, full-scale field evaluation is usually required to confirm the laboratory results obtained for a given crude blend and desalter combination.

Although caustic soda is often used to neutralize acidic components, as with any chemical treatment, caution is advised. Uncontrolled addition of caustic soda to the desalter can have a detrimental effect. An excess of caustic can result in the formation of soap due to, for instance, the presence of fatty acids (as occur in high-acid crudes). Soap stabilizes the oil–water mixture and obstructs the separation process due to the production of emulsions that are difficult to break. When this is the case, contaminants pass out of the desalter with the feedstock and cause fouling in downstream processes.

Another process parameter that can play an important role in both neutralizing acids as well as promoting demulsification is the relative acidity/alkalinity. Carefully monitoring pH in the desalter water effluent allows for efficient dosing of caustic soda or acid which may result in significant cost savings. Maintaining the pH of the mixture within a specified range (that is most likely feedstock dependent) assists in chemical demulsification.

Opportunity crude oils and high-acid crude oils, as well as heavy crude oils, extra heavy oils, and tar sand bitumen, are often appealing feedstocks for refineries due to their lower cost (Speight, 2011b, 2014a). The acceptance of the opportunity crude oils, the high-acid crude oils, and various types of heavy (viscous) feedstocks is improving as refinery processes evolve but the asphaltene constituents present in viscous feedstocks can significantly affect the key performance indicators when they become destabilized and agglomerate to the extent where precipitation can occur in the desalter. It is therefore advisable that each crude oil to be included in a blend be subjected to the dewatering and desalting processes separately prior to the blending operation.

Asphaltene particles can stabilize emulsions, causing desalter performance and oil carry-under problems, and can contribute to accelerated fouling in crude unit preheat exchangers. Also, asphaltene separation (precipitation from the feedstock) can be troublesome for both the preheat exchangers and the desalting process. Asphaltene precipitation typically occurs when incompatible crude oils are blended together and a consequence is the fouling of preheat exchangers. The application of an antifoulant in the cold preheat exchanger can help reduce the fouling, but can also potentially stabilize an emulsion in the desalter (McDaniels and Olowu, 2016).

In the modern refinery which handles feedstock blends of a variety of crude oils, efficient operation of the desalter is a key element of feedstock processing. Continued operation of the desalter requires continuous monitoring of the unit. Conventional analytical methods can be used on a day-to-day basis to properly monitor desalter performance. However, the variability seen in analytical characteristics of the heavy feedstock properties and any such feedstock blends (Chapter 3) requires increased vigilance in the frequency of desalter performance measurements, in order to respond quickly to unit disturbances.

For example, the level of filterable solids levels is a crucial monitoring parameter with many heavy feedstocks and measuring the filterable solids content must be a frequent (perhaps even a daily) occurrence. Back-up action plans must also be in place to respond to any upsets that might occur when solids levels are particularly high, or when feedstock instability occurs, due to the instability of the asphaltene (and resin) constituents (Asomaning and Watkinson, 2000; Stark et al., 2002; Asomaning, 2003; Stark and Asomaning, 2003). Such events can be reduced if a laboratory program is in place that provides valuable information about the corrosion behavior and the fouling potential of the blends.

This type of program would lead to the development of a blending index (Chapter 3) which would assist the refinery to determine the compatibility of various feedstocks accepted by the refinery, enhance prediction of the behavior of specific feedstock blends, and prevent blending of incompatible feedstocks. This would also enable the refinery to select the most appropriate best set of desalting conditions and chemical treatment methods to handle various blends of the crude oils accepted by the refinery. The characteristics of many refinery feedstocks (especially the opportunity crude oil, the high-acid crude oils, and the viscous feedstocks) can include high solids levels, unstable asphaltene derivatives, non-extractable chloride derivatives, and considerable variability in one or more of these parameters for a given grade of feedstock. Efficient operation of the dewatering–desalting operations as well as suitable chemical treatment programs and enhanced desalter monitoring are all keys to successful feedstock processing.

6.3   CORROSION AND FOULING

In the current context, corrosion is a phenomenon (initiated by the use of untreated feedstocks) that involves the deterioration of equipment (Speight, 2014b). Furthermore, corrosion processes not only influence the chemical properties of a metal or metal alloy but also generate changes in the physical properties and the mechanical behavior.

There are numerous sources of corrosion in refining operations, but they can be divided into three groups which are (1) corrosion from crude oil components, (2) corrosion from chemicals used in refining processes, and (3) environmental corrosion. To comprehend the corrosion phenomena at refineries, it is necessary to first comprehend the physicochemical properties of crude oil and natural gas.

Crude oil is a liquid mixture of hydrocarbons ranging from a volatile hydrocarbon compound to a non-volatile hydrocarbon compound. Although crude oil is not corrosive, impurities and components containing nitrogen, sulfur, and oxygen can be found in crude oil and can be corrosive. These impurities can be found in crude oil as liquids, solids, gases, and microorganisms. The same compounds are present in all crude oils, but in different proportions. Natural gas in crude oil is primarily a mixture of nitrogen, carbon dioxide (CO_2), hydrogen sulfide (H_2S), and water which can cause corrosion to the equipment if not removed at the most appropriate point, i.e. at the dewatering and desalting stage. The total acid number (TAN), total sulfur content, water, salt content, and microorganisms are all factors that influence the corrosiveness of crude oil. The combination of these components will corrode metallic equipment at various locations throughout the refinery, resulting in various types of corrosion such as localized pitting corrosion, erosion corrosion, stress corrosion cracking, hydrogen embrittlement, and intergranular corrosion. Furthermore, the presence of mixed corrosive gases (hydrogen sulfide, H_2S, carbon dioxide, CO_2, and naphthenic acid derivatives) adjacent to boiled water in the desalination unit is a major cause of corrosion in the internal surface of the steel tube material.

Of particular relevance in the current context is the potential for the occurrence of acidic corrosion which is typically (but not always) due to the presence of naphthenic acids which is a generic name used for all of the organic acids present in crude oils, especially high-acid crude oils (Chapter 1). Most of these acids are represented by simple chemical formulas – such as $R(CH_2)_nCOOH$, where R is a cyclopentane ring or cyclohexane ring and n is typically greater than 12. Any such acids in the refinery feedstock must be removed before the feedstock enters the refinery proper (i.e. the distillation units and other downstream units).

Thus, in the current context, the importance of dewatering and desalting of refinery feedstocks should not be minimized in any way. Any refinery feedstock (i.e. fossil fuel feedstocks and any alternate feedstocks, such as liquids produced from coal, liquids produced from oil shale, and bio-oil) must be subjected to pretreatment. The penalties for not subjecting refinery feedstocks to pretreat as they enter the refinery is the occurrence of corrosion and/or fouling.

A variety of corrosive substances occur in refineries (Table 6.1) and corrosion occurs in various forms in the refining process, such as pitting corrosion from acidic water, hydrogen embrittlement, and stress corrosion cracking from sulfide attack. Carbon steel is used for the majority (>80%) of refinery components and is resistant to the most common forms of corrosion, particularly from hydrocarbon impurities at temperatures below 205°C (400°F); however, other corrosive chemicals and environments prevent the use of carbon steel in all refinery units. Common replacement materials are low-alloy steel containing chromium and molybdenum with high-chromium stainless steel used for more corrosive environments. Nickel, titanium, and copper alloys are used in the most problematic areas where extremely high temperatures and/or very corrosive chemicals are present.

6.3.1 Corrosion

Corrosion is the deterioration of a material as a result of its interaction of the material with the surroundings and it can occur at any point or at any time during crude oil and natural gas processing. Although this definition is applicable to any type of material, it is typically reserved for metallic alloys. Furthermore, corrosion processes not only influence the chemical properties of a metal or metal alloy but also generate changes in the physical properties and the mechanical behavior. Corrosion can be represented by relatively simple chemical reactions and there are various forms of corrosion, some of which involve redox reactions and many of which do not involve redox reactions (Speight, 2014b).

Prior to the use of crude oil in the refinery, the crude oil feedstock (or the blend of two or more crude oils) may be stored in aboveground steel storage tanks (the tank farm), which act as a buffer between feedstock supply and demand for the refinery. The storage tanks can be as high as 50 and 100 feet or more wide but tank size depends on the typical daily demand of crude oil, cycle time, safety stock, and working capacity – cycle time is the time between the production and the delivery of a particular product.

The tanks are available in many shapes: vertical and horizontal cylindrical, open top and closed top, flat bottom, cone bottom, slope bottom, and dish bottom. Large tanks typically have vertical cylindrical shapes or rounded corners that transition from vertical side wall to bottom profile, in order to safely withstand hydrostatically induced pressure of contained liquid. Many tanks are covered with fixed or floating roofs to avoid evaporation and to minimize risk of fire. Aboveground storage tanks (AST) differ from underground storage tanks (UST) in terms of various regulations

TABLE 6.1
Corrosive Substances that Occur in Refinery Feedstocks[a]

Ammonia

Nitrogen in feedstocks combines with hydrogen to form ammonia (NH_3)
Used for neutralization
May combine with other elements to form corrosive compounds, such as ammonium chloride (NH_4Cl)

Carbon

Not corrosive
At high-temperature results in carburization that causes embrittlement
Can result in reduced corrosion resistance in some alloys

Carbon Dioxide

Occurs in steam reforming of hydrocarbon in hydrogen plants
May also be formed, to some extent, in catalytic cracking
Combines with moisture to form carbonic acid (H_2CO_3)

Chloride Derivatives

Present in the form of salts (such as magnesium chloride and calcium chloride)
Originates from crude oil, catalysts, and cooling water

Cyanides

Generated in the cracking of high-nitrogen feedstocks
When present, corrosion rates are likely to increase

Hydrogen

Not typically a corrosive
Can lead to blistering and embrittlement of steel
Readily combines with other elements to produce corrosive compounds

Hydrogen Chloride

Formed by the hydrolysis of magnesium chloride ($MgCl_2$) and calcium chloride ($CaCl_2$)
May occur in overhead (vapor) streams
On condensation in the presence of water, forms corrosive hydrochloric acid

Naphthenic Acid Derivatives

A collective name for organic acids found in crude oils
Structural aspects have not been fully characterized
Known to be responsible for acidic corrosion

Oxygen

Originates in crude oils, aerated water, or packing gland leaks
Aerial oxygen is used in furnace combustion and catalyst (FCC) regeneration units
Causes oxidation and scaling of metal surfaces

Phenol Derivatives

Found primarily in sour water strippers but also in some crude oils
Can contribute to acidic corrosion

Polythionic Acid Derivatives

Oxoacids which have a straight chain of sulfur atoms and have the chemical formula $S_n(SO_3H)_2$, n>2
Formed by the interaction of sulfides, moisture, and oxygen
Often formed by various chemical interactions when equipment is shut down

(Continued)

TABLE 6.1 (*Continued*)

Corrosive Substances that Occur in Refinery Feedstocks[a]

Sulfur

Occurs in crude oils, especially in the more viscous feedstocks

Causes high-temperature sulfidation of metals

Combines with other elements to form corrosive products, such as sulfide derivatives, sulfate derivatives, sulfurous acid derivatives, sulfuric acid derivative, and polythionic acid derivatives

Sulfuric Acid

Used as a catalyst in alkylation plants

Formed in process streams that contain sulfur trioxide (SO_3) and water: $SO_3 + H_2O \rightarrow H_2SO_4$.

[a] *The substances occur in all of the potential refinery feedstocks including the alternate feedstocks (Chapter 1) and are listed alphabetically and not in any order of preferences; the occurrence of such substance will be feedstock and process dependent.*

that are applied to such units. Most container tanks for handling liquids during transportation are designed to handle varying degrees of pressure. However, regardless of the tank shape, size, or function, corrosion can and will occur.

In fact, crude oil pretreatment actually commences in the storage tanks. This involves adding specially formulated chemicals to crude receipts entering refinery crude storage tanks. When added ahead of tankage, these chemicals have more time to diffuse to solids particles, asphaltene constituents, and other emulsion-stabilizing materials at the oil/water interface surrounding brine droplets. This provides more effective phase separation and solids removal in the desalting vessel. For more modern desalting operations, the primary goals for crude pretreatment programs include reduced hydrocarbon contamination of desalter effluent water, reduced emulsion layer buildup, better dehydration, and the potential to reduce fouling in downstream units.

The majority of storage tanks are built to withstand varying degrees of temperature and pressure but corrosion can occur regardless of geometry, size, or function of the tanks. The exterior of the tanks is typically coated with a thermally insulating material. However, one issue that must be addressed in relation to corrosion in storage tanks is biocorrosion.

Microorganisms are unavoidable in crude oil stored in storage tanks. Contamination of crude oil is unavoidable, and it occurs immediately after it is pumped from the reservoir. Aerobic and anaerobic microorganisms both colonize crude oil storage facilities. The presence of water and microbes in the storage tank can cause biocorrosion, which can considerably increase the rate of corrosion of a crude oil. Furthermore, microorganisms need water as a source of energy to drive their metabolism, and nutrients provide essential building materials (such as carbon, nitrogen, phosphorus, and trace metals) for cell renewal and growth, which is readily available at the bottom of the tank.

To combat corrosion, a liner should be included on the internal floor and walls of storage tanks to prevent contact between the metal shell and any accumulated water or sediment. In fact, storage tanks suffer corrosion only if water is allowed to accumulate at the bottom with or without the sediment. Metal tanks in contact with soil and containing crude oil products must be protected from corrosion; the most effective and common corrosion control technique for steel in contact with soil is cathodic protection.

After dewatering and desalting, the next step in the refining process is the separation of crude oil into various fractions or straight-run cuts by distillation in atmospheric and vacuum towers. The main fractions (*cuts*) obtained have specific boiling point ranges and can be classified in the order of decreasing volatility into gases, light distillates, middle distillates, gas oils, and residuum. The concentration of certain constituents by the distillation process can cause corrosion, sediment fouling, and affect flow rates (Speight, 2014a, b). A properly designed distillation column can reduce the effects of these consequences, but corrosion and other effects are very prominent in reducing separation efficiency in the column.

Prior to entering the distillation unit, desalted crude feedstock is preheated and then flows to a direct-fired crude charge heater where it is introduced into the distillation column at pressures slightly above atmospheric pressure and at temperatures ranging from 345°C to 370°C (650°F–700°F) where the lower boiling constituents flash into vapor. As the hot vapor rises in the tower, its temperature is reduced. Heavy fuel oil or residuum is taken from the bottom and, at successively higher points on the tower, various major products are drawn off (Parkash, 2003; Gary et al., 2007; Speight, 2014a, 2017; Hsu and Robinson, 2017).

The acidic species in the feedstock can be quantified by the TAN of the crude oil which is expressed in terms of milligrams of potassium hydroxide per gram (mg KOH/g) (Speight, 2014b, 2015). The TAN is not specific to particular acid constituents but rather refers to all possible acidic components of the oil; although it usually refers to (organic) naphthenic acids, mineral acids such as hydrogen sulfide (H_2S), hydrogen cyanide (HCN), and carbon dioxide (CO_2) also significantly contribute to corrosion of equipment. High-acid crude oils are crude oils which typically have a TAN in excess of 0.5. It is preferable that these acids be removed from the refinery feedstock during the dewatering/desalting stages of the feedstock pretreatment (Chapter 2).

However, despite an efficient desalting operation, corrosion-causing agents can still appear in the various post-desalting streams. For example, in the distillation unit, acid gases are formed of which hydrogen sulfide is notorious. Steam, which is injected into the crude tower to improve the fractionation, condenses in the upper part of the unit. The hydrogen sulfide dissolves in the condensate and forms a weak acid which is known to cause stress corrosion cracking in the top section of the tower and in the overhead condenser. This may lead to frequent renewal of tubing of the condenser, and in severe cases, to replacement of the entire crude tower top.

In addition, the desalter wastewater (brine) and the aqueous condensate from the overhead reflux drums of main fractionating column contain appreciable levels of hydrogen sulfide as does the hotwell water from the vacuum unit. The common practice is to strip hydrogen sulfide from the sour water (sour water stripping) before disposal. Fuel gas and the liquefied petroleum gas (LPG) stream, which may contain hydrogen sulfide, are (should be) subjected to an amine wash to remove hydrogen sulfide before the caustic wash. The sour water, fuel gas, and rich amine containing hydrogen sulfide are potential sources for corrosion.

In the present context, acid corrosion is typically (but not always) due to the presence of constituents in crude oil known as naphthenic acids, which is a generic name used for all of the organic acids present in crude oils and most of which arise as biochemical markers of crude oil origin and various maturation processes. In fact, naphthenic acid derivatives are of particular interest when high-acid crude oils are used as refinery feedstocks and must go through the dewatering and desalting operations.

Most of these acids are represented by simple chemical formulas – such as $R(CH_2)_nCOOH$, where R is a cyclopentane ring or cyclohexane ring and n is typically greater than 12 – and a multitude of other acidic organic constituents are also present in crude oil(s) but not all of the species have been fully analyzed and identified (Speight, 2014a).

Naphthenic acid corrosion is one of the serious long-known problems in the crude oil refining industry (Slavcheva, 1998, 1999; Speight, 2014b). The rates of naphthenic acid corrosion typically increase with the increase in TAN and also increase with the increase in temperature. In the crude oil refining industry and the gas processing industry, factors associated with the composition and behavior of the feedstocks (such as temperature, TAN, fluid velocity, gas velocity, and pipeline/reactor material) work simultaneously and cumulatively. In order to obtain the credible corrosion rate, comprehensive analysis and estimation of feedstock should be done taking into account not only the temperature but also any other relevant factors.

Thus, naphthenic acid corrosion in the presence of hydrogen sulfide can be represented as follows:

$$Fe + 2RCOOH \rightarrow Fe(RCOO)_2 + H_2$$

$$Fe + H_2S \rightarrow FeS + H_2$$

The first reaction produces the oil-soluble iron (iron naphthenate) and the second reaction inhibits soluble iron production – each reaction is dependent on the functionality of the sulfur compounds. In short, the amount of sulfur in crude oil (determined as total sulfur, S% w/w) is not necessarily related to reactivity – for example, hydrogen sulfide is very reactive toward iron, producing a protective layer of iron sulfide (FeS) that can prevent further corrosion by acidic species.

Naphthenic acid corrosion is differentiated from sulfidic corrosion by the nature of the corrosion (pitting and impingement) and by the severe (naphthenic acid) corrosion at high velocity in distillation units. Feedstock heaters, furnaces, transfer lines, feed and reflux sections of columns, atmospheric and vacuum columns, heat exchangers, and condensers are among the types of equipment subject to this type of corrosion. Furthermore, isolated deep pits in partially passivated areas and/ or impingement attack in essentially passivation-free areas are typical of naphthenic acid corrosion. In many cases, even very highly stable alloys (i.e. 12 Cr, AISI types 316 and 317) have been found to exhibit sensitivity to corrosion under these conditions.

Attempts to mitigate corrosion will continue to use a complex system of monitoring, preventative repairs, and careful material selection. Monitoring methods include both off-line checks taken during maintenance and online monitoring. Off-line checks measure corrosion after it has occurred, telling the engineer when equipment must be replaced based on the historical information he has collected.

Blending of refinery feedstocks (with the inherent danger of phase separation and incompatibility) will continue to be used to mitigate the effects of corrosion as well as blending, inhibition, materials upgrading, and process control. Blending will be used to reduce the naphthenic acid content of the feed, thereby reducing corrosion to an acceptable level. However, while blending of heavy and light crude oils can change shear stress parameters and might also help reduce corrosion, there is also the potential for incompatibility of heavy (viscous, high density) crude oils and light (low-density) crude oils (Chapters 1 and 3).

Also, because of the high content of alkane derivatives in some tight oils, these oils can have major effects on blended refinery feedstock, especially if high asphaltene crude oils are a part of the blend. The low molecular weight alkane derivatives are used in the standard methods to precipitate an asphaltene fraction from asphaltene-containing crude oils (Speight, 2014a, 2015). Blending tight oils with other feedstocks runs the risk of causing incompatibility in the form of a principal of the asphaltene fraction. Also, because of the presence of high molecular weight alkanes in the tight crude oils (Chapter 1), there is a risk of separation of waxes when the constituents of the blend have not been fully identified. Either event (i.e. asphaltene precipitation or the separation of wax derivatives) can cause serious fouling in the refinery system.

When alternate feedstocks are under consideration, it must be acknowledged that corrosion incidents in the fractionation area of direct coal liquefaction pilot plants have been severe. The Wilsonville, Alabama Solvent Refined Coal (SRC)-1 pilot plant operators quickly identified chloride, a minor component of the feed coal, as a contributing factor to this corrosion. They also learned that adding sodium carbonate to the coal feed was an effective way to provide a short-term measure to reduce corrosion; however, processing conditions would not permit the addition of alkali salts in a commercial plant (Barnett et al., 1983).

While the composition of coal liquids and of shale oil varies according to the source (and even within sources for the same area), some of the common characteristics can lead to significant disruptions across the refining supply chain – from transportation from the production site to shipment from the refinery. The high content of phenol derivatives in many coal liquids may result in incompatibility issues with crude oil, and the paraffinic nature and low sulfur content of shale oil can lead to cold flow, water separation, and lubricity issues in distillate products (Scouten, 1990; Lee, 1991, 1996). Some light-ends products may experience corrosion problems, and heavy fuel oil stability can be negatively impacted by blending shale oil residual materials with asphaltenic stocks.

Shale oil can range from condensates to more viscous crudes (35°–60° API) and it has a poor wetting ability on steel and other surfaces, primarily due to the low concentration of polar compounds.

The oil volatility can have an effect on the solubility of the acid gases in the hydrocarbon phase, which can change the fugacity of carbon dioxide and hydrogen sulfide in the vapor phase as the production stream flows from the reservoir through the well. Other factors complicating the refining of shale oil include the entrainment of solids in the oil. Also, waxy solids are often observed in shale oil, which can precipitate when the temperature cools below the wax appearance temperature (WAT) of the oil and can occur both downhole and on the surface.

Shale oils are highly paraffinic (with many featuring waxes melting above 93°C (200°F)) and can consequently create wax deposits that can foul transportation modes, storage tanks, and process units. When the light paraffinic shale oil is blended with heavy, asphaltenic crude oil, the resulting blend may experience asphaltene instability, creating sludge/deposits that reduce tank capacity in crude tanks, stabilize emulsions in the desalter unit, and foul process equipment. Refiners typically blend a mix of moderate-to-high sulfur crudes with higher TAN crudes to help reduce the risk of naphthenic acid corrosion. Replacing these moderate/high sulfur crudes with lower sulfur shale crudes can lead to an increase in the risk of naphthenic acid corrosion. Shale oil contain hydrogen sulfide (H_2S) – a natural corrosive and a deadly gas that presents significant health and safety issues during transportation.

Also, when shale oil is treated with scavengers to curb the presence of hydrogen sulfide, the resulting tramp amines can affect economics throughout the refining process. When present in the crude blend, tramp amines can partition into the oil phase at the desalter. Once past the desalter, they can react with hydrogen chloride (HCl) in the atmospheric column and overhead system to deposit as corrosive salts.

As the use of non-fossil fuel feedstocks becomes increasingly widespread, issues of material compatibility with these biofuels arise. The corrosion behavior of industrial metals and alloys, such as carbon and stainless steels, is well understood in traditional fossil fuel environments, and methods of mitigating and preventing corrosion of steel tanks, pipes, engine parts, and other items have been developed. When the environment in contact with these metal surfaces is changed from fossil fuel to biofuel, the nature of the metal–environment interaction can change, leading to corrosion or degradation of expensive industrial equipment, sometimes leading to catastrophic failures.

Biorefining is a versatile concept that promotes the sustainable production of a portfolio of bio-based products including biomolecules, biomaterials, bioenergy, and energy vectors from various biomass resources and residues. Early development of the so-called first-generation biorefineries that were essentially concentrating on mass production of biodiesel or bioethanol has shown a number of limitations in the related value chains in terms of sustainability issues. Among those issues, a number of corrosion problems have been reported and identified, at least partially understood and sometimes solved. We can quote, for example, the corrosion of some metallic components of engines due to heat-driven conversion of ethanol into acidic species before combustion, the high-temperature corrosion in biomass burning furnaces due to the presence of alkali salts as well as stress corrosion cracking in ethanol carbon steel tanks and in pipelines. However, no global assessment of corrosion issues in future biorefineries has been performed so far, leaving significant interrogation where research efforts have to be deployed to accompany their sustainable implementation.

For example, the failure of steel tanks and pipes has been attributed to stress corrosion cracking caused by biofuels such as ethanol. The corrosion and cracking of steels are mostly caused by minor chemical constituents and impurities within the fuels. Specifically, water and oxygen contained in the non-fossil fuels play a large role in the corrosivity of these materials. Similarly, bio-oil produced from biomass is significantly more corrosive compared to traditional fossil fuel-based liquids. As-produced bio-oil typically contains a substantial amount of oxygen, primarily as component of water, carboxylic acid derivatives, phenol derivatives, ketone derivatives, and aldehyde derivatives. Depending on the source of the bio-oil, the production process, and the resulting chemical constituents, bio-oil can be significantly more corrosive than fossil fuel-based liquids. The extent and type of corrosion of different bio-oils depend on the chemical constituents. Water, which occurs in all bio-oils as an impurity, plays an important role in the overall corrosivity. Thus, it is necessary to understand the detrimental interactions of bio-oil with industrial materials so that the existing fuel infrastructure and equipment in contact with these materials can be properly maintained and new structures can be designed with appropriate materials.

6.3.2 Fouling

Fouling, as it pertains to crude oil refineries, refers to deposit formation, encrustation, deposition, scaling, scale formation, slagging, and sludge formation which has an adverse effect on operations (Deshannavar et al., 2010). It is, in fact, caused by the accumulation of incompatible material (Chapter 3) within a processing unit or on the solid surfaces of the unit to the detriment of function. For example, when it does occur during refinery operations, the major effects include (1) loss of heat transfer as indicated by charge outlet temperature decrease and pressure drop increase, (2) blocked process pipes, (3) under-deposit corrosion and pollution, and (4) localized hot spots in reactors, all of which culminate in production losses and increased maintenance costs. In addition, the term macrofouling is often used to generally describe the blockage of tubes and pipes, while the term microfouling is generally used to describe scaling on the walls of tubes and pipes. Again, the outcome is a loss of efficiency and output to the refinery as well as a series of potential environmental-related issues.

Fouling during crude oil production, transportation, or refining can occur in a variety of processes, either inadvertently when the separation is detrimental to the process or intentionally (such as in the deasphalting process or in the dewaxing process). Thus, separation of solids occurs whenever the solvent characteristics of the liquid phase are no longer adequate to maintain polar and/or high molecular weight constituents in solution. Examples of such occurrences are (1) the separation of asphaltene constituents, which occurs when the paraffin nature of the liquid medium increases, (2) the separation of waxes, which occurs when there is a drop in temperature or increase in the aromaticity of the liquid medium, and (3) the formation of sludge or sediment in a reactor, which occurs when the solvent characteristics of the liquid medium change so that asphaltene constituents (and, in some cases, resin constituents) or wax materials separate, and coke formation, which occurs at high temperatures and commences when the solvent power of the liquid phase is not sufficient to maintain the potential sludge or sediment precursors in solution.

Typically, the fouling material consists of organic materials and/or inorganic materials that are deposited by the feedstock as a result of the instability or incompatibility of the deposited materials from one crude oil with another during and/or shortly after a blending operation. Blending is one of the typical operations that a refinery must pursue not only to prepare a product to meet sales specifications but also to blend different crudes and heavy feedstocks to produce a refinery. Although simple in principle, the blending operation must be performed with care and diligence because refineries frequently accept heavy feedstocks as part of their refinery slate. Lack of attention to the properties of the individual feedstocks prior to the blending operations can lead to asphaltene precipitation or phase separation (fouling) due to incompatibility of different components of the blend (Chapter 3). It is advisable for the refiner to be able to predict the potential for incompatibility by determining not only the appropriate components for the blend but also the ratio of the individual crude oils and any heavy crude oil that are in the blend.

The compatibility of crude oils is generally evaluated by colloidal stability based on bulk composition or asphaltene precipitation (Schermer et al., 2004; Speight, 2014b). Typically, the test methods are used to evaluate oil stability at ambient conditions but applying the data to the potential for fouling under the actual conditions found in heat transfer equipment must be done with caution. Fouling is dependent not only on the conditions of asphaltene separation fluid and the stability of the crude oil/heavy feedstock system but also on flow conditions and other parameters (Chapters 3 and 6). Fouling is concerned with not only asphaltene precipitation. In addition, fouling can also be a consequence of corrosion in a unit when deposits of inorganic solids become evident (Speight, 2014b). With the influx of opportunity crudes, high-acid crudes, heavy crude oils, extra heavy crude oils, and tar sand bitumen into refineries, incompatibility phenomena and fouling phenomena are more diverse and are becoming more common (Chapters 1 and 3).

In the crude oil industry, the components that may be subject to fouling and the corresponding effects of fouling are (1) the production zone of crude oil reservoirs and oil wells, which is reflected by a decrease in production with time through the formation of plugs which can lead to the complete cessation of flow, (2) pipes and flow channels which result in reduced flow, increased pressure drop, increased upstream pressure, slugging in two-phase flow, and flow blockage, (3) heat-exchanger surfaces, which result in a reduction in thermal efficiency along with decreased heat flux, increased temperature on the hot side, decreased temperature on the cold side, and under-deposit corrosion, (4) injection/spray nozzles (e.g., a nozzle spraying a fuel into a furnace or a reactor) in which the incorrect amount of feedstock is injected, and (5) within a reactor due to uncontrollable chemical and/or physical reaction. In addition, there is macrofouling and microfouling.

Macrofouling is caused by coarse matter from either organic, biological, or inorganic origin. Such substances foul the surfaces of heat exchangers and may cause deterioration of the relevant heat transfer coefficient as well as flow blockages.

Microfouling is somewhat more complex and several distinctive events can be identified: (1) particulate fouling, which is the accumulation of particles on a surface, (2) chemical reaction fouling, such as decomposition of organic matter on heating surfaces, (3) solidification fouling, which occurs when components of a flowing fluid with a high-melting point freeze onto a subcooled surface, (4) corrosion fouling, which is caused by corrosion, (5) biofouling, which frequently follows biocorrosion and is caused by the action of bacteria or algae, and (6) composite fouling, which refers to fouling that involves more than one foulant or fouling mechanism.

Fouling is caused by the presence of particulate matter in the fluid is a common form of fouling and it can be defined as the process in which particles in the process stream deposit onto heat-exchanger surfaces. These particles include those that were originally carried by the feedstock before entering the heat exchanger and those that were formed in the heat exchanger itself as a result of various reactions, aggregation, and flocculation. Particulate fouling increases with particle concentration, and typically, particles greater than $1\,\mu m$ size lead to significant fouling problems.

Corrosion fouling is deposit formation as a result of the corrosion of the substrate metal on heat transfer surfaces or reactor surfaces. In this type of fouling, corrosion occurs first and initiates fouling, whereas under-deposit corrosion occurs as an after-effect of fouling. Corrosion fouling is dependent on several factors such as (1) thermal resistance, (2) surface roughness, (3) composition of the substrate, and (4) composition of the feedstock. In particular, impurities present in the feedstock stream (which have not been removed in the dewatering–desalting process) will contribute to the onset of corrosion – examples of these impurities are hydrogen sulfide (H_2S), ammonia (NH_3), and hydrogen chloride (HCl).

Fouling of a surface through the formation of deposits does not always develop steadily with time. There may be an induction period when the surface is new or very clean and the foulant does not accumulate immediately. After the induction period, the fouling rate increases. On the other hand, there is also negative fouling which occurs when relatively small amounts of deposit can improve heat transfer, relative to clean surface, and can give an appearance of a negative fouling rate and negative total amount of the foulant. After the initial period of surface roughness control or surface roughness adjustment, the fouling rate may become positive.

In asymptotic fouling, the fouling rate decreases with time, until it finally reaches zero, and at this point, the deposit thickness remains constant with time. This often occurs when the deposits are relatively soft or poorly adherent in areas of fast flow or turbulent flow and is usually assigned to the point at which the deposition rate is equal to the deposit removal rate. However, accelerating fouling is almost the opposite since the fouling rate increases with time and the rate of deposit buildup accelerates until it becomes transport limited. This type of fouling can develop when fouling increases the surface roughness or when the deposit surface exhibits higher chemical propensity to fouling than the pure underlying metal.

Fouling can be caused by a number of different mechanisms, which include (1) particles in the feedstock, (2) particle formation, (3) corrosion fouling, (4) coking, (5) aggregation and flocculation, (6) phase separation, (7) particle deposition, (8) deposit growth, aging, and hardening, and (9) auto-retardation and erosion or removal. In addition to these stages, the rate of fouling and the prediction of a fouling factor must also be considered (Hirschberg et al., 1984; Speight, 2014b).

Particles in the crude oil feedstocks originate from poor desalter performance or have been entrained during the distillation process. These particles are, for the most part, insoluble inorganic particles, such as dirt, silt and sand particles, as well as other inorganic salts, such as sodium chloride, calcium chloride, and magnesium chloride (which arise from poor desalter performance), corrosion products (iron sulfide and rust), and catalyst particles or fines. There may also contain some organic particles that may have been formed during their storage or transport if, for example, the fetlock was exposed to oxygen. In particular, streams from refinery process units such as vacuum distillation, visbreaking, and cokers may have more particulates and metals than straight-run products due to the heavier feedstocks that are processed in these units. Feedstocks purchased from other refiners may also be suspect because of the increased transportation time and exposure to oxygen leading to higher levels of particulate matter as a result of various chemical reactions in addition to the higher potential for contract with corroded system and/or the potential from the feedstock itself to cause corrosion (Speight, 2014c).

These particles, wherever the source, can be sub-categorized into the following classes: (1) basic sediment and (2) filterable solids. The amount of filterable solids in the feedstock (reported in pounds per barrel % w/w) may be determined by filtration of the feedstock and it is possible to evaluate the potential for fouling by indicating the type of materials that could contribute to fouling if allowed to pass into the heat exchanger. Such particles in the feedstock at amounts in excess of 1 (1 pound per thousand barrels, also often shown as 1 ptb) lead to significant fouling problems. The effect of the particles on fouling can be diminished considerably (if not avoided) by solid–liquid filtration, sedimentation, and centrifugation or by any of various fluid cleaning devices. However, particles that need to be considered as serious in terms of foulant production are those that are not filterable (such as the micron-sized clay minerals in crude oil) and are likely to pass through the desalter and proceed into the heat exchanger prior to the distillation unit.

By the way of explanation and in the context of this text, sedimentation is the accumulation of solids that results from poor dewatering and desalting operations, and these solids are deposited in low velocity areas in process equipment. The equipment can include heat exchangers, tower distributors, distillation trays, random packing, and structured packing. If the feedstock contains suspended solids – such as salts, metal oxides, catalyst fines, asphaltene particles, and coke fines – sedimentation can occur on the mass transfer surface (sedimentation fouling). Precipitation and crystallization of dissolved salts can occur when process conditions become supersaturated, especially at mass transfer surfaces. Ammonia salt deposition resulting from both water vaporization and direct solid deposition from the gas phase is a common refining problem.

In some cases, the deposit may adhere strongly to the surface and be self-limiting in that as a deposit becomes thicker, it is more likely to be removed by the fluid flow and thus achieve some asymptotic average value over time. Sedimentation fouling is strongly affected by fluid velocity and less so by temperature, although a deposit can bake on a surface and become very difficult to remove.

Certain salts such as calcium sulfate are less soluble in warm water than in cold water, and if such a stream encounters a surface at a temperature above that corresponding to saturation for the dissolved salt, the salt will crystallize on the surface. Typically, crystallization will begin at specially active points (nucleation sites) such as area where corrosion has occurred, and after a considerable induction period, it will spread to cover the entire surface. The buildup of the foulant will continue as long as the surface in contact with the fluid has a temperature above saturation. In addition, solidification fouling can occur due to cooling below the solidification temperature of a dissolved component, such as solidification and separation of wax from crude oil.

Corrosion fouling and deposit formation due to corrosion of the substrate metal on heat transfer surfaces or reactor surfaces can occur as a result of poor or inadequate dewatering and desalting operations. In this type of fouling, corrosion occurs first and initiates fouling, whereas under-deposit corrosion occurs as an after-effect of fouling. Corrosion fouling is dependent on several factors such as thermal resistance, surface roughness, composition of the substrate, and composition of the feedstock. In particular, impurities present in the feedstock stream can greatly contribute to the onset of corrosion – examples of these impurities are hydrogen sulfide, ammonia, and hydrogen chloride.

In many crude oils (especially the heavy feedstocks such as heavy crude oil, extra heavy crude oil, tar sand bitumen, and the refinery-produced residua), sulfur- and nitrogen-containing compounds are common contaminants that are decomposed during refining to hydrogen sulfide and ammonia, respectively. Chloride derivatives which may be found in feedstocks (and originate from the formation brine which is co-produced with crude oil) are converted to hydrogen chloride in, for example, the distillation section of the refinery:

$$RCl + [H] \rightarrow HCl + R$$

In addition to entering the refinery from the production wells, chloride derivatives may also be derived from various chemicals used during oil production such as recovery enhancement chemicals and solvents used to clean tankers, barges, trucks, and pipelines. As the crude oil is processed, some of these chemicals and solvents, which are thermally stable and not soluble in water, pass overhead in the main tower of the atmospheric distillation unit along with the naphtha fraction (boiling range 0°C–200°C; 32°F–390°F). Further processing of the naphtha (a gasoline blend stock or solvent precursor) causes generation and release of the hydrogen chloride.

In a hydrogen sulfide environment, the sulfur reacts with any exposed iron to form iron sulfide compounds – this can occur in both the hot and colder sections of a refinery unit. Although corrosion is typically expected, the iron sulfide can form a complex protective scale or lattice on the base metal, which inhibits further corrosion (Speight, 2014c). In such a case, the corrosion rate would be minimal if no other impurities were present in the system to interact with the sulfide lattice (which is then unable to remain in equilibrium with the unit environment), and however, corrosion not only continues but is accelerated.

Although often ignored in favor of discussions about the sulfur-containing and nitrogen-containing contaminants and their contribution to fouling, hydrogen chloride is also an important contributor to fouling, particularly corrosion fouling. By the way of clarification, hydrogen chloride itself is a much lesser problem and typically does not foul equipment or corrode the carbon steel. However, chloride corrosion and the ensuing fouling take place when hydrogen chloride, ammonia, and water all interact in the colder sections of a unit and cause serious damage – the extent of the damage depends on the concentration (of the chloride ions) and is directly dependent on pH, with the corrosion rate increasing rapidly with pH decrease to a more acidic environment. Hydrogen chloride is corrosive when it comes in contact with free water, i.e. water that is not in the vapor phase. Hydrogen chloride is highly soluble in water, and in a free water environment, any hydrogen chloride present in the vapor or hydrocarbon liquid will be quickly absorbed by the water, thus decreasing the pH down.

Another form of fouling that is particularly noteworthy when a blended feedstock is employed as the refinery feedstock is phase separation (also called sedimentation fouling when gravity is the controlling force as might occur during the dewatering/desalting operation) which is the separation of solid particles from a feedstock stream and the eventual deposition of the particles onto a heat-exchanger surface or within a reactor, especially in the distillation unit. This may be a result of a combination of chemical and physical processes that can be assigned to various constituents of the feedstock, particularly the asphaltene constituents, which when thermally charged move into a region of instability and phase separate (Chapter 3) (Speight, 1994, 2014a).

Suspended particles such as sand, silt, clay, and iron oxide (which occur in refinery feedstocks (Chapter 1) (Speight 2014a) may separate from the feedstock if they are beyond a limiting size (which is system dependent). This phenomenon can often be prevented by pre-filtration at the time of the dewatering/desalting operation. Sedimentation fouling is strongly affected by fluid velocity, and suspended particles in the process fluids will deposit in low-velocity regions, especially where the velocity changes quickly, such as in pipe elbows (provided the elbow does not cause turbulent flow), heat-exchanger water boxes, on the shell side of the heat exchanger, or in the reactor.

Despite the operation of a desalting unit, small amounts of inorganic salts dissolved in water, such as magnesium, chloride, and calcium chloride, may remain in the crude oil that is sent to the distillation section of the refinery. However, there are many factors that can impair desalter performance, including high crude density and viscosity, which makes separation more difficult, and high salt content. When a high concentration of naphthenic acid derivatives is present in the crude oil, the likelihood of forming a rag layer (stable water in an oil emulsion), which can lead to corrosion and fouling, increases.

The following section contains sub-sections (presented alphabetically) that give an indication of the corrosion events that can occur when the performance of the desalter is below the necessary standard and corrosive-causing agents are allowed to pass into the refinery.

6.4 TYPES OF CORROSION AND FOULING

The potential for dealing with corrosion and fouling should occur early in the refining process with the identification of the types of corrosion. Even though an efficient desalting operation is important for corrosion and fouling control, there are several strategies to address the corrosion-causing contaminants in the refinery feedstock before submitting the feedstock to the desalter. In addition, tank farm management is also essential since the tank farm serves as the first stop for holding crude oil once it reaches the refinery by pipeline, truck, rail, or ship.

Crude oil in the tank farm will likely contain water with dissolved salts and solids. Crude tanks should be drained of water before the oil is charged to the unit. This addresses several problems: it reduces the salt contents in the raw crude charge and helps to avoid slugs of water to the desalter. Water slugs will cause loss of level control and potentially send excess water containing salts into the atmospheric distillation tower which could increase tower pressure and back out the feedstock charge or perhaps blow out the tower trays, causing a forced shutdown. Crude tank switches should be performed over 2 hours or so if possible in order to minimize abrupt changes in the quality of the crude oil.

Thus, combating or preventing corrosion and fouling is achieved by a system of (1) identifying, (2) monitoring, and (3) preventing corrosion. Monitoring methods include both off-line checks taken during maintenance and on-line monitoring. Off-line checks measure corrosion after it has occurred, telling the engineer when equipment must be replaced based on the historical information he has collected (preventative management). In addition, on-line monitoring systems have made substantial differences in the way corrosion is detected and mitigated. The systems also allow immediate responses to process changes to control corrosion mechanisms, allowing corrosion to be minimized while also maximizing production output. In an ideal situation, having on-line corrosion information that is accurate and real-time will allow conditions that cause high corrosion rates to be identified and reduced (predicative management) (Speight, 2014b).

By the way of recall (Chapter 3), fouling, as it pertains to crude oil refineries, is deposit formation, encrustation, deposition, scaling, scale formation, slagging, and sludge formation which has an adverse effect on operations. Corrosion and fouling can result in the formation of degradation products and other undesirable changes in the original properties of crude oil products. More specifically, fouling is due to the accumulation of incompatible material within a processing unit or on the solid surfaces of the unit, which can be a direct outcome of corrosion. Hence, the two effects (corrosion and foiling) are often associated with each other and both contribute to the detriment of the processing unit function.

The types of corrosion – each of which leads to fouling through the deposition of insoluble products from the corrosion process – that commonly occur in the refinery are presented (alphabetically) in the following sub-sections. In each case, the potential for fouling also exists.

6.4.1 Acidic Corrosion

The process of separating crude oil from brine results in the formation of hydrogen chloride. When sodium chloride and magnesium chloride dissolve in water, hydrolysis at 150°C–205°C (300°F–400°F) produces hydrogen chloride which reacts with iron:

$$Fe + 2HCl \rightarrow FeCl_2 + H_2$$

When the temperature is less than the dew point (100°C, 212°F) at the top of the distillation column, hydrogen chloride can come into contact with water and cause corrosion. When hydrogen sulfide is present, it results in even more corrosion:

$$FeCl_2 + H_2S \rightarrow FeS + HCl$$

Hydrogen chloride causes both general and localized corrosion and it is very aggressive to most common materials of construction across a wide range of concentrations. Damage in refineries is most often associated with dew point corrosion, in which vapors containing water and hydrogen chloride condense from the overhead stream of a distillation, fractionation, or stripping tower. The first water droplets that condense can be highly acidic (low pH) and can promote high corrosion rates.

Therefore, it is essential that the dewatering and desalting system be monitored carefully and that there should be no acidic effluents present that can lead to the corrosion mentioned above.

Of particular interest here is the potential for corrosion caused by the presence of naphthenic acids derivatives in high-acid crude oils, and to mitigate the effects of these acids, it is essential that these acid derivatives be removed at the dewatering–desalting stage. In fact, naphthenic acid derivatives are among the most corrosive agents in crude oil with numerous effects on refinery equipment. To address the problem of this type of corrosion, it is necessary to first understand the chemical behavior of these constituents (Al-Moubaraki and Obot, 2021).

Naphthenic acid derivatives are aliphatic acids with cycloalkene rings, an alkyl chain, and a terminal carboxylic acid group (Chapter 1). The varying structure of the naphthenic acid derivatives is related to (1) the structure of the constituents of the source material and (2) the biodegradation mechanism, which occurs as the oil field matures (Speight, 2014a, b, c; Al-Moubaraki and Obot, 2021). At room temperature, naphthalene acids are not corrosive but the corrosivity becomes more aggressive at temperatures where refining takes place (typically at temperatures >200°C, >390°F).

The corrosivity of the naphthenic acid (NA) derivatives is related to the TAN, which is a measure of acidity and is expressed in milligrams of potassium hydroxide (KOH) that is required to neutralize the acids in one gram of crude oil (Speight, 2015). The corrosion rate may also be affected by the sulfur and NA concentrations in crude oil (Al-Moubaraki and Obot, 2021). As a result of the direct reaction between the naphthenic acid derivatives and iron, iron naphthenates are formed. Simultaneously, hydrogen sulfide (H_2S) reacts with the iron, resulting in the formation of solid iron sulfide (FeS), which is then deposited on the metal surface. The hydrogen sulfide also reacts with the iron naphthenate to form iron sulfide and regenerate naphthenic acid. Thus:

$$Fe + 2RCOOH \rightarrow Fe(RCOO)_2 + H_2$$

$$Fe + H_2S \rightarrow FeS + H_2$$

$$Fe(RCOO)_2 + H_2S \rightarrow FeS + 2RCOOH$$

While these simple equations present a very simplified representation of corrosion by naphthenic acid derivative, corrosion by these constituents of crude oil is more complicated and is influenced by a variety of factors, including (1) crude oil type, (2) temperature, (3) flow velocity, (4) alloy type, and (5) sulfur content.

Naphthenic acid corrosion often occurs in the same place as high-temperature sulfur attack such as heater tube outlets, transfer lines, column flash zones, and pumps. Furthermore, naphthenic acids alone or in combination with other organic acids such as phenols can cause corrosion at temperatures as low as 65°C (150°F) up to 420°C (790°F). Crude oils with a TAN higher than 0.5 and crude oil fractions with a TAN higher than 1.5 are considered to be potentially corrosive between the temperature of 230°C and –400°C (450°F–750°F).

The rate of corrosion by naphthenic acid derivatives typically increases with an increase in TAN and also increases with an increase in temperature. Naphthenic acid corrosion is differentiated from sulfidic corrosion by the nature of the corrosion (pitting and impingement) and by the severe (naphthenic acid) corrosion at high velocity in distillation units. Feedstock heaters, furnaces, transfer lines, feed and reflux sections of columns, atmospheric and vacuum columns, heat exchangers, and condensers are among the types of equipment subject to this type of corrosion. Furthermore, isolated deep pits in partially passivated areas and/or impingement attack in essentially passivation-free areas are typical of naphthenic acid corrosion. In many cases, even very highly stable alloys (i.e. 12 Cr, AISI types 316 and 317) have been found to exhibit sensitivity to corrosion under these conditions.

Corrosion by naphthenic acids typically has a localized pattern, particularly at areas of high velocity and, in some cases, where condensation of concentrated acid vapors can occur in crude distillation units. The attack is also described as lacking corrosion products. Damage is in the form of unexpected high corrosion rates on alloys that would normally be expected to resist sulfidic corrosion (particularly steels with more than 9% w/w chromium). In some cases, even very highly alloyed materials (i.e. 12% w/w chromium, type 316 stainless steel (SS) and type 317 SS), and in severe cases, even 6% w/w Mo stainless steel have been found to exhibit sensitivity to corrosion under these conditions.

However, not all acidic species in crude oil are derivatives of carboxylic acids (RCOOH) and the acidic constituents (especially phenol derivatives, C_6H_5OH, which do not contain the carboxylic acid function, -COOH) may also be resistant to high temperatures. For example, acidic species appear in the vacuum residue after having been subjected to the inlet temperatures of an atmospheric distillation tower and a vacuum distillation tower. In addition, for the acid species that are volatile, naphthenic acids are most active at their boiling point and the most severe corrosion generally occurs on condensation from the vapor phase back to the liquid phase.

Also, during naphthenic crude processing, corrosion at high temperature is mitigated by injecting either phosphate-based ester additives or sulfur-based additives, which provide an adherent layer that does not corrode or erode due to the effect of naphthenic acids. It has been suggested and partially proven that corrosion during processing of high-acid crude oils is a lower risk if the sulfur content is high – the relationship between the acid number and the amount of sulfur is not fully understood but it does appear that the presence of sulfur-containing constituents has an inhibitive effect (Piehl, 1988).

In terms of high-acid crude oils, corrosion is predominant at temperatures higher than 180°C (355°F), where shear stress on pipe walls is significant. Such corrosion problems in the high-temperature section of the atmospheric distillation column are normally mitigated by dosing phosphate ester-based or sulfur (S)-based inhibitors at certain critical locations inside the process units. The inhibitors are dosed in process streams with the help of injection quills.

The first phase of an engineered solution is to perform a comprehensive high TAN impact assessment of a crude unit processing a target high TAN blend under defined operating conditions. An important part of any solution system is the design and implementation of a comprehensive corrosion monitoring program. Effective corrosion monitoring helps confirm which areas of the unit require a corrosion-mitigation strategy and provides essential feedback on the impact of any mitigation steps taken.

For example, the reliability of the refinery equipment during the processing of high-acid crude oils is of the utmost importance. Hardware changes – such as upgrading materials construction from carbon steel (CS) and alloy steel to stainless steel (SS) 316/317, which contains molybdenum and is significantly resistant to naphthenic acid corrosion – are complicated tasks that require large capital investment and a long turnaround for execution. Alternatives to hardware changes are corrosion mitigation with additives and corrosion monitoring with the application of inspection technologies and analytical tests.

With a complete understanding of the unit operating conditions, crude oil and distillate properties, unit metallurgies and equipment performance history, a probability of failure analysis can be performed for those areas which would be susceptible to naphthenic acid corrosion. Each process circuit is assigned a relative failure probability rating based on the survey data and industry experience.

Mitigation of process corrosion includes blending, inhibition, materials upgrading, and process control. Blending may be used to reduce the naphthenic acid content of the feed, thereby reducing corrosion to an acceptable level. Blending of heavy and light crudes can change shear stress parameters and might also help reduce corrosion. Blending is also used to decrease the level of sulfur content in the feed in an effort to somewhat limit naphthenic acid corrosion, but this strategy is not always successful (Chapter 3).

Corrosion inhibitors are often the most economical choice for mitigation of naphthenic acid corrosion. Effective inhibition programs can allow refiners to defer or avoid capital-intensive alloy upgrades, especially where high TAN crudes are not processed on a full-time basis.

Injection of corrosion inhibitors may provide protection for specific fractions that are known to be particularly severe. Monitoring needs to be adequate in this case to check on the effectiveness of the treatment. Process control changes may provide adequate corrosion control if there is a possibility of reducing charge rate and temperature. For long-term reliability, upgrading the construction materials is the best solution. Above 290°C (555°F), with very low naphthenic acid content, cladding with chromium (Cr) steels (5%–12% w/w Cr) is recommended for crudes of greater than 1% w/w sulfur when no operating experience is available. When hydrogen sulfide is evolved, an alloy containing a minimum of 9% w/w chromium is preferred. In contrast to high-temperature sulfidic corrosion, low-alloy steels containing up to 12% w/w Cr do not seem to provide benefits over carbon steel in naphthenic acid service. Type 316 stainless steel (>2.5% w/w molybdenum) or Type 317 stainless steel (>3.5% w/w molybdenum) is often recommended for cladding of vacuum and atmospheric distillation columns.

Another form of the mitigation of corrosion by naphthenic acids is neutralization or esterification of the acids at the point of the dewatering–desalting process. For example, caustic treatment can be used to neutralize acidic components, but the disadvantage is that non-soluble emulsions, product loss, and pollution may be the end result.

The conversion of the naphthenic acid derivatives to naphthenic acid esters is also an option (Wang et al., 2007, 2011; Wang and Watkinson, 2011; Quiroga-Becerra et al., 2012; Al-Moubaraki and Obot, 2021). This type of process could be installed as a batch system in addition to the dewatering–desalting units.

Another option for mitigation of corrosion due to the presence of naphthenic acids is the decarboxylation of the acids (Zhang et al., 2006; Al-Moubaraki and Obot, 2021). In this process, the acidic carboxyl group would be removed from the naphthenic acids by thermal decarboxylation that can occur in the refinery distillation column; although temperatures in the column can reach 400°C (750°F), the majority of the reaction would take place at a temperature on the order of 150°C–300°C (300°F–570°F).

6.4.2 Carbon Dioxide Corrosion

Corrosion by carbon dioxide (also known as sweet corrosion in contrast to the sour corrosion caused by hydrogen sulfide) is a common type of corrosion in the oil and gas industry (Usman and Ali, 2018; Obot et al., 2020; Al-Moubaraki and Obot, 2021). However, carbon dioxide is only corrosive when dissolved in water to form carbonic acid which (as dissolved carbon dioxide) reacts with iron. Thus:

$$Fe + CO_2 + H_2O \rightarrow FeCO_3 + H_2$$

The iron carbonate is in the form of a passive film that forms on carbon steel and low alloy metals under specific conditions. This film will prevent corrosion on the metal surface but on the other hand the passive film will degrade under stress or in high-velocity streams, making it susceptible to pits and cracks.

The dissolution of the metal is the primary cause of uniform or general corrosion. Pitting corrosion occurs at low fluid flow velocities, and this type of corrosion is enhanced by high temperatures and high partial pressures of carbon dioxide.

6.4.3 Carburization

Carburization is not a common occurrence in most refining operations because of the relatively low tube temperatures of most refinery fired heaters. However, it can and does occur in those higher temperature process heaters (e.g., coking units), where its control in 9Cr-Mo tubes has been somewhat successful through the use of an aluminum vapor diffusion coating. Carburization can also occur during an upset which results in the exposed material being heated to unacceptably high temperature, where it is expected to occur within the range of normal operating conditions; the broader approach has been to use Type 304H for temperatures up to about 815°C (1500°F). There is no advantage in using either of the stabilized grades since any unreacted titanium or niobium from the original melt would be quickly tied up. Type 310 or Alloy 800H may be used for temperatures up to about 1010°C (1850°F). For the most part, refinery application of the latter two alloys for this purpose is confined to hydrogen reformer furnaces. Unfortunately, the 300 series stainless steel alloys, including Type 310, are subject to sigma phase embrittlement in the temperature range where they have useful carburization resistance. Alloy 800H would be a better choice.

The most effective element in controlling carburization is nickel in combination with chromium. Silicon also has a strong effect and aluminum in excess of 3.5%–4% w/w is also beneficial. Unfortunately, the presence of much more than 2% w/w silicon adversely affects the rupture strength and weldability of both wrought and cast heat resistant alloys. Aluminum, in concentrations greater than 2%–2.5% w/w, has an adverse effect on ductility and fabrication properties that are essential for piping, tubing, and pressure vessels.

Carburization is far more common in the petrochemical industry than in refining. The most common occurrence is in the radiant and shield sections of ethylene cracking furnaces. Carburization is a serious problem in these furnaces because of the high tube metal temperatures up to 1150°C (2100°F) and high carbon potential associated with the ethane, propane, naphtha, and other hydrocarbon feedstocks which are cracked. However, it also occurs, albeit less frequently and with less severity, in reforming operations and in other processes handling hydrocarbon streams or certain ratios of $CO/CO_2/H_2$ gas mixtures at high temperatures.

The form and severity of decoking operations appear to play important roles in the rate of carburization. High-temperature decoking with low quantities of steam are thought to accelerate carburization. Likewise, steam–air decoking appears to be more deleterious than steam alone. Appropriate metallurgy can be used to reduce but rarely totally eliminate carburization. The most important characteristic of a successful alloy is its ability to form and maintain a stable, protective oxide film. Chromium oxide is considered to be such a film. However, it is not sufficiently stable at higher operating temperatures and low oxygen partial pressures. Alumina or silica are much better. Unfortunately, the addition of aluminum or silicon to the heat resistant alloys in quantities to develop full protection involves tradeoffs in strength, aged ductility, and/or weldability that are often unacceptable. Viable alloys are generally restricted to about 2% of either element.

6.4.4　Cooling Water Corrosion

Also, problems exist in every cooling water system in the oil refining industry: (1) corrosion, (2) inorganic deposits containing carbonate scale, (3) corrosion products of iron, phosphates, silicates, and some others, and (4) biofouling (microbial contamination). On-line corrosion and deposit monitoring systems used in the cooling water system allow monitoring the general corrosion of carbon steel (or any other alloy) at ambient temperature (non-heated steel surface) and at the drop temperature in the heat exchanger (heated steel surface), as well as pitting tendency for heated and non-heated surfaces and heat transfer resistance – the quantitative value of inorganic and organic deposits (fouling).

One of the major contributors to corrosion is the pH value of process water; pH measurements in oil refinery service have acquired a bad reputation due to their poor ability to measure in the aggressive environment they have to contend with. When the correct equipment is chosen, however, in-line pH measurement and control facilities have proven to be of great value in reducing plant-wide corrosion and the consumption of chemicals such as pH control reagents and corrosion inhibitors. This results not only in significant cost savings but also in increased earnings through increased on-stream process time.

6.4.5　Crude Oil Quality

Crude oil quality is an important aspect of corrosion that is often not recognized as much as other causes (Chapter 6). Crude oil value – to a refinery – is based on the expected yield and value of the products value, less the operating costs expected to be incurred to achieve the desired yield. With the growing popularity of heavy feedstocks, opportunity crudes, and high-acid crudes, one of the greatest challenges the refining industry faces is ensuring that the quality of crude oil received is equivalent to the purchased quality (value acquired is equal to value expected).

Furthermore, when multiple crude oils are processed as a blend, the complexity of the crude delivery system increases, making it more difficult to minimize differences between purchased quality and refinery-receipt quality. Shipping crude oil through multiple pipelines and redistribution storage tanks – a reality faced by most inland refiners – results in the delivered crude oil being a composite of the many crude oils. Thus, the resultant composite blend may vary significantly from the expected purchased quality, and the sources of quality problems are much more difficult to estimate.

Issues regarding crude oil properties occur regardless of whether the dominant crude slate is comprised of domestic crude delivered by pipeline or foreign crude delivered via waterborne transportation. In both cases, using simple categories such as gravity or sulfur does not provide an accurate measure of a particular crude oil value, and monitoring only gravity and sulfur does not provide adequate safeguards for the integrity of the crude oil while it is in transit. More sophisticated analyses (with an analysis for constituents likely to cause corrosion and correlated to refinery performance) can provide a comprehensive estimate of quality value to a specific refinery. This analysis needs to give equal weight to quality consistency, improved yield, reduced operating expenses, and the compatibility of the crude oil feedstock to the refinery processing hardware, where appropriate.

In addition, a combination of (1) aging refineries, (2) greater fluid corrosiveness, and (3) tightening of health, safety, security, and environment requirements has made corrosion management a key consideration for refinery operators. The prevention of corrosion damage through live monitoring provides a real-time picture of how the refinery is coping with the high demands placed upon it by corrosive fluids. This information can assist in risk management assessments.

6.4.6 TEMPERATURE-DEPENDENT CORROSION

There are two forms of temperature-dependent corrosion in a refinery and these are (1) low-temperature corrosion and (2) high-temperature corrosion. The approximate rule of thumb is that the corrosion rate of a metal doubles for every 10°C (18°F) increase in temperature but there will be exception to this rule.

For example, the rule is based on the situation when the corrosion rate is controlled by a chemical reaction, such as dilute sulfuric acid attacking carbon steel which, for the chemical reaction, may be true; but if the corrosion rate is controlled by other factors, such as the presence of oxygen in the corrosive environment, the rule of thumb may fail.

Typically, an increase in temperature does contribute to an increase in the number of active centers of corrosion on a metal surface and accelerates the development of corrosion processes but the exact relationship of the corrosion rate to the temperature is subject to much speculation.

6.4.6.1 Low-Temperature Corrosion

Low-temperature corrosion (where the temperature is <100°C, <212°F) occurs in (1) the presence of electrolytes, usually aqueous solutions of electrolytes and water, (2) dissolved corrosive gases (hydrogen sulfide, hydrogen chloride, and ammonia), or (3) dissolved salts (such as sodium chloride and sodium sulfate).

Thus, category includes corrosion in aqueous amine solutions (in amine treating units). The majority of low-temperature corrosion problems in oil refineries and petrochemical plants are caused by inorganic compounds rather than processed hydrocarbons, such as hydrogen chloride, hydrogen sulfide, ammonia, sulfuric acid, sodium hydroxide, sodium carbonate, and dissolved oxygen. Organic acids that are contained or sometimes formed in the crude oil, corrosion in the overhead of distillation systems, and amines used as neutralizers may all promote low-temperature corrosion.

There are two principal sources of compounds causing corrosion which are (1) contaminants contained in crude oil (such as air, water, hydrogen chloride, hydrogen sulfide, and ammonia) and (2) process chemicals including solvents, neutralizers, and catalysts. In addition, the use of aqueous solutions of amines for the absorption of acid gases such as carbon dioxide and hydrogen sulfide at the recovery site can result in the formation of corrosive heat stable amine salts. Temperature, pressure, flow regime, and media are the primary factors influencing the low-temperature corrosion rate and intensity.

6.4.6.2 High-Temperature Corrosion

High-temperature corrosion (where the temperature is >200°C, >390°F) involves corrosion by non-electrolytes (usually gaseous hydrogen, hydrogen sulfide, hydrogen sulfide, naphthenic acids, and hot ash corrosion).

High-temperature crude corrosion is a complex problem. There are at least three corrosion mechanisms: (1) furnace tubes and transfer lines where corrosion is dependent on velocity and vaporization, and is accelerated by naphthenic acid, (2) vacuum column where corrosion occurs at the condensing temperature, is independent of velocity, and increases with naphthenic acid concentration, and (3) side-cut piping where corrosion is dependent on naphthenic acid content and is inhibited somewhat by sulfur compounds in the refinery feedstock (Chapters 1 and 6).

To mitigate high-temperature corrosion, a high-temperature corrosion inhibitor is dosed into the process streams, normally at a concentration of 3–15 ppm w/w. The additive is mixed with the suitable process stream, typically at a ratio of 1:30, and the mix is injected into the cooled stream (<100°C, <212°F) with specially designed corrosion-mitigation quills, since the inhibitor is highly corrosive to alloy steel and even stainless steel. The dosage limit of the phosphorus-based inhibitor is calculated from the allowable phosphorus level in the products – products such as the vacuum gas oil that is used as the feedstock for a hydrocracking unit or as the feedstock for a fluid catalytic cracking unit.

Alloys intended for high-temperature applications are designed to have the capability of forming protective oxide scales. Alternatively, where the alloy has ultra-high temperature strength capabilities (which is usually synonymous with reduced levels of protective scale-forming elements), it must be protected by a specially designed coating. Oxides that effectively meet the criteria for protective scales listed above and can be formed on practical alloys are limited to chromia (Cr_2O_3), alumina (Al_2O_3), and silicon dioxide (SO_2). In the pure state, alumina exhibits the slowest transport rates for metal and oxygen ions and so should provide the best oxidation resistance.

A useful concept in assessing the potential high-temperature oxidation behavior of an alloy is that of the reservoir of scale-forming element contained by the alloy in excess of the minimum level (approximately 20% w/w for iron-chromium alloys at 1000°C (1832°F)). The more likely the service conditions are to cause repeated loss of the protective oxide scale, the greater the reservoir of scale-forming element required in the alloy for continued protection. Extreme cases of this concept result in chromizing or aluminizing to enrich the surface regions of the alloy or in providing an external coating rich in the scale-forming elements.

6.4.7 SULFIDIC CORROSION

In addition to the corrosive properties of high-acid crude oils, sulfur may be present in crude oil as hydrogen sulfide (H_2S), as compounds (such as mercaptan derivatives, RSH, sulfide derivatives, RSR, disulfide derivatives, RSSR, and thiophene derivatives), or as elemental sulfur. Each crude oil has different amounts and types of sulfur compounds but, generally, the proportion, stability, and complexity of the compounds are greater in heavier crude oils. Hydrogen sulfide is a primary contributor to corrosion in refinery processing units. Other corrosive substances are elemental sulfur and mercaptan derivatives.

This type of corrosion can manifest itself as general (uniform corrosion), localized pitting corrosion, and hydrogen embrittlement. The latter will result in the appearance of a crack at the metal, releasing H_2S gas into the environment, causing environmental damage and serious health issues for workers. Temperature is an important factor in corrosion by hydrogen sulfide, with severe corrosion occurring in the distillation unit if the temperature exceeds 205°C (400°F).

Heavy feedstocks (of the types that are relevant to this chapter, i.e. heavy crude oil, extra heavy crude oil, and tar sand bitumen) contain inorganic salts such as sodium chloride, magnesium chloride, and calcium chloride in suspension or dissolved in entrained water (brine). These salts must be removed or neutralized before processing to prevent catalyst poisoning, equipment corrosion, and fouling. Salt corrosion is caused by the hydrolysis of some metal chlorides to hydrogen chloride (HCl) and the subsequent formation of hydrochloric acid when crude is heated. Hydrogen chloride may also combine with ammonia to form ammonium chloride (NH_4Cl), which causes fouling and corrosion.

The sections of the process susceptible to corrosion include (but may not be limited to) preheat exchanger (due to the presence of hydrogen chloride, HCl, and hydrogen sulfide, H_2S), preheat furnace and bottoms exchanger (due to the presence of hydrogen sulfide, H_2S, and other sulfur-containing compounds), atmospheric tower and vacuum furnace (due to the presence of hydrogen sulfide, H_2S, other sulfur-containing compounds, and organic acids), vacuum tower (due to the presence of hydrogen sulfide, H_2S, and organic acids), and overhead (due to the presence of hydrogen sulfide, H_2S, hydrogen chloride, HCl, and water). Where sour crudes are processed, severe corrosion can occur in furnace tubing and in both atmospheric and vacuum towers where metal temperatures exceed 230°C (450°F). Wet hydrogen sulfide will also cause cracks in steel. When processing high-nitrogen crudes, nitrogen oxides can form in the flue gases of furnaces and these oxides are corrosive to steel when cooled to low temperatures (nitric and nitrous acids are formed) in the presence of water.

One type of sulfidic corrosion is the corrosion caused by the presence of hydrogen sulfide in the feedstock which can occur as general (uniform) corrosion, localized pitting corrosion, and hydrogen embrittlement (Qu et al., 2005, 2006; Yépez, 2005). The latter will result in the appearance of a crack at the metal, thereby releasing hydrogen sulfide gas into the environment, causing environmental damage and serious health issues for workers. Temperature is an important factor in corrosion by hydrogen sulfide, with severe corrosion occurring in the atmospheric distillation unit if the temperature exceeds 205°C (400°F).

Any of the 18Cr-8Ni stainless steel grades can be used to control sulfidation. However, it is best to use the stabilized grades mentioned earlier. Some sensitization is unavoidable if exposure in the sensitizing temperature range is continuous or long term. Stainless steel equipment subjected to such exposure and to sulfidation corrosion should be treated with a 2% w/w soda ash solution or an ammonia solution immediately upon shutdown to avoid the formation of polythionic acid $[S_n(SO_3H)_2$, where n >2] which can cause severe intergranular corrosion and stress cracking.

Polythionic acid is an oxoacid which has a straight chain of sulfur atoms and has the chemical formula $S_n(SO_3H)_2$ (n > 2):

$$
\begin{array}{ccc}
& O & & & O & \\
& \parallel & & & \parallel & \\
HO - & S & - S_X - & S & - OH \\
& \parallel & & & \parallel & \\
& O & & & O &
\end{array}
$$

Examples are dithionic acid ($H_2S_2O_6$), trithionic acid ($H_2S_3O_6$), tetrathionic acid ($H_2S_4O_6$), and pentathionic acid ($H_2S_5O_6$).

Vessels for high-pressure hydrotreating and other heavy crude fraction upgrading processes (e.g., hydrocracking) are usually made of one of the Cr-Mo alloys. To control sulfidation, they are internally clad with one of the 300 series stainless steels by roll or explosion bonding or by weld overlay. In contrast, piping, exchangers, and valves exposed to high-temperature hydrogen–hydrogen sulfide environments are usually made of solid 300 series stainless steel alloys. In some designs, Alloy 800H has been used for piping and headers. In others, centrifugally cast HF-modified piping has been used. High nickel alloys are rarely used in refinery or petrochemical plants in hydrogen–hydrogen sulfide environments because of their susceptibility to the formation of deleterious nickel sulfide. They are particularly susceptible to this problem in reducing environments. As a general rule, it is recognized that the higher the nickel in the alloy, the more susceptible the material to corrosion.

Vapor diffusion aluminum coatings have been used with carbon, Cr-Mo, and stainless steels to help control sulfidation and reduce scaling. For the most part, this has been restricted to smaller components. Aluminum metal spray coatings have also been used but not widely or very successfully.

6.5   CORROSION AND FOULING MANAGEMENT

Corrosion control is an ongoing, dynamic process in the prevention of metal deterioration by three general ways: (1) change the feedstock environment in which corrosion occurs, (2) change the character of the feedstock that leads to corrosion and fouling, or (3) place a barrier between the corrosion-forming feedstock and the environment or process that gives rise to corrosion and fouling. The material that corrodes does not have to be metal – but *is* a metal or an alloy of metals in most cases – and the metal does not have to be steel, but, because of the strength and cheapness of this material, it usually is steel or an alloy of steel. Again, the environment is, in most cases, the atmosphere, water, or the earth and is an important contributor to corrosion chemistry.

For practical purposes, corrosion in refineries can be classified into low-temperature corrosion and high-temperature corrosion. Low-temperature corrosion is considered to occur below approximately 260°C (500°F) in the presence of water. Carbon steel can be used to handle most hydrocarbon streams in this temperature range, except where aqueous corrosion by inorganic contamination, such as hydrogen chloride or hydrogen sulfide, necessitates selective application of more resistant alloys. High-temperature corrosion is considered to take place above approximately 260°C (500°F). The presence of water is not necessary, because corrosion occurs by the direct reaction between metal and environment.

The major cause of low-temperature (and, for that matter, high-temperature) refinery corrosion is the presence of contaminants in crude oil as it is produced. Although some contaminants are removed during preliminary treating at the wellhead fields as well as during dewatering and desalting, they can still appear in refinery tankage along with contaminants picked up in pipelines or marine tankers. However, in most cases, the actual corrosives are formed during initial refinery operations. For example, potentially corrosive hydrogen chloride evolves in crude preheat furnaces from relatively benign calcium chloride ($CaCl_2$) and magnesium chloride ($MgCl_2$) entrained in crude oil. Mitigation of the problems related to low-temperature corrosion is associated with adequate cleaning of the crude oil and removal of corrosive contaminates at the time of, or immediately after, formation.

The pH stabilization technique can be used for corrosion control in wet gas pipelines when no or very little formation water is transported in the pipeline. This technique is based on precipitation of protective corrosion product films on the steel surface by adding pH-stabilizing agents to increase the pH of the water phase in the pipeline. This technique is very well suited for use in pipelines where glycol is used as hydrate preventer, as the pH stabilizer will be regenerated together with the glycol – thus, there is very little need for replenishment of the pH stabilizer.

Some of the ways adapted today to overcome the effects of corrosion are, for example, use of specific types of metal to be longstanding in spite of the effects of corrosion. Carbon steel is used for the majority of refinery equipment requirements as it is cost-efficient and withstands most forms of corrosion due to hydrocarbon impurities below a temperature of 205°C (400°F); however, it is not used universally as it is not able to resist other chemicals and environments. Other kinds of metals used are low alloys of steel containing chromium and molybdenum and stainless steel containing high concentrations of chromium for excessively corrosive environments. More durable metals such as nickel, titanium, and copper alloys are used for the most corrosive areas of the plant which are mostly exposed to the highest of temperatures and the most corrosive of chemicals.

Many problems of correct use of corrosion control measures (for example, injection of chemicals such as inhibitors, neutralizers, and biocides) may be solved by means of corrosion monitoring methods (Jambo et al., 2002; Speight, 2014b). For example, hydrocarbons containing water vapors, hydrogen chloride, and hydrogen sulfide leave the atmospheric distillation column at 130°C (265°F). This mixture becomes very corrosive when cooled below the dew point temperature of 100°C (212°F). In order to prevent high acidic corrosion in the air cooler and condensers, neutralizers and corrosion inhibitor are injected in the overhead of the distillation column. In addition, corrosion monitoring equipment such as weight loss coupons and electrical resistance probes should be installed in several places, including the naphtha pump-around and kerosene pump-around lines. The electrical resistance probes show the corrosion situation continuously. The weight loss should be changed every few months, in order to compare with the results of the electrical resistance probes and to examine the danger of chloride attack (pitting corrosion). The more points in the unit used for corrosion monitoring, the better and more efficient is the corrosion coverage.

In summary, control of corrosion requires (1) evaluation of the potential corrosion risks, (2) consideration of control options – principally inhibition as well as material selection, (3) monitoring whole life cycle suitability, (4) life cycle costing to demonstrate economic choice, and (5) diligent quality assurance (QA) at all stages (Van Den Berg et al., 2003).

Information is the key to effective corrosion and fouling management since it is on the basis of this information that ongoing adjustments to corrosion control are made. Information is valid data.

Thus, to make effective corrosion management decisions on a day-to-day basis, the monitoring data must be valid. This is not simply a requirement for the probes to be operating correctly. It requires that they be placed in the most appropriate places, i.e. at those points where the corrosion controlling activity might be expected to work but also where it might equally be expected to be least effective, e.g., remote from the inhibitor injection point.

In many cases, specially designed traps are introduced into a plant so that corrosion probes may be inserted. These often produce their own microenvironment, which is atypical of the plant itself and provides little hope of effective entry for an inhibitor. Data from a probe in such a location are unlikely to be relevant to corrosion management elsewhere in the system. Invalid data lead to ineffective corrosion management.

From time to time, a corrosion management program should be reviewed at both the strategic and the tactical level. In human affairs, things change. The management of a facility will always be alive to current market trends and it may be necessary to revise the management objectives from time to time. Since the corrosion management program was constructed to meet the objectives of an earlier plant management plan, it will be necessary to review the program and possibly alter it. Likewise, the pace of technological change is rapid compared to the anticipated lifetime of most facilities. Thus, newer, more effective, cheaper means of achieving the same ends may emerge, and indeed, it may be possible to adopt them in place of existing tactics within the corrosion management program. Thus, the program is not a fixed blueprint, but a means to an end that must be reviewed and revised to meet the current management objective.

Corrosion cannot be ignored for it will not go away. However, there is little merit in controlling corrosion simply because it occurs, and none in ignoring it completely. The consequences of corrosion must always be considered. If the consequence of corrosion can be tolerated, it is perfectly acceptable to take no action to control it. If the consequences are unacceptable, steps must be taken to manage them throughout the facility's life at a level that is acceptable. To manage is not simply to control.

Good corrosion management aims to maintain, at a minimum life cycle cost, the levels of corrosion within predetermined acceptable limits. This requires that, where appropriate, corrosion control measures be introduced and their effectiveness ensured by judicious, and not excessive, corrosion monitoring and inspection. Good corrosion management serves to support the general management plan for a facility. Since the latter changes as market conditions, for example, change, the corrosion management plan must be responsive to that change. The perceptions of the consequences and risk of a given corrosion failure may change as the management plan changes. Equally, some aspects of the corrosion management strategy may become irrelevant. Changes in the corrosion management plan must, inevitably, follow. For example, crude blending is the most common solution to high TAN crude processing. But blending can only be effective if proper care is taken to control crude oil and distillate acid numbers to proper threshold levels.

In order to manage corrosion, not only detection of corrosion is necessary but also an assessment of the extent of the corrosion. Assessment of the extent of damage depends on inspection, or on an estimation of the accumulation of damage based on a model for damage accumulation, or both. Sound planning of inspections is critical so that the areas inspected are those where damage is expected to accumulate and the inspection techniques used will provide reliable estimates of the extent of damage. If the extent of the damage is known or can be estimated, a reduced strength can be ascribed to the component and its adequacy to perform safely can be calculated.

The general procedures for estimating fitness-for-service are outlined in the American Petroleum Institute (API) Recommended Practice 579 – Fitness-for-Service, which provides assessment procedures for various types of defects to be expected in pressurized equipment in the refinery and the steps are (1) the identification of flaws and damage mechanisms, (2) the identification of the applicability of the assessment procedures applicable to the particular damage mechanism, (3) the identification of the requirements for data for the assessment, (4) the evaluation of the acceptance of the component in accordance with the appropriate assessment techniques and procedures, (5) the remaining life evaluation, which may include the evaluation of appropriate inspection intervals to monitor the growth of damage or defects, (6) the remediation if required, (7) in-service monitoring where a remaining life or inspection interval cannot be established, and (8) the documentation, providing appropriate records of the evaluation made.

Creep damage can be assessed by various procedures including those described earlier. Life estimates can also be made based on the predicted life at the temperature and stress that are involved by subtracting the calculated life used up and making an allowance for loss of thickness by oxidation or other damage. The growth of cracks in components operating at high temperature that is detected can be estimated using established predictive methods. Additionally, various examples of simplified methods to predict safe life in petrochemical plant containing cracks have been published, for example, in a reformer furnace.

There is also the need for the management of slop oil which, by definition, is crude oil that is emulsified with water and solids rendering it a waste stream and is typically kept in evaporation ponds, sludge pits, storage tanks, and permitted commercial disposal facilities. In some cases, the slop oil and waste oil will be blended into fresh crude oil feedstock(s) as a form of reprocessing but this is a practice that is not common because the slop oil and waste oil contain surfactants. The premise behind the desalting process is to wash crude oil with water to remove contaminants and then separate the contaminated water from the dry, clean oil. A surfactant will reduce the surface tension between two fluids (in the case of desalting, the oil and water), and they are allowed to mix together and become difficult to separate.

The best practices for treating slop oil are (1) remove surfactant-laden solids via centrifuge before sending slop to the crude tank, (2) break any emulsion that has formed in the slop and separate the oil from the water before sending the oil to crude charge, (3) bypass crude charge and feed slop to another part of the refinery, such as the coking unit, (4) inject the slop oil directly and continuously into the crude charge line at a very low rate, typically <1% of the crude charge rate, and (5) have a dedicated slop tank and test the slop for BS&W and filterable solids before charging it to the appropriate refinery unit (McDaniels and Olowu, 2016).

REFERENCES

Al-Moubaraki, A.H., and Obot, I.B. 2021. Corrosion challenges in petroleum refinery operations: Sources, mechanisms, mitigation, and future outlook. *Journal of Saudi Chemical Society*, 25: 101370. https://www.researchgate.net/publication/355678568_Corrosion_Challenges_in_Petroleum_Refinery_Operations_Sources_Mechanisms_Mitigation_and_Future_Outlook.

Andersen, S.I., and Pederson, C. 1999. Thermodynamics of asphaltene stability by flocculation titration. *Proceedings of AICHE International conference on Petroleum Behavior and Fouling, 3rd International Symposium on the Thermodynamics of Asphaltenes and Heavy Oils.* American Institute of Chemical Engineers, New York.

Asomaning, S. 2003. Test methods for determining asphaltene stability in crude oils. *Petroleum Science and Technology*, 21: 581–590.

Asomaning, S., and Watkinson A.P. 2000. Petroleum stability and heteroatom species effects in fouling of heat exchangers by asphaltenes. *Heat Transfer Engineering*, 21: 10–16.

Barletta, A., and White, S. 2007. Crude overhead system design considerations. *Digital Refining*. http://www.digitalrefining.com/article/1000111,Crude_overhead_system_design_considerations.html#.VJhrUV4AA.

Barnett, J.W. 1988. Desalters can remove more than salts and sediment. *Oil & Gas Journal*, 86: 43.

Barnett, W.P., Sagüés, A.A., and Baumert, K.L. 1983. Coal liquids distillation tower corrosion: Correlation of on-stream monitoring with process variables at the wilsonville, alabama, SRC-1 pilot plant. *Fuel Processing Technology*, 8(1): 53–64.

Bhatia, S., and Sharma, D.K. 2006. Emerging role of biorefining of heavier crude oils and integration of biorefining with crude oil refineries in the future. *Petroleum Science and Technology*, 24(10): 1125–1159.

Blanco, F. and Hopkinson, B. 1983. Experience with Naphthenic Acid corrosion in Refinery Distillation Process Units. Paper No. 99. Proceedings. CORROSION 83. NACE International, Houston, TX.

Chambers, B., Srinivasan, S., Yap, K.M., and Yunovich, M. 2011. Corrosion in crude distillation unit overhead operations: A comprehensive review. *Proceedings of Corrosion 2011, NACE International*, Houston, TX. March 13–17.

Coletti, F., and Macchietto, S. 2009. Refinery pre-heat train network simulation undergoing fouling assessment of energy efficiency and carbon emission. *Proceedings of International Conference on Heat Exchanger Fouling and Cleaning VIII – 2009*, H. Müller-Steinhagen, M.R. Malayeri, and A.P. Watkinson (Editors). Schladming, Austria, June 14–19.

Coletti, F., Macchietto, S., and Polley, G.T. 2010. Effects of fouling on performance of retrofitted heat exchanger networks: A thermo- hydraulic based analysis. *Proceedings of 20th European Symposium on Computer Aided Process Engineering – ESCAPE20*, S. Pierucci and G. Buzzi Ferraris (Editors). Elsevier, Amsterdam, Netherlands

Collins, T., and Barletta, A. 2012. Desalting heavy Canadian crudes. Digital Refining. http://www.digitalrefining.com/article/1000566,Desalting_heavy_Canadian_crudes.html#.VJiTC14AA.

Costa, A.L.H., Tavares, V.B.G., Queiroz, E.M., Pessoa, F.L.P., Liporace, F.S., Oliveira, S.G., and Borges, J.L. 2011a. Parameter estimation of fouling models in crude preheat. *Proceedings of International Conference on Heat Exchanger Fouling and Cleaning – 2011*, Crete Island, Greece, M.R. Malayeri, H. Muller-Steinhagen and A.P. Watkinson (Editors), June 5–10, pp. 37–46.

Costa, A.L.H., Tavares, V.B.G., Queiroz, E.M., Pessoa, F.L.P., Liporace, F.S., Oliveira, S.G., and Borges, J.L. 2011b. Analysis of the environmental and economic impact of fouling in crude preheat trains for petroleum distillation. *Proceedings. International Conference on Heat Exchanger Fouling and Cleaning – 2011*, Crete Island, Greece, M.R. Malayeri, H. Muller-Steinhagen and A.P. Watkinson (Editors), June 5–10, pp. 47–49. www.heatexchanger-fouling.com; accessed December 12, 2014.

Crittenden, B.D., Kolaczkowski, S.T., and Downey, L.L. 1992. Fouling of crude oil preheat exchangers. *Chemical Engineering Research and Design*, 70: 547–557.

De Croocq, D. 1997. Major scientific and technical challenges about development of new processes in refining and petrochemistry. *Revue Institut Français de Pétrole*, 52(5): 469–489.

Derakhshesh, M., Eaton, P., Newman, B., Hoff, A., Mitlin, D., and Gray, M.R. 2013. Effect of asphaltene stability on fouling at delayed coking process furnace conditions. *Energy & Fuels*, 27: 1856–1864.

Derungs, W.A. 1956. Naphthenic acid corrosion – An old enemy of the petroleum industry. *Corrosion*, 12(2): 41.

Deshannavar, U.B., Rafeen, M.S., Ramasamy, M., and Subbarao, D. 2010. Crude oil fouling: A review. *Journal of Applied Sciences*, 10: 3167–3174.

Dickakian, G.B., and Seay, S. 1988. Asphaltene precipitation primary crude exchanger fouling mechanism. *Oil and Gas Journal*, 86: 47–50.

Dos Santos Liporace, F., and de Oliveira, S.G. 2005. Real time fouling diagnosis and heat exchanger performance. *Proceedings of ECI Symposium Series, Volume RP2, 6th International Conference on Heat Exchanger Fouling and Cleaning – Challenges and Opportunities*, H. Müller-Steinhagen, M.R. Malayeri, and A.P. Watkinson (Editors). Engineering Conferences International, Kloster Irsee, Germany, June 5–10.

Gary, J.H., Handwerk, G.E., and Kaiser, M.J. 2007. *Petroleum Refining: Technology and Economics*, 5th Edition. CRC Press, Taylor & Francis Group, Boca Raton, FL.

Georgiadis, M.C., Papageorgiou, L.G., and Macchietto, S. 2000. Optimal cleaning policies in heat exchanger networks under rapid fouling. *Industrial and Engineering Chemistry Research*, 39: 441–454.

Gutzeit, J. 1977. Naphthenic acid corrosion in oil refineries. *Materials Performance*, 33(10): 24.

Gutzeit, J. 2000. Effect of organic chloride contamination of crude oil on refinery corrosion. *Proceedings, CORROSION/2000*. NACE International, Houston TX.

Gutzeit, J. 2008. Controlling Crude Unit Overhead Corrosion by Improved Desalting. Hydrocarbon Processing, February: page 119.

Hirschberg, A., deJong, L.N.J., Schipper, B.A. 1984. Influence of Temperature and Pressure on Asphaltene Flocculation. SPE Journal, 24(3): 283–293.

Hsu, C.S., and Robinson, P.R. (Editors). 2017. *Handbook of Petroleum Technology*. Springer International Publishing AG, Cham.

Jambo, H.C.M., Freitas, D.S., and Ponciano, J.A.C. 2002. Ammonium hydroxide injection for overhead corrosion control in a crude distillation unit. *Proceedings of International Corrosion Congress*, Granada, Spain.

Kremer, L.N. 2000. Challenges to desalting heavy crude oil. *Proceedings of International Conference on Refinery Processing*, 2000, AIChE Spring National Meeting, Atlanta, Georgia, March 5–9.

Kremer, L.N. 2006a. Crude oil management: Reduce operating problems while processing opportunity crudes. *Proceedings of International Conference on Refinery Processing*. AIChE Spring National Meeting, Orlando, FL, April 23–27.

Kremer, L.N. 2006b. Controlling quality variations in the feed to desalters. *Proceedings of International Conference on Refinery Processing*. AIChE Spring National Meeting, Orlando, FL, April 23–27.

Kronenberger, D.L. 1984. Paper No. 128. Corrosion Problems Associated with the Desalting Difficulties of Maya and Other Heavy Crudes. Proceedings. CORROSION/84. NACE International, Houston, TX.

Lee, S. 1991. *Oil Shale Technology*. CRC Press, Taylor & Francis Group, Boca Raton, FL.

Lee, S. 1996. *Alternative Fuels*. CRC Press, Taylor & Francis Group, Boca Raton, FL.

Lindemuth, P.M., Lessard, R.B., and Lozynski, M. 2001. Improve desalter operations. *Hydrocarbon Processing*, 80: 67.

Macchietto, S., Hewitt, C.F., Coletti, F., Crittenden, B.D., Dugwell, D.R., Galindo, A., Jackson, G., Kandiyoti, R., Kazarian, S.G., Luckham, P.F., Matar, O.K., Millan-Agorio, M., Miller, E.A., Paterson, W., Pugh, S.J., Richardson, S.M., and Wilson, D.I. 2009. Fouling in crude oil preheat trains: A systematic solution to an old problem. *Proceedings of International Conference on Heat Exchanger Fouling and Cleaning VIII – 2009*, H. Muller-Steinhagen, M.R. Malayeri and A.P. Watkinson (Editors), Schladming, Austria, June 14–19.

McDaniels, J., and Olowu, W. 2016. Removing contaminants from crude oil. *Digital Refining*, PTQ, Q1. https://www.digitalrefining.com/article/1001246/removing-contaminants-from-crude-oil#.YmgTBdrMJPY.

Murphy, G., and Campbell, J. 1992. Fouling in refinery heat exchangers. In: *Fouling Mechanisms*, M. Bohnet (Editor). GRETH Seminar, Grenoble, pp. 249–261.

Mushrush, G.W., and Speight, J.G. 1995. *Petroleum Products: Instability and Incompatibility*. Taylor & Francis, Washington, DC.

Mushrush, G.W., and Speight, J.G. 1998. Instability and incompatibility of petroleum products. In: *Petroleum Chemistry and Refining*, J.G. Speight (Editor). Taylor & Francis, Washington, DC, Chapter 8, 1998.

Obot, I.B. Onyeachu, I.B., Umoren, S.A., Quraishi, M.A., Sorour, A.A., Chen, T., Aljeaban, N., and Wang, Q. 2020. High temperature sweet corrosion and inhibition in the oil and gas industry: Progress, challenges and future perspectives. *Journal of Petroleum Science and Engineering*, 185: 106469. https://www.sciencedirect.com/science/article/abs/pii/S0920410519308903.

Parkash, S. 2003. *Refining Processes Handbook*. Gulf Professional Publishing, Elsevier, Amsterdam.

Piehl, R.L. 1988. Naphthenic acid corrosion in crude distillation units. *Materials Performance*, 44(1): 37.

Polley, G.T., Morales-Fuentes, A., and Wilson, D.I. 2009. Simultaneous consideration of flow and thermal effects of fouling in crude oil preheat trains. *Heat Transfer Engineering*, 30(10–11): 815–821.

Polley, G.T., Wilson, D.I. and Pugh, S. 2000. Designing crude oil pre-heat trains with fouling mitigation. *Proceedings of 3rd International Conference on Refining Procedures*. American Institute of Chemical Engineers (AIChE), New York, pp. 519–523.

Polley, G.T., Wilson, D.I., Yeap, B.L., and Pugh, S.J. 2002. Evaluation of laboratory crude oil threshold fouling for application to refinery preheat trains. *Applied Thermal Engineering*, 22: 777–788.

Pruneda, EF., Escobedo, E.R.B., and Vazquez, F.J.G. 2005. Optimum temperature in the electrostatic desalting of maya crude oil. *Journal of Mexican Chemical Society*, 49(1): 14–19.

Qu, D.R., Zheng, Y.G., Jing, H.M., Jiang, X., and Ke, W. 2005. Erosion-corrosion of Q235 and 5Cr1/2Mo steels in oil with naphthenic acid and/or sulfur compound at high temperature. *Materials and Corrosion*, 56(8): 533–541.

Qu, D.R., Zheng, Y.G., Jing, H.M., Yao, Z.M., and Ke, W. 2006. High temperature naphthenic acid corrosion and sulfidic corrosion of Q235 and 5Cr1/2Mo steels in synthetic refining media. *Corrosion Science*, 48: 1960–1985.

Quiroga-Becerra, H., Mejía-Miranda, C., Laverde-Cataño, D., Hernández-López, M., and Gómez-Sánchez, M. 2012. A kinetic study of esterification of naphthenic acids from a colombian heavy crude oil. *CTyF – Ciencia, Tecnologia y Futuro*, 4: 21–31. https://www.researchgate.net/publication/260765970_A_kinetic_study_of_esterification_of_naphthenic_acids_from_a_Colombian_Heavy_Crude_Oil.

Rodriguez, C., and Smith, R. 2007. Optimization of operating conditions for mitigating fouling in heat exchanger networks. *Chemical Engineering Research and Design*, 85: 839–851.

Saleh, Z.S., Sheikholeslami, R. and Watkinson, A.P. 2004. Fouling characteristics of two crude oils used in Australia. *Proceedings of Chemical Conference*, Sydney, NSW, Australia.

Saleh, Z.S., Sheikholeslami, R. and Watkinson, A.P. 2005a. Fouling characteristics of a light australian crude oil. *Heat Transfer Engineering*, 26(1): 15–22.

Saleh, Z.S., Sheikholeslami, R., and Watkinson, A.P. 2005b. Blending effects on fouling of four crude oils. *Proceedings of ECI Symposium Series, Volume RP2, 6th International Conference on Heat Exchanger Fouling and Cleaning – Challenges and Opportunities*, H. Müller-Steinhagen, M.R. Malayeri, and A.P. Watkinson (Editors). Engineering Conferences International, Kloster Irsee, Germany, June 5–10, pp. 37–46.

Schermer, W.E.M., Melein, P.M.J., and Van den Berg, F.G.A. 2004. Simple techniques for evaluation of crude oil compatibility. *Petroleum Science and Technology*, 22: 1045–1054.

Scouten, C.S. 1990. Oil shale. In *Fuel Science and Technology Handbook.* Marcel Dekker Inc., New York, Chapters 25–31, pp. 795–1053.

Slavcheva, E., Shone, B., and Turnbull, A. 1998. Factors Controlling Naphthenic Acid Corrosion. Paper No. 98579. Proceedings. CORROSION/98. NACE International, Houston, TX.

Slavcheva, E., Shone, B., and Turnbull, A. 1999. Review of naphthenic acid corrosion in oil refining. *British Corrosion Journal,* 34(2): 125–131.

Sloley, A.W. 2013a. Mitigate fouling in crude unit overhead Part 1. *Hydrocarbon Processing,* 92(9): 73–81.

Sloley, A.W. 2013b. Mitigate fouling in crude unit overhead Part 2. *Hydrocarbon Processing,* 92(11): 73–75.

Speight, J.G. 1994. Chemical and physical studies of petroleum asphaltenes. In: *Asphaltenes and Asphalts, I. Developments in Petroleum Science, 40,* T.F. Yen and G.V. Chilingarian (Editors), Elsevier, Amsterdam, Chapter 2.

Speight, J.G. 2005. Natural bitumen (Tar Sands) and heavy oil. In Coal, Oil Shale, Natural Bitumen, Heavy Oil and Peat, from *Encyclopedia of Life Support Systems (EOLSS),* Developed under the Auspices of the UNESCO, EOLSS Publishers, Oxford, UK. http://www.eolss.net.

Speight, J.G. 2011a. *An Introduction to Petroleum Technology, Economics, and Politics.* Scrivener Publishing, Beverly, MA.

Speight, J.G. 2011b. *The Refinery of the Future.* Gulf Professional Publishing, Elsevier, Oxford, UK.

Speight, J.G. 2015. *Handbook of Petroleum Product Analysis,* 2nd Edition. John Wiley & Sons Inc., Hoboken, NJ.

Speight, J.G. 2017. *Handbook of Petroleum Refining.* CRC Press, Taylor & Francis Group, Boca Raton, FL.

Speight, J.G. 2019. *Handbook of Petrochemical Processes.* CRC Press, Taylor & Francis Group, Boca Raton, FL.

Speight, J.G. 2020c. *Refinery of the Future,* 2nd Edition. Gulf Professional Publishing, Elsevier, Cambridge, MA.

Speight, J.G. 2021a. *Refinery Feedstocks.* CRC Press, Taylor & Francois Group, Boca Raton, FL.

Speight, J.G. 2021b. *Chemistry and Technology of Alternate Fuels.* World Scientific Publishing Co., PTE Ltd., Singapore and Hackensack, NJ.

Speight, J.G., and Islam, M.R. 2016. *Peak Energy – Myth or Reality.* Scrivener Publishing, Beverly, MA.

Stark, J.L., and Asomaning, S. 2003. Crude oil blending effects on asphaltene stability in refinery fouling. *Petroleum Science and Technology,* 21(3&4): 569–579.

Stark, J.L., Nguyen, J., and Kremer, L.N. 2002. Crude stability as related to desalter upsets. *Proceedings of 5th International Conference on Refinery Processing.* AIChE Spring National Meeting. New Orleans, LA, March 11–14.

US DOE 2004a. Strategic Significance of America's Oil Shale Reserves, I. Assessment of Strategic Issues, March. http://www.fe.doe.gov/programs/reserves/publications.

US DOE 2004b. Strategic Significance of America's Oil Shale Reserves, II. Oil Shale Resources, Technology, and Economics, March. http://www.fe.doe.gov/programs/reserves/publications.

US DOE 2004c. America's Oil Shale: A Roadmap for Federal Decision Making. USDOE Office of US Naval Petroleum and Oil Shale Reserves.

Usman, B.J., and Ali, S.A. 2018. Carbon dioxide corrosion inhibitors: A review. *Arabian Journal for Science and Engineering,* 34: 1–22. https://link.springer.com/article/10.1007/s13369-017-2949-5.

Van den Berg, F.G.A., Kapusta, S.D., Ooms, A.C., and Smith, A.J. 2003. Fouling and compatibility of crudes as the basis for a new crude selection strategy. *Petroleum Science and Technology,* 21: 557–568.

Wang, Y.Z., Sun, X.Y., Liu, Y.P., and Liu, C.G. 2007. Removal of naphthenic acids from a diesel fuel by esterification. *Energy Fuels,* 21: 941–943.

Wang, W., and Watkinson, A.P. 2011. Iron sulfide and coke fouling from sour oils: Review and initial experiments. *Proceedings of International Conference on Heat Exchanger Fouling and Cleaning – 2011,* Crete Island, Greece, M.R. Malayeri, H. Muller-Steinhagen and A.P. Watkinson (Editors), June 5–10, pp. 23–30. www.heatexchanger-fouling.com.

Wang, C., Wang, Y., Chen, J., Sun, X., Liu, Z., and Wan, Q., Dai, Y., and Zheng, W. 2011. High temperature naphthenic acid corrosion of typical steels. *Canadian Journal on Mechanical Sciences and Engineering,* 2(2): 23–29.

Wiehe, I.A., Kennedy, R.J., and Dickakian, G. 2001. Fouling of nearly incompatible oils. *Energy & Fuels,* 15: 1057–1058.

Yeap, B.L., Wilson, D.I., Polley G.T., and Pugh, S.J. 2004. Mitigation of crude oil refinery heat exchanger fouling through retrofits based on thermo-hydraulic fouling models. *Chemical Engineering Research and Design,* 82(A1): 53–71.

Yépez, O. 2005. Influence of different sulfur compounds on corrosion due to naphthenic acid. *Fuel,* 84(1): 97–104.

Zhang, A., Ma, Q., Wang, K., Liu, X., Shuler, P., and Tang, Y. 2006. Naphthenic acid removal from crude oil through catalytic decarboxylation on magnesium oxide. *Applied Catalysis A: General,* 303(1): 103–109.

Glossary

ABN separation: a method of fractionation by which petroleum is separated into acidic, basic, and neutral constituents.

Absorber: see Absorption tower.

Absorption gasoline: gasoline extracted from natural gas or refinery gas by contacting the absorbed gas with an oil and subsequently distilling the gasoline from the higher boiling components.

Absorption oil: oil used to separate the heavier components from a vapor mixture by absorption of the heavier components during close contact of the oil and vapor; used to recover natural gasoline from wet gas.

Absorption plant: a plant for recovering the condensable portion of natural or refinery gas, by absorbing the higher boiling hydrocarbons in an absorption oil, followed by separation and fractionation of the absorbed material.

Absorption tower: a tower or column which promotes contact between a rising gas and a falling liquid so that part of the gas may be dissolved in the liquid.

Acetone-benzol process: a dewaxing process in which acetone and benzol (benzene or aromatic naphtha) are used as solvents.

Acid catalyst: a catalyst having acidic character; the alumina minerals are examples of such catalysts.

Acid deposition: acid rain; a form of pollution depletion in which pollutants, such as nitrogen oxides and sulfur oxides, are transferred from the atmosphere to soil or water; often referred to as atmospheric self-cleaning. The pollutants usually arise from the use of fossil fuels.

Acid gas removal: a process for the removal of hydrogen sulfide, other sulfur species, and some carbon dioxide from syngas by absorption in a solvent with subsequent solvent regeneration and production of a hydrogen sulfide-rich (H_2S-rich) stream for sulfur recovery.

Acid number: a measure of the reactivity of petroleum with a caustic solution and it is given in terms of milligrams of potassium hydroxide that are neutralized by one gram of petroleum.

Acid rain: the precipitation phenomenon that incorporates anthropogenic acids and other acidic chemicals from the atmosphere to the land and water (see Acid deposition).

Acid sludge: the residue left after treating petroleum oil with sulfuric acid for the removal of impurities; a black, viscous substance containing the spent acid and impurities.

Acid treating: a process in which unfinished petroleum products, such as gasoline, kerosene, and lubricating oil stocks, are contacted with sulfuric acid to improve their color, odor, and other properties.

Acidity: the capacity of an acid to neutralize a base such as a hydroxyl ion (OH^-).

Acidizing: a technique for improving the permeability of a reservoir by injecting acid.

Acoustic log: see Sonic log.

Acre-foot: a measure of bulk rock volume where the area is one acre and the thickness is one foot.

Additive: a material added to another (usually in small amounts) in order to enhance desirable properties or to suppress undesirable properties; different additives, even when added for identical purposes, may be incompatible with each other, for example, react and form new compounds. See also Incompatibility and Instability.

Add-on control methods: the use of devices that remove refinery process emissions after they are generated but before they are discharged to the atmosphere.

Adsorption: transfer of a substance from a solution to the surface of a solid resulting in relatively high concentration of the substance at the place of contact; see also Chromatographic adsorption.

Adsorption gasoline: natural gasoline obtained by the adsorption process from wet gas.

After-flow: flow from the reservoir into the wellbore that continues for a period after the well has been shut in; it can complicate the analysis of a pressure transient test.

Afterburn: the combustion of carbon monoxide (CO) to carbon dioxide (CO_2); usually in the cyclones of a catalyst regenerator.

Air-blown asphalt: asphalt produced by blowing air through residua at elevated temperatures.

Air injection: an oil recovery technique using air to force oil from the reservoir into the wellbore.

Airlift Thermofor catalytic cracking: a moving-bed continuous catalytic process for conversion of heavy gas oils into lighter products; the catalyst is moved by a stream of air.

Air pollution: the discharge of toxic gases and particulate matter introduced into the atmosphere, principally as a result of human activity.

Air separation unit (ASU): a plant that separates oxygen and nitrogen from air, usually by cryogenic distillation.

Air sweetening: a process in which air or oxygen is used to oxidize lead mercaptan derivatives (RSH) to disulfide derivatives (RSSR) instead of using elemental sulfur.

Air toxics: hazardous air pollutants.

Albertite: a black, brittle, natural hydrocarbon possessing a conchoidal fracture and a specific gravity of approximately 1.1.

Alcohol: the family name of a group of organic chemical compounds composed of carbon, hydrogen, and oxygen. The molecules in the series vary in chain length and are composed of a hydrocarbon plus a hydroxyl group. Alcohol includes methanol and ethanol.

Alicyclic hydrocarbon: a compound containing only carbon and hydrogen which has a cyclic structure (e.g., cyclohexane); also collectively called naphthenes.

Aliphatic hydrocarbon: a compound containing only carbon and hydrogen which has an open-chain structure (e.g., ethane, butane, octane, butene) or a cyclic structure (e.g., cyclohexane).

Aliquot: that quantity of material of proper size for measurement of the property of interest; test portions may be taken from the gross sample directly, but often preliminary operations such as mixing or further reduction in particle size are necessary.

Alkali treatment: see Caustic wash.

Alkali wash: see Caustic wash.

Alkaline: a high pH usually of an aqueous solution; aqueous solutions of sodium hydroxide, sodium orthosilicate, and sodium carbonate are typical alkaline materials used in enhanced oil recovery.

Alkaline flooding: see EOR process.

Alkalinity: the capacity of a base to neutralize the hydrogen ion (H^+).

Alkanes: hydrocarbons that contain only single carbon–hydrogen bonds. The chemical name indicates the number of carbon atoms and ends with the suffix "ane".

Alkenes: hydrocarbons that contain carbon–carbon double bonds. The chemical name indicates the number of carbon atoms and ends with the suffix "ene".

Alkylate: the product of an alkylation process.

Alkyl groups: a group of carbon and hydrogen atoms that branch from the main carbon chain or ring in a hydrocarbon molecule. The simplest alkyl group, a methyl group, is a carbon atom attached to three hydrogen atoms.

Alkylate bottoms: residua from fractionation of alkylate; the alkylate product which boils higher than the aviation gasoline range; sometimes called heavy alkylate or alkylate polymer.

Alkylation: in the petroleum industry, a process by which an olefin (e.g., ethylene) is combined with a branched-chain hydrocarbon (e.g., *iso*-butane); alkylation may be accomplished as a thermal or catalytic reaction.

Alpha-scission: the rupture of the aromatic carbon–aliphatic carbon bond that joins an alkyl group to an aromatic ring.

Alumina (Al_2O_3): used in separation methods as an adsorbent and in refining as a catalyst.

American Society for Testing and Materials (ASTM): the official organization in the United States for designing standard tests for petroleum and other industrial products.

Amine washing: a method of gas cleaning whereby acidic impurities such as hydrogen sulfide and carbon dioxide are removed from the gas stream by washing with an amine (usually an alkanolamine).

Anaerobic digestion: decomposition of biological wastes by microorganisms, usually under wet conditions, in the absence of air (oxygen), to produce a gas comprising mostly methane and carbon dioxide.

Analytical equivalence: the acceptability of the results obtained from different laboratories; a range of acceptable results.

Analyte: the chemical for which a sample is tested or analyzed. *Antibody* A molecule having chemically reactive sites specific for certain other molecules.

Aniline point: the temperature, usually expressed in °F, above which equal volumes of a petroleum product are completely miscible; a qualitative indication of the relative proportions of paraffins in a petroleum product which are miscible with aniline only at higher temperatures; a high aniline point indicates low aromatics.

Annual removals: the net volume of growing stock trees removed from the inventory during a specified year by harvesting, cultural operations such as timber stand improvement, or land clearing.

Antibody: a molecule having chemically reactive sites specific for certain other molecules.

Antiknock: resistance to detonation or pinging in spark-ignition engines.

Antiknock agent: a chemical compound such as tetraethyl lead which, when added in small amount to the fuel charge of an internal-combustion engine, tends to lessen knocking.

Antistripping agent: an additive used in an asphaltic binder to overcome the natural affinity of an aggregate for water instead of asphalt.

API gravity: a measure of the *lightness* or *heaviness* of petroleum which is related to density and specific gravity.

$$°API = (141.5/sp\ gr\ @\ 60°F) - 131.5$$

Apparent bulk density: the density of a catalyst as measured, usually loosely compacted in a container.

Apparent viscosity: the viscosity of a fluid, or several fluids flowing simultaneously, measured in a porous medium (rock), and subject to both viscosity and permeability effects; also called effective viscosity.

Aquifer: a subsurface rock interval that will produce water; often the underlay of a petroleum reservoir.

Areal sweep efficiency: the fraction of the flood pattern area that is effectively swept by the injected fluids.

Aromatic hydrocarbon: a hydrocarbon characterized by the presence of an aromatic ring or condensed aromatic rings; benzene and substituted benzene, naphthalene and substituted naphthalene, phenanthrene and substituted phenanthrene, as well as the higher condensed ring systems; compounds that are distinct from those of aliphatic compounds or alicyclic compounds.

Aromatics: see Aromatic hydrocarbon.

Aromatization: the conversion of non-aromatic hydrocarbons to aromatic hydrocarbons by (1) rearrangement of aliphatic (noncyclic) hydrocarbons into aromatic ring structures and (2) dehydrogenation of alicyclic hydrocarbons (naphthenes).

Arosorb process: a process for the separation of aromatic derivatives from non-aromatic derivatives by adsorption on a gel from which they are recovered by desorption.

Asphalt: the non-volatile product obtained by distillation and treatment of an asphaltic crude oil; a manufactured product.

Asphalt cement: asphalt especially prepared to ensure quality and consistency for direct use in the manufacture of bituminous pavements.

Asphalt emulsion: an emulsion of asphalt cement in water containing a small amount of emulsifying agent.

Asphaltene (asphaltenes): the brown to black powdery material produced by treatment of petroleum, petroleum residua, or bituminous materials with a low-boiling liquid hydrocarbon, e.g., pentane or heptane; it is soluble in benzene (and other aromatic solvents), carbon disulfide, and chloroform (or other chlorinated hydrocarbon solvents).

Asphalt flux: an oil used to reduce the consistency or viscosity of hard asphalt to the point required for use.

Asphalt primer: a liquid asphaltic material of low viscosity that, when applied to a non-bituminous surface, waterproofs the surface and prepares it for further construction.

Asphaltene association factor: the number of individual asphaltene species which associate in non-polar solvents as measured by molecular weight methods; the molecular weight of asphaltenes in toluene divided by the molecular weight in a polar non-associating solvent, such as dichlorobenzene, pyridine, or nitrobenzene.

Asphaltic pyrobitumen: see Asphaltoid.

Asphaltic road oil: a thick, fluid solution of asphalt; usually a residual oil; see also Non-asphaltic road oil.

Asphaltite: a variety of naturally occurring, dark brown to black, solid, non-volatile bituminous material that is differentiated from bitumen primarily by a high content of material insoluble in n-pentane (asphaltene) or other liquid hydrocarbons.

Asphaltoid: a group of brown to black, solid bituminous materials of which the members are differentiated from asphaltites by their infusibility and low solubility in carbon disulfide.

Asphaltum: see Asphalt.

Associated molecular weight: the molecular weight of asphaltenes in an associating (non-polar) solvent, such as toluene.

Atmospheric residuum: a residuum obtained by distillation of crude oil under atmospheric pressure and which boils above 350°C (660°F).

Atmospheric equivalent boiling point (AEBP): a mathematical method of estimating the boiling point at atmospheric pressure of non-volatile fractions of petroleum.

Attainment area: a geographical area that meets NAAQS for criteria air pollutants (see also Non-attainment area).

Attapulgus clay: see Fuller's earth.

Autofining: a catalytic process for desulfurizing distillates.

Average particle size: the weighted average particle diameter of a catalyst.

Aviation gasoline: any of the special grades of gasoline suitable for use in certain airplane engines.

Aviation turbine fuel: see Jet fuel.

Back mixing: the phenomenon observed when a catalyst travels at a slower rate in the riser pipe than the vapors.

BACT: best available control technology.

Baghouse: a filter system for the removal of particulate matter from gas streams; it is so called because of the similarity of the filters to coal bags.

Bank: the concentration of oil (oil bank) in a reservoir that moves cohesively through the reservoir.

Bari-Sol process: a dewaxing process which employs a mixture of ethylene dichloride and benzol as the solvent.

Barrel (bbl): the unit of measure used by the petroleum industry; equivalent to approximately 42 US gallons or approximately 34 (33.6) Imperial gallons or 159 L; 7.2 barrels are equivalent to one ton of oil (metric).

Barrel of oil equivalent (boe): the amount of energy contained in a barrel of crude oil, i.e. approximately 6.1 GJ (5.8 million Btu), equivalent to 1700 kWh.

Base number: the quantity of acid, expressed in milligrams of potassium hydroxide per gram of sample that is required to titrate a sample to a specified end-point.

Base stock: a primary refined petroleum fraction into which other oils and additives are added (blended) to produce the finished product.

Basic nitrogen: nitrogen (in petroleum) which occurs in pyridine form.

Basic sediment and water (BS&W, BSW): the material which collects in the bottom of storage tanks, usually composed of oil, water, and foreign matter; also called bottoms or bottom settlings.

Batch blending: a blending process (also called the in-tank process) in which specific (calculated and measured) volumes of different crude oils that have been stored in separate tanks are loaded into a blending tank where they are mixed (by a mechanical stirring operation).

Battery: a series of stills or other refinery equipment operated as a unit.

Baumé gravity: the specific gravity of liquids expressed as degrees on the Baumé (°Bé) scale; for liquids lighter than water:

Sp gr 60°F = 140/(130 + °Bé)

For liquids heavier than water:

Sp gr 60°F = 145/(145 − °Bé)

Bauxite: mineral matter used as a treating agent; hydrated aluminum oxide formed by the chemical weathering of igneous rock.

Bbl: see Barrel.

Bell cap: a hemispherical or triangular cover placed over the riser in a (distillation) tower to direct the vapors through the liquid layer on the tray; see Bubble cap.

Bender process: a chemical treating process using lead sulfide catalyst for sweetening light distillates by which mercaptan derivatives (RSH) are converted to disulfide derivatives (RSSR) by oxidation.

Bentonite: montmorillonite (a magnesium-aluminum silicate); used as a treating agent.

Benzene: a colorless aromatic liquid hydrocarbon (C_6H_6).

Benzin: a refined light naphtha used for extraction purposes.

Benzine: an obsolete term for light petroleum distillates covering the gasoline and naphtha range; see Ligroine.

Benzol: the general term which refers to commercial or technical (not necessarily pure) benzene; also the term used for aromatic naphtha.

Beta-scission: the rupture of a carbon–carbon bond; two bonds removed from an aromatic ring.

Billion: 1×10^9

Biocide: any chemical capable of killing bacteria and bio-organisms.

Biochemical conversion: the use of fermentation or anaerobic digestion to produce fuels and chemicals from organic sources.

Biodiesel: a fuel derived from biological sources that can be used in diesel engines instead of petroleum-derived diesel; through the process of transesterification, the triglycerides in the biologically derived oils are separated from the glycerin, creating a clean-burning, renewable fuel.

Bioenergy: useful, renewable energy produced from organic matter – the conversion of the complex carbohydrates in organic matter to energy; organic matter may either be used directly as a fuel, processed into liquids and gases, or be a residual of processing and conversion.

Bioethanol: ethanol produced from biomass feedstocks; includes ethanol produced from the fermentation of crops, such as corn, as well as cellulosic ethanol produced from woody plants or grasses.

Biofuels: a generic name for liquid or gaseous fuels that are not derived from petroleum-based fossil fuels or contain a proportion of non-fossil fuel; fuels produced from plants, crops such as sugar beet and rape seed oil, or reprocessed vegetable oils or fuels made from gasified biomass; fuels made from renewable biological sources and include ethanol, methanol, and biodiesel; sources include, but are not limited to, corn, soybeans, flaxseed, rapeseed, sugarcane, palm oil, raw sewage, food scraps, animal parts, and rice.

Biogas: a combustible gas derived from decomposing biological waste under anaerobic conditions. Biogas normally consists of 50%–60% methane. See also Landfill gas.

Biogenic: material derived from bacterial or vegetation sources.

Biological lipid: any biological fluid that is miscible with a non-polar solvent. These materials include waxes, essential oils, and chlorophyll.

Biological oxidation: the process by which bacteria consume organic matter in an oxidative manner and produce gases as a result.

Biomass: any organic matter that is available on a renewable or recurring basis, including agricultural crops and trees, wood and wood residues, plants (including aquatic plants), grasses, animal manure, municipal residues, and other residue materials. Biomass is generally produced in a sustainable manner from water and carbon dioxide by photosynthesis. There are three main categories of biomass –primary, secondary, and tertiary.

Biomass to liquid (BTL): the process of converting biomass to liquid fuels. Hmm, that seems painfully obvious when you write it out.

Biopolymer: a high molecular weight carbohydrate produced by bacteria.

Biopower: the use of biomass feedstock to produce electric power or heat through direct combustion of the feedstock, through gasification and then combustion of the resultant gas, or through other thermal conversion processes. Power is generated with engines, turbines, fuel cells, or other equipment.

Biorefinery: a facility that processes and converts biomass into value-added products. These products can range from biomaterials to fuels such as ethanol or important feedstocks for the production of chemicals and other materials.

Bitumen: also, on occasion, referred to as native asphalt and extra heavy oil; a naturally occurring material that has little or no mobility under reservoir conditions and which cannot be recovered through a well by conventional oil well production methods including currently used enhanced recovery techniques; current methods involve mining for bitumen recovery.

Bituminous: containing bitumen or constituting the source of bitumen.

Bituminous rock: see Bituminous sand.

Bituminous sand: a formation in which the bituminous material (see Bitumen) is found as a filling in veins and fissures in fractured rock or impregnating relatively shallow sand, sandstone, and limestone strata; a sandstone reservoir that is impregnated with a heavy, viscous black petroleum-like material that cannot be retrieved through a well by conventional production techniques.

Black acid(s): a mixture of sulfonates found in acid sludge which are insoluble in naphtha, benzene, and carbon tetrachloride; very soluble in water but insoluble in 30% sulfuric acid; in the dry, oil-free state, the sodium soaps are black powders.

Black liquor: solution of lignin residue and the pulping chemicals used to extract lignin during the manufacture of paper.

Black oil: any of the dark-colored oils; a term now often applied to heavy oil.

Black soap: see Black acid.

Black strap: the black material (mainly lead sulfide) formed in the treatment of sour light oils with doctor solution and found at the interface between the oil and the solution.

Blended crude: a mixture of crude oils, blended in the pipeline to create a crude with specific physical properties. For example, heavy crude oil, extra heavy crude oil, and tar sand bitumen cannot flow from the field to the refinery in their original state, and at normal surface temperatures, they are blended with lighter crude oils primarily to reduce the viscosity, thereby enabling transportation to a refinery. In addition, the blend can provide a blended refinery feedstock crude oil that has significantly higher value than the raw heavy feedstocks in which case the blend is usually constructed so that the value of the overall blended volume is greater than the summed value of the initial volumes of the individual heavy and light components of the blend.

Blending: in the current context, the process of mixing two or more crude oils without producing a chemical reaction or a physical reaction (i.e. phase separation of asphaltene constituents) to create a composite feedstock that is still suitable for refining.

Blown asphalt: the asphalt prepared by air blowing a residuum or an asphalt.

Bogging: a condition that occurs in a coking reactor when the conversion to coke and light ends is too slow causing the coke particles to agglomerate.

Boiling point: a characteristic physical property of a liquid at which the vapor pressure is equal to that of the atmosphere and the liquid is converted to a gas.

Boiling range: the range of temperature, usually determined at atmospheric pressure in standard laboratory apparatus, over which the distillation of oil commences, proceeds, and finishes.

Bone dry: having zero percent moisture content. Wood heated in an oven at a constant temperature of 100°C (212°F) or above until its weight stabilizes is considered bone dry or oven dry.

Bottled gas: usually butane or propane, or butane–propane mixtures, liquefied and stored under pressure for domestic use; see also Liquefied petroleum gas.

Bottoming cycle: a cogeneration system in which steam is used first for process heat and then for electric power production.

Bottoms: the liquid which collects in the bottom of a vessel (tower bottoms, tank bottoms) either during distillation; also the deposit or sediment formed during storage of petroleum or a petroleum product; see also Residuum and Basic sediment and water.

Bright stock: refined, high-viscosity lubricating oils usually made from residual stocks by processes such as a combination of acid treatment or solvent extraction with dewaxing or clay finishing.

British thermal unit: see Btu.

Bromine number: the number of grams of bromine absorbed by 100 g of oil which indicates the percentage of double bonds in the material.

Brown acid: oil-soluble petroleum sulfonates found in acid sludge which can be recovered by extraction with naphtha solvent. Brown-acid sulfonates are somewhat similar to mahogany sulfonates but are more water soluble. In the dry, oil-free state, the sodium soaps are light-colored powders.

Brown soap: see Brown acid.

Brønsted acid: a chemical species which can act as a source of protons.

Brønsted base: a chemical species which can accept protons.

BS&W: see Basic sediment and water.

BTEX: benzene, toluene, ethylbenzene, and the xylene isomers.

Btu (British thermal unit): the energy required to raise the temperature of one pound of water by one degree Fahrenheit.

Bubble cap: an inverted cup with a notched or slotted periphery to disperse the vapor in small bubbles beneath the surface of the liquid on the bubble plate in a distillation tower.

Bubble plate: a tray in a distillation tower.

Bubble point: the temperature at which incipient vaporization of a liquid in a liquid mixture occurs, corresponding with the equilibrium point of 0% vaporization or 100% condensation.

Bubble tower: a fractionating tower so constructed that the vapors rising through it pass up through layers of condensate on a series of plates or trays (see Bubble plate); the vapor passes from one plate to the next above by bubbling under one or more caps (see Bubble cap) and out through the liquid on the plate where the less volatile portions of vapor condense in bubbling through the liquid on the plate, overflow to the next lower plate, and ultimately back into the reboiler thereby effecting fractionation.

Bubble tray: a circular, perforated plates having the internal diameter of a bubble tower, set at specified distances in a tower to collect various fractions produced during distillation.

Buckley–Leverett method: a theoretical method of determining frontal advance rates and saturations from a fractional flow curve.

Bumping: the knocking against the walls of a still occurring during distillation of petroleum or a petroleum product which usually contains water.

Bunker: a storage tank.

Bunker C oil: see No. 6 Fuel oil.

Burner fuel oil: any petroleum liquid suitable for combustion.

Burning oil: an illuminating oil, such as kerosene (kerosine), suitable for burning in a wick lamp.

Burning point: see Fire point.

Burning-quality index: an empirical numerical indication of the likely burning performance of a furnace or heater oil; derived from the distillation profile and the API gravity, and generally recognizing the factors of paraffin character and volatility.

Burton process: an older thermal cracking process in which oil was cracked in a pressure still and any condensation of the products of cracking also took place under pressure.

Butane dehydrogenation: a process for removing hydrogen from butane to produce butenes and, on occasion, butadiene.

Butane vapor-phase isomerization: a process for isomerizing n-butane to *iso*-butane using aluminum chloride catalyst on a granular alumina support and with hydrogen chloride as a promoter.

Butanol: though generally produced from fossil fuels, this four-carbon alcohol can also be produced through bacterial fermentation of alcohol.

C_1, C_2, C_3, C_4, C_5 fractions: a common way of representing fractions containing a preponderance of hydrocarbons having 1, 2, 3, 4, or 5 carbon atoms, respectively, and without reference to hydrocarbon type.

CAA: clean Air Act; this act is the foundation of air regulations in the United States.

Calcining: heating a metal oxide or an ore to decompose carbonates, hydrates, or other compounds often in a controlled atmosphere.

Capillary forces: interfacial forces between immiscible fluid phases, resulting in pressure differences between the two phases.

Capillary number: N_c, the ratio of viscous forces to capillary forces which is equal to viscosity times velocity divided by interfacial tension.

Carbene: the pentane- or heptane-insoluble material that is insoluble in benzene or toluene but which is soluble in carbon disulfide (or pyridine); a type of rifle used for hunting bison.

Carboid: the pentane- or heptane-insoluble material that is insoluble in benzene or toluene and which is also insoluble in carbon disulfide (or pyridine).

Carbonate washing: processing using a mild alkali (e.g., potassium carbonate) process for emission control by the removal of acid gases from gas streams.

Carbon dioxide augmented waterflooding: injection of carbonated water, or water and carbon dioxide, to increase water flood efficiency; see Immiscible carbon dioxide displacement.

Carbon dioxide miscible flooding: see EOR process.

Carbon-forming propensity: see Carbon residue.

Carbonization: the conversion of an organic compound into char or coke by heat in the substantial absence of air; often used in reference to the destructive distillation (with simultaneous removal of distillate) of coal.

Carbon monoxide (CO): a lethal gas produced by incomplete combustion of carbon-containing fuels in internal combustion engines. It is colorless and odorless.

Carbon-oxygen log: information about the relative abundance of elements such as carbon, oxygen, silicon, and calcium in a formation; usually derived from pulsed neutron equipment.

Carbon rejection: upgrading processes in which coke is produced, e.g., coking.

Carbon residue: the amount of carbonaceous residue remaining after thermal decomposition of petroleum, a petroleum fraction, or a petroleum product in a limited amount of air; also called the *coke-* or *carbon-forming propensity*; often prefixed by the terms Conradson or Ramsbottom in reference to the inventor of the respective tests.

Carbon sink: a geographical area whose vegetation and/or soil soaks up significant carbon dioxide from the atmosphere. Such areas, typically in tropical regions, are increasingly being sacrificed for energy crop production.

CAS: chemical Abstract Service.

Cascade tray: a fractionating device consisting of a series of parallel troughs arranged on stair-step fashion in which liquid from the tray above enters the uppermost trough and liquid thrown from this trough by vapor rising from the tray below impinges against a plate and a perforated baffle, and liquid passing through the baffle enters the next longer of the troughs.

Casinghead gas: natural gas which issues from the casinghead (the mouth or opening) of an oil well.

Casinghead gasoline: the liquid hydrocarbon product extracted from casinghead gas by one of three methods: compression, absorption, or refrigeration; see also Natural gasoline.

Catagenesis: the alteration of organic matter during the formation of petroleum that may involve temperatures in the range 50°C (120°F) to 200°C (390°F); see also Diagenesis and Metagenesis.

Catalyst: a chemical agent which when added to a reaction (process) will enhance the conversion of a feedstock without being consumed in the process.

Catalyst selectivity: the relative activity of a catalyst with respect to a particular compound in a mixture or the relative rate in competing reactions of a single reactant.

Catalyst stripping: the introduction of steam, at a point where spent catalyst leaves the reactor, in order to strip, i.e. remove, deposits retained on the catalyst.

Catalytic activity: the ratio of the space velocity of the catalyst under test to the space velocity required for the standard catalyst to give the same conversion as the catalyst being tested; usually multiplied by 100 before being reported.

Catalytic cracking: the conversion of high-boiling feedstocks into lower-boiling products by means of a catalyst which may be used in a fixed bed or fluid bed.

Cat cracking: see Catalytic cracking.

Catalytic reforming: rearranging hydrocarbon molecules in a gasoline boiling-range feedstock to produce other hydrocarbons having a higher antiknock quality; isomerization of paraffins, cyclization of paraffins to naphthenes; and dehydrocyclization of paraffins to aromatics.

Catforming: a process for reforming naphtha using a platinum-silica-alumina catalyst which permits relatively high space velocities and results in the production of high-purity hydrogen.

Caustic consumption: the amount of caustic lost from reacting chemically with the minerals in the rock, the oil, and the brine.

Caustic wash: the process of treating a product with a solution of caustic soda to remove minor impurities; often used in reference to the solution itself.

Ceresin: a hard, brittle wax obtained by purifying ozokerite; see Microcrystalline wax and Ozokerite.

Cetane index: an approximation of the cetane number calculated from the density and mid-boiling point temperature; see also Diesel index.

Cetane number: a number indicating the ignition quality of diesel fuel; a high cetane number represents a short ignition delay time.

CFR: code of Federal Regulations; Title 40 (40 CFR) contains the regulations for protection of the environment.

Characterization factor: the UOP characterization factor K, defined as the ratio of the cube root of the molal average boiling point, T_B, in degrees Rankine (°R = °F + 460), to the specific gravity at 60°F/60°F:

$$K = \left(T_B\right)^{1/3} / \text{sp gr}$$

The value ranges from 12.5 for paraffin stocks to 10.0 for the highly aromatic stocks; also called the Watson characterization factor.

Cheesebox still: an early type of vertical cylindrical still designed with a vapor dome.

Chelating agents: complex-forming agents having the ability to solubilize heavy metals.

Chemical flooding: see EOR process.

Chemical octane number: the octane number added to gasoline by refinery processes or by the use of octane number improvers such as tetraethyl lead.

Chemical waste: any solid, liquid, or gaseous material discharged from a process and that may pose substantial hazards to human health and environment.

Chlorex process: a process for extracting lubricating oil stocks in which the solvent used is Chlorex (ß-ß dichlorodiethyl ether).

Chromatographic adsorption: selective adsorption on materials such as activated carbon, alumina, or silica gel; liquid or gaseous mixtures of hydrocarbons are passed through the adsorbent in a stream of diluent, and certain components are preferentially adsorbed.

Chromatographic separation: the separation of different species of compounds according to their size and interaction with the rock as they flow through a porous medium.

Chromatography: a method of separation based on selective adsorption; see also Chromatographic adsorption.

Clarified oil: the heavy oil which has been taken from the bottom of a fractionator in a catalytic cracking process and from which residual catalyst has been removed.

Clarifier: equipment for removing the color or cloudiness of oil or water by separating the foreign material through mechanical or chemical means; may involve centrifugal action, filtration, heating, or treatment with acid or alkali.

Clastic: composed of pieces of pre-existing rock.

Clay: silicate minerals that also usually contain aluminum and have particle sizes that are less than 0.002 μm; used in separation methods as an adsorbent and in refining as a catalyst.

Clay contact process: see Contact filtration.

Clay refining: a treating process in which vaporized gasoline or other light petroleum product is passed through a bed of granular clay such as fuller's earth.

Clay regeneration: a process in which spent coarse-grained adsorbent clay minerals from percolation processes are cleaned for reuse by deoiling the clay minerals with naphtha, steaming out the excess naphtha, and then roasting in a stream of air to remove carbonaceous matter.

Clay treating: see Gray clay treating.

Clay wash: light oil, such as kerosene (kerosine) or naphtha, used to clean fuller's earth after it has been used in a filter.

Cleanup: a preparatory step following extraction of a sample media designed to remove components that may interfere with subsequent analytical measurements.

Closed-loop biomass: crops grown, in a sustainable manner, for the purpose of optimizing their value for bioenergy and bioproduct uses. This includes annual crops such as maize and wheat, and perennial crops such as trees, shrubs, and grasses such as switch grass.

Cloud point: the temperature at which paraffin wax or other solid substances begin to crystallize or separate from the solution, imparting a cloudy appearance to the oil when the oil is chilled under prescribed conditions.

Coal: an organic rock.

Coalescence: the union of two or more droplets to form a larger droplet and, ultimately, a continuous phase.

Coal tar: the specific name for the tar produced from coal.

Coal tar pitch: the specific name for the pitch produced from coal.

Coarse materials: wood residues suitable for chipping, such as slabs, edgings, and trimmings.

COFCAW: an EOR process that combines forward combustion and water flooding.

Cogeneration: an energy conversion method by which electrical energy is produced along with steam generated for EOR use.

Coke: a gray to black solid carbonaceous material produced from petroleum during thermal processing; characterized by having a high carbon content (95%+ by weight) and a honeycomb type of appearance and is insoluble in organic solvents.

Coke drum: a vessel in which coke is formed and which can cut oil from the process for cleaning.

Coke number: used, particularly in Great Britain, to report the results of the Ramsbottom carbon residue test, which is also referred to as a coke test.

Coker: the processing unit in which coking takes place.

Coking: a process for the thermal conversion of petroleum in which gaseous, liquid, and solid (coke) products are formed.

Cold pressing: the process of separating wax from oil by first chilling (to help form wax crystals) and then filtering under pressure in a plate and frame press.

Cold settling: processing for the removal of wax from high-viscosity stocks, wherein a naphtha solution of the waxy oil is chilled and the wax crystallizes out of the solution.

Color stability: the resistance of a petroleum product to color change due to light, aging, etc.

Combined cycle: a combustion (gas) turbine equipped with a heat recovery steam generator that produces steam for the steam turbine; power is produced from both the gas and steam turbines – hence the term combined cycle.

Combustible liquid: a liquid with a flash point in excess of 37.8°C (100°F) but below 93.3°C (200°).

Combustion zone: the volume of reservoir rock wherein petroleum is undergoing combustion during enhanced oil recovery.

Composition: the general chemical makeup of petroleum.

Completion interval: the portion of the reservoir formation placed in fluid communication with the well by selectively perforating the wellbore casing.

Composition map: a means of illustrating the chemical makeup of petroleum using chemical and/or physical property data.

Con Carbon: see Carbon residue.

Condensate: a mixture of light hydrocarbon liquids obtained by condensation of hydrocarbon vapors: predominately butane, propane, and pentane with some heavier hydrocarbons and relatively little methane or ethane; see also Natural gas liquids.

Conductivity: a measure of the ease of flow through a fracture, perforation, or pipe.

Conformance: the uniformity with which a volume of the reservoir is swept by injection fluids in area and vertical directions.

Connate water: water trapped in the pores of a rock during formation of the rock – also described as fossil water. The chemistry of connate water can change in composition throughout the history of the rock; connate water can be dense and saline compared with seawater. On the other hand, formation water or interstitial water is water found in the pore spaces of a rock, but might not have been present when the rock was formed.

Conradson carbon residue: see Carbon residue.

Contact filtration: a process in which finely divided adsorbent clay is used to remove color bodies from petroleum products.

Contaminant: a substance that causes deviation from the normal composition of an environment.

Continuous contact coking: a thermal conversion process in which petroleum-wetted coke particles move downward into the reactor in which cracking, coking, and drying take place to produce coke, gas, gasoline, and gas oil.

Continuous contact filtration: a process to finish lubricants, waxes, or special oils after acid treating, solvent extraction, or distillation.

Conventional crude oil (conventional petroleum): crude oil that is pumped from the ground and recovered using the energy inherent in the reservoir; also recoverable by application of secondary recovery techniques.

Conventional recovery: primary and/or secondary recovery.

Conversion: the thermal treatment of petroleum which results in the formation of new products by the alteration of the original constituents.

Conversion cost: the cost of changing a production well to an injection well or some other change in the function of an oilfield installation.

Conversion factor: the percentage of feedstock converted to light ends, gasoline, other liquid fuels, and coke.

Copper sweetening: processes involving the oxidation of mercaptan derivatives (RSH) to disulfide derivatives (RSSR) by oxygen in the presence of cupric chloride ($CuCl_2$).

Cord: a stack of wood comprising 128 cubic feet ($3.62\,m^3$); standard dimensions are $4 \times 4 \times 8$ feet, including air space and bark. One cord contains approx. 1.2 U.S. tons (oven-dry) = 2400 pounds = 1089 kg.

Core floods: laboratory flow tests through samples (cores) of porous rock.

Co-surfactant: a chemical compound, typically alcohol that enhances the effectiveness of a surfactant.

Cp (centipoise): a unit of viscosity.

Craig-Geffen-Morse method: a method for predicting oil recovery by water flood.

Cracked residua: residua that have been subjected to temperatures above 350°C (660°F) during the distillation process.

Cracking: the thermal processes by which the constituents of petroleum are converted to lower molecular weight products.

Cracking activity: see Catalytic activity.

Cracking coil: equipment used for cracking heavy petroleum products consisting of a coil of heavy pipe running through a furnace so that the oil passing through it is subject to high temperature.

Cracking still: the combined equipment furnace, reaction chamber, and fractionator for the thermal conversion of heavier feedstocks to lighter products.

Cracking temperature: the temperature (350°C; 660°F) at which the rate of thermal decomposition of petroleum constituents becomes significant.

Criteria air pollutants: air pollutants or classes of pollutants regulated by the Environmental Protection Agency; the air pollutants are (including VOCs): ozone, carbon monoxide, particulate matter, nitrogen oxides, sulfur dioxide, and lead.

Cracking: a secondary refining process that uses heat and/or a catalyst to break down high molecular weight chemical components into lower molecular weight products which can be used as blending components for fuels.

Cropland: total cropland includes five components: cropland harvested, crop failure, cultivated summer fallow, cropland used only for pasture, and idle cropland.

Cropland pasture: land used for long-term crop rotation. However, some cropland pasture is marginal for crop uses and may remain in pasture indefinitely. This category also includes land that was used for pasture before crops reached maturity and some land used for pasture that could have been cropped without additional improvement.

Cross-linking: combining of two or polymer molecules by the use of a chemical that mutually bonds with a part of the chemical structure of the polymer molecules.

Crude assay: a procedure for determining the general distillation characteristics (e.g., distillation profile, *q.v.*) and other quality information of crude oil.

Crude oil: see Petroleum.

Crude scale wax: the wax product from the first sweating of the slack wax.

Crude still: distillation equipment in which crude oil is separated into various products.

Cull tree: a live tree, 5.0 inches in diameter at breast height (d.b.h.) or larger that is non-merchantable for saw logs now or prospectively because of rot, roughness, or species. (See definitions for rotten and rough trees.)

Cultivated summer fallow: cropland cultivated for one or more seasons to control weeds and accumulate moisture before small grains are planted.

Cumene: a colorless liquid [$C_6H_5CH(CH_3)_2$] used as an aviation gasoline blending component and as an intermediate in the manufacture of chemicals.

Cut point: the boiling-temperature division between distillation fractions of petroleum.

Cutback: the term applied to the products from blending heavier feedstocks or products with lighter oils to bring the heavier materials to the desired specifications.

Cutback asphalt: asphalt liquefied by the addition of a volatile liquid such as naphtha or kerosene which, after application and on exposure to the atmosphere, evaporates leaving the asphalt.

Cutting oil: an oil to lubricate and cool metal-cutting tools; also called cutting fluid or cutting lubricant.

Cycle stock: the product taken from some later stage of a process and recharged (recycled) to the process at some earlier stage.

Cyclic steams injection: the alternating injection of steam and production of oil with condensed steam from the same well or wells.

Cyclization: the process by which an open-chain hydrocarbon structure is converted to a ring structure, e.g., hexane to benzene.

Cyclone: a device for extracting dust from industrial waste gases. It is in the form of an inverted cone into which the contaminated gas enters tangential from the top; the gas is propelled down a helical pathway, and the dust particles are deposited by means of centrifugal force onto the wall of the scrubber.

Deactivation: reduction in catalyst activity by the deposition of contaminants (e.g., coke, metals) during a process.

Dealkylation: the removal of an alkyl group from aromatic compounds.

Deasphaltened oil: the fraction of petroleum after the asphaltene constituents have been removed.

Deasphaltening: removal of a solid powdery asphaltene fraction from petroleum by the addition of the low-boiling liquid hydrocarbons such as n-pentane or n-heptane under ambient conditions.

Deasphalting: the removal of the asphaltene fraction from petroleum by the addition of a low-boiling hydrocarbon liquid such as n-pentane or n-heptane; more correctly, the removal of asphalt (tacky, semi-solid) from petroleum (as occurs in a refinery asphalt plant) by the addition of liquid propane or liquid butane under pressure.

Debutanization: distillation to separate butane and lighter components from higher boiling components.

Decant oil: the highest boiling product from a catalytic cracker; also referred to as slurry oil, clarified oil, or bottoms.

Decarbonizing: a thermal conversion process designed to maximize coker gas-oil production and minimize coke and gasoline yields; operated at essentially lower temperatures and pressures than delayed coking.

Decoking: removal of petroleum coke from equipment such as coking drums; hydraulic decoking uses high-velocity water streams.

Decolorizing: removal of suspended, colloidal, and dissolved impurities from liquid petroleum products by filtering, adsorption, chemical treatment, distillation, bleaching, etc.

De-ethanization: distillation to separate ethane and lighter components from propane and higher boiling components; also called de-ethanation.

Degradation: the loss of desirable physical properties of EOR fluids, e.g., the loss of viscosity of polymer solutions.

Dehydrating agents: substances capable of removing water (drying, q.v.) or the elements of water from another substance.

Dehydrocyclization: any process by which both dehydrogenation and cyclization reactions occur.

Dehydrogenation: the removal of hydrogen from a chemical compound; for example, the removal of two hydrogen atoms from butane to make butene(s) as well as the removal of additional hydrogen to produce butadiene.

Delayed coking: a coking process in which the thermal reaction are allowed to proceed to completion to produce gaseous, liquid, and solid (coke) products.

Demethanization: the process of distillation in which methane is separated from the higher boiling components; also called demethanation.

Demulsifiers: a group of chemicals or surfactants used to separate water content in the water-in-oil and oil-in-water emulsions usually at low concentrations. See Emulsion.

Density: the mass (or weight) of a unit volume of any substance at a specified temperature; see also Specific gravity.

Deoiling: reduction in quantity of liquid oil entrained in solid wax by draining (sweating) or by a selective solvent; see MEK deoiling.

Depentanizer: a fractionating column for the removal of pentane and lighter fractions from a mixture of hydrocarbons.

Depropanization: distillation in which lighter components are separated from butanes and higher boiling material; also called depropanation.

Desalting: the removal of mineral salts (mostly chlorides) from crude oils; the first refining process applied to crude oil which removes salt, water, and solid particles that would otherwise lead to operational problems during refining such as corrosion, fouling of equipment, or poisoning of catalysts.

Desorption: the reverse process of adsorption whereby adsorbed matter is removed from the adsorbent; also used as the reverse of absorption.

Desulfurization: the removal of sulfur or sulfur compounds from a feedstock.

Detergent oil: lubricating oil possessing special sludge-dispersing properties for use in internal combustion engines.

Dewaxing: see Solvent dewaxing.

Devolatilized fuel: smokeless fuel; coke that has been reheated to remove all of the volatile materials.

Diagenesis: the concurrent and consecutive chemical reactions which commence the alteration of organic matter (at temperatures up to 50°C (120°F)) and ultimately result in the formation of petroleum from the marine sediment; see also Catagenesis and Metagenesis.

Diagenetic rock: rock formed by conversion through pressure or chemical reaction from a rock, e.g., sandstone is a diagenetic.

Diesel cycle: a repeated succession of operations representing the idealized working behavior of the fluids in a diesel engine.

Diesel engine: Named for the German engineer Rudolph Diesel, this internal combustion, compression ignition engine works by heating fuels and causing them to ignite; can use either petroleum or bio-derived fuel.

Diesel fuel: fuel used for internal combustion in diesel engines; usually that fraction which distills after kerosene.

Diesel index: an approximation of the cetane number of diesel fuel calculated from the density and aniline point.

DI = (aniline point (°F) × API gravity)100

Diesel knock: the result of a delayed period of ignition is long and an accumulation of diesel fuel in the engine.

Differential strain analysis: measurement of thermal stress relaxation in a recently cut well.

Digester: an airtight vessel or enclosure in which bacteria decomposes biomass in water to produce biogas.

Direct-injection engine: a diesel engine in which fuel is injected directly into the cylinder.

Dispersion: a measure of the convective flow of fluids in a reservoir.

Displacement efficiency: the ratio of the amount of oil moved from the zone swept by the reprocess to the amount of oil present in the zone prior to start of the process.

Distribution coefficient: a coefficient that describes the distribution of a chemical in reservoir fluids, usually defined as the equilibrium concentrations in the aqueous phases.

Distillate: any petroleum product produced by boiling crude oil and collecting the vapors produced as a condensate in a separate vessel, for example, gasoline (light distillate), gas oil (middle distillate), or fuel oil (heavy distillate).

Distillation: the primary refining process which uses high temperature to separate crude oil into vapor and fluids (bulk products) which can then be fed into a variety of processes to produce saleable products.

Distillation curve: see Distillation profile.

Distillation loss: the difference, in a laboratory distillation, between the volume of liquid originally introduced into the distilling flask and the sum of the residue and the condensate recovered.

Distillation range: the difference between the temperature at the initial boiling point and at the end point, as obtained by the distillation test.

Distillation profile: the distillation characteristics of petroleum or petroleum products showing the temperature and the percent distilled.

Doctor solution: a solution of sodium plumbite used to treat gasoline or other light petroleum distillates to remove mercaptan sulfur; see also Doctor test.

Doctor sweetening: a process for sweetening gasoline, solvents, and kerosene by converting mercaptan derivatives (RSH) to disulfide derivatives (RSSR) using sodium plumbite (Na_2PbO_2) and sulfur.

Doctor test: a test used for the detection of compounds in light petroleum distillates which react with sodium plumbite; see also Doctor solution.

Domestic heating oil: see No. 2 Fuel Oil.

Donor solvent process: a conversion process in which hydrogen donor solvent is used in place of or to augment hydrogen.

Downcomer: a means of conveying liquid from one tray to the next below in a bubble tray column.

Downdraft gasifier: a gasifier in which the product gases pass through a combustion zone at the bottom of the gasifier.

Downhole steam generator: a generator installed downhole in an oil well to which oxygen-rich air, fuel, and water are supplied for the purposes of generating steam for it into the reservoir. Its major advantage over a surface steam generating facility is the losses to the wellbore and surrounding formation are eliminated.

Drying: removal of a solvent or water from a chemical substance; also referred to as the removal of solvent from a liquid or suspension.

Dropping point: the temperature at which grease passes from a semi-solid to a liquid state under prescribed conditions.

Dry gas: a gas which does not contain fractions that may easily condense under normal atmospheric conditions.

Dry point: the temperature at which the last drop of petroleum fluid evaporates in a distillation test.

Dualayer distillate process: a process for removing mercaptan derivatives (RSH) and oxygenated compounds from distillate fuel oils and similar products, using a combination of treatment with concentrated caustic solution and electrical precipitation of the impurities.

Dualayer gasoline process: a process for extracting mercaptan derivatives (RSH) and other objectionable acidic compounds from petroleum distillates; see also Dualayer solution.

Dualayer solution: a solution which consists of concentrated potassium or sodium hydroxide containing a solubilizer; see also Dualayer gasoline process.

Dubbs cracking: an older continuous, liquid-phase thermal cracking process formerly used.

Dutch oven furnace: one of the earliest types of furnaces, having a large, rectangular box lined with firebrick (refractory) on the sides and top; commonly used for burning wood.

Dykstra-Parsons coefficient: an index of reservoir heterogeneity arising from permeability variation and stratification.

E85: an alcohol fuel mixture containing 85% ethanol and 15% gasoline by volume, and the current alternative fuel of choice of the U.S. government.

Ebullated bed: a process in which the catalyst bed is in a suspended state in the reactor by means of a feedstock recirculation pump which pumps the feedstock upward at sufficient speed to expand the catalyst bed at approximately 35% above the settled level.

Edeleanu process: a process for refining oils at low temperature with liquid sulfur dioxide (SO_2) or with liquid sulfur dioxide and benzene; applicable to the recovery of aromatic concentrates from naphtha and heavier petroleum distillates.

Effective viscosity: see Apparent viscosity.

Effluent: any contaminating substance, usually a liquid, which enters the environment via a domestic industrial, agricultural, or sewage plant outlet.

Electric desalting: a continuous process to remove inorganic salts and other impurities from crude oil by settling out in an electrostatic field.

Electrical precipitation: a process using an electrical field to improve the separation of hydrocarbon reagent dispersions. May be used in chemical treating processes on a wide variety of refinery stocks.

Electrofining: a process for contacting a light hydrocarbon stream with a treating agent (acid, caustic, doctor, etc.), then assisting the action of separation of the chemical phase from the hydrocarbon phase by an electrostatic field.

Electrolytic mercaptan process: a process in which aqueous caustic solution is used to extract mercaptan derivatives (RSH) from refinery streams.

Electrostatic precipitators: devices used to trap fine dust particles (usually in the size range $30–60\,\mu m$) that operate on the principle of imparting an electric charge to particles in an incoming air stream and which are then collected on an oppositely charged plate across a high voltage field.

Eluate: the solutes, or analytes, moved through a chromatographic column (see *Elution*).

Eluent: solvent used to elute sample.

Elution: a process whereby a solute is moved through a chromatographic column by a solvent (liquid or gas) or eluent.

Emissions: substances discharged into the air during combustion.

Emission control: the use gas cleaning processes to reduce emissions.

Emission standard: the maximum amount of a specific pollutant permitted to be discharged from a particular source in a given environment.

Emulsion: a two-phase system comprising two liquids which are not homogeneous when mixed. In the emulsion, one of the liquids is the constantly dispersed (the dispersed phase) as globules in the second phase (the continuous phase), such as oil in water. See Demulsifiers.

Emulsion breaking: the settling or aggregation of colloidal-sized emulsions from suspension in a liquid medium.

End-of-pipe emission control: the use of specific emission control processes to clean gases after production of the gases.

Energy: the capacity of a body or system to do work, measured in joules (SI units); also the output of fuel sources.

Energy balance: the difference between the energy produced by a fuel and the energy required to obtain it through agricultural processes, drilling, refining, and transportation.

Energy crops: crops grown specifically for their fuel value; include food crops, such as corn and sugarcane, and nonfood crops, such as poplar trees and switch grass.

Energy-efficiency ratio: a number representing the energy stored in a fuel as compared to the energy required to produce, process, transport, and distribute that fuel.

Energy from biomass: the production of energy from biomass.

Engler distillation: a standard test for determining the volatility characteristics of a gasoline by measuring the per cent distilled at various specified temperatures.

Enhanced oil recovery (EOR): petroleum recovery following recovery by conventional (i.e. primary and/or secondary) methods.

Enhanced oil recovery (EOR) process: a method for recovering additional oil from a petroleum reservoir beyond that economically recoverable by conventional primary and secondary recovery methods. EOR methods are usually divided into three main categories: (1) *chemical flooding:* injection of water with added chemicals into a petroleum reservoir. The chemical processes include surfactant flooding, polymer flooding, and alkaline flooding; (2) *miscible flooding:* injection into a petroleum reservoir of a material that is miscible, or can become miscible, with the oil in the reservoir. Carbon dioxide, hydrocarbons, and nitrogen are used; (3) *thermal recovery:* injection of steam into a petroleum reservoir, or propagation of a combustion zone through a reservoir by air or oxygen-enriched air injection. The thermal processes include steam drive, cyclic steam injection, and in situ combustion.

Entrained bed: a bed of solid particles suspended in a fluid (liquid or gas) at such a rate that some of the solid is carried over (entrained) by the fluid.

Entrainer: the separating agent used to enhance the separation of closely boiling compounds in azeotropic distillation or in extractive distillation.

EPA: Environmental Protection Agency.

Ester: a compound formed by the reaction between an organic acid and an alcohol; ethoxylated alcohols (i.e. alcohols having ethylene oxide functional groups attached to the alcohol molecule).

Ethanol (ethyl alcohol, alcohol, or grain-spirit): a clear, colorless, flammable oxygenated hydrocarbon; used as a vehicle fuel by itself (E100 is 100% ethanol by volume), blended with gasoline (E85 is 85% ethanol by volume), or as a gasoline octane enhancer and oxygenate (10% by volume); formed during fermentation of sugars; used as an intoxicant and as a fuel.

Evaporation: a process for concentrating non-volatile solids in a solution by boiling off the liquid portion of the waste stream.

Expanding clays: clays that expand or swell on contact with water, e.g., montmorillonite.

Explosive limits: the limits of percentage composition of mixtures of gases and air within which an explosion takes place when the mixture is ignited.

Extract: the portion of a sample preferentially dissolved by the solvent and recovered by physically separating the solvent.

Extractive distillation: the separation of different components of mixtures which have similar vapor pressures by flowing a relatively high-boiling solvent, which is selective for one of the components in the feed, down a distillation column as the distillation proceeds; the selective solvent scrubs the soluble component from the vapor.

Fabric filters: filters made from fabric materials and used for removing particulate matter from gas streams (see Baghouse).

Facies: one or more layers of rock that differs from other layers in composition, age, or content.

FAST: fracture-assisted steamflood technology.

Fast pyrolysis: a pyrolysis process that involves heating a feedstock, such as biomass, rapidly (2 seconds) at temperatures ranging in the order of 350°C– 650°C (650°F–1200°F).

Fat oil: the bottom or enriched oil drawn from the absorber as opposed to lean oil.

Faujasite: a naturally occurring silica-alumina (SiO_2-Al_2O_3) mineral.

FCC: fluid catalytic cracking.

FCCU: fluid catalytic cracking unit.

Feedstock: petroleum as it is fed to the refinery; a refinery product that is used as the raw material for another process; biomass used in the creation of a particular biofuel (e.g., corn or sugarcane for ethanol, soybeans or rapeseed for biodiesel); the term is also generally applied to raw materials used in other industrial processes.

Fermentation: conversion of carbon-containing compounds by microorganisms for production of fuels and chemicals such as alcohols, acids, or energy-rich gases.

Ferrocyanide process: a regenerative chemical treatment for mercaptan removal using caustic-sodium ferrocyanide reagent.

Fiber products: products derived from fibers of herbaceous and woody plant materials; examples include pulp, composition board products, and wood chips for export.

Field-scale: the application of EOR processes to a significant portion of a field.

Filtration: the collection of solids using an impassable barrier while allowing liquids to pass.

Fine materials: wood residues not suitable for chipping, such as planer shavings and sawdust.

Fingering: the formation of finger-shaped irregularities at the leading edge of a displacing fluid in a porous medium which move out ahead of the main body of fluid.

Fire point: the lowest temperature at which, under specified conditions in standardized apparatus, a petroleum product vaporizes sufficiently rapidly to form an air–vapor mixture above its surface which burns continuously when ignited by a small flame.

First contact miscibility: see Miscibility.

Fischer–Tropsch process: a process for synthesizing hydrocarbons and oxygenated chemicals from a mixture of hydrogen and carbon monoxide.

Fixed bed: a stationary bed (of catalyst) to accomplish a process (see Fluid bed).

Five-spot: an arrangement or pattern of wells with four injection wells at the corners of a square and a producing well in the center of the square.

Flammability range: the range of temperature over which a chemical is flammable.

Flammable: a substance that will burn readily.

Flammable liquid: a liquid having a flash point below 37.8°C (100°F).

Flammable solid: a solid that can ignite from friction or from heat remaining from its manufacture, or which may cause a serious hazard if ignited.

Flash point: the lowest temperature to which the product must be heated under specified conditions to give off sufficient vapor to form a mixture with air that can be ignited momentarily by a flame.

Flash vaporization: a simple unit operation in which a heated liquid mixture is throttled through a valve in order to vaporize the liquid mixture. The vaporization is done in order to separate the constituents of the liquid mixture. Flash vaporization can be seen as a single stage.

Flexible-fuel vehicle (flex-fuel vehicle): a vehicle that can run alternately on two or more sources of fuel; includes cars capable of running on gasoline and gasoline/ethanol mixtures, as well as cars that can run on both gasoline and natural gas.

Flexicoking: a modification of the fluid coking process insofar as the process also includes a gasifier adjoining the burner/regenerator to convert excess coke to a clean fuel gas.

Flocculants: chemicals that flocculate the water droplets and facilitate coalescence.

Flocculation threshold: the point at which constituents of a solution (e.g., asphaltene constituents or coke precursors) will separate from the solution as a separate (solid) phase.

Floc point: the temperature at which wax or solids separate as a definite floc.

Flood, flooding: the process of displacing petroleum from a reservoir by the injection of fluids.

Flue gas: gas from the combustion of fuel, the heating value of which has been substantially spent and which is, therefore, discarded to the flue or stack.

Flue gases: the gaseous products of the combustion process mostly comprised of carbon dioxide, nitrogen, and water vapor.

Fluid: a reservoir gas or liquid.

Fluid bed: the use of an agitated bed of inert granular material to accomplish a process in which the agitated bed resembles the motion of a fluid.

Fluid-bed: a bed (of catalyst) that is agitated by an upward passing gas in such a manner that the particles of the bed simulate the movement of a fluid and have the characteristics associated with a true liquid; cf. Fixed bed.

Fluid catalytic cracking: cracking in the presence of a fluidized bed of catalyst.

Fluid coking: a continuous fluidized solids process that cracks feed thermally over heated coke particles in a reactor vessel to gas, liquid products, and coke.

Fluidized bed combustion: a process used to burn low-quality solid fuels in a bed of small particles suspended by a gas stream (usually air that will lift the particles but not blow them out of the vessel). Rapid burning removes some of the offensive byproducts of combustion from the gases and vapors that result from the combustion process.

Fluidized-bed boiler: a large, refractory-lined vessel with an air distribution member or plate in the bottom, a hot gas outlet in or near the top, and some provisions for introducing fuel; the fluidized bed is formed by blowing air up through a layer of inert particles (such as sand or limestone) at a rate that causes the particles to go into suspension and continuous motion.

Fly ash: particulate matter produced from mineral matter in coal that is converted during combustion to finely divided inorganic material and emerges from the combustor in the gases.

Foots oil: the oil sweated out of slack wax; named from the fact that the oil goes to the foot, or bottom, of the pan during the sweating operation.

Forest land: land at least 10% stocked by forest trees of any size, including land that formerly had such tree cover and that will be naturally or artificially regenerated; includes transition zones, such as areas between heavily forested and non-forested lands that are at least 10% stocked with forest trees and forest areas adjacent to urban and built-up lands; also included are pinyon-juniper and chaparral areas; minimum area for classification of forest land is 1 acre.

Forest health: a condition of ecosystem sustainability and attainment of management objectives for a given forest area; usually considered to include green trees, snags, resilient stands growing at a moderate rate, and endemic levels of insects and disease.

Forest residues: material not harvested or removed from logging sites in commercial hardwood and softwood stands as well as material resulting from forest management operations such as precommercial thinnings and removal of dead and dying trees.

Formation: an interval of rock with distinguishable geologic characteristics.

Formation volume factor: the volume in barrels that one stock tank barrel occupies in the formation at reservoir temperature along with the solution gas that is held in the oil at reservoir pressure.

Fossil fuel resources: a gaseous, liquid, or solid fuel material formed in the ground by chemical and physical changes (diagenesis, q.v.) in plant and animal residues over geological time; natural gas, petroleum, coal, and oil shale.

Fouling: the name applied to the disposition of solids or phase separation which typically commences with the introduction into the refinery of a feedstock (crude oil or the blends of crude oils) that is incompatible with (in the current context) the other components of a blend of feedstocks.

Fractional composition: the composition of petroleum as determined by fractionation (separation) methods.

Fractional distillation: the separation of the components of a liquid mixture by vaporizing and collecting the fractions, or cuts, which condense in different temperature ranges.

Fractional flow: the ratio of the volumetric flow rate of one fluid phase to the total fluid volumetric flow rate within a volume of rock.

Fractional flow curve: the relationship between the fractional flow of one fluid and its saturator during simultaneous flow of fluids through rock.

Fracture: a natural or man-made crack in a reservoir rock.

Fracturing: the break-apart of reservoir rock by applying very high fluid pressure at the rock face.

Fractionating column: a column arranged to separate various fractions of petroleum by a single distillation and which can be tapped at different points along its length to separate various fractions in the order of their boiling points.

Fractionation: the separation of petroleum into the constituent fractions using solvent or adsorbent methods; chemical agents such as sulfuric acid may also be used.

Frasch process: a process formerly used for removing sulfur by distilling oil in the presence of copper oxide.

Fuel cell: a device that converts the energy of a fuel directly to electricity and heat without combustion.

Fuel cycle: the series of steps required to produce electricity. The fuel cycle includes mining or otherwise acquiring the raw fuel source, processing and cleaning the fuel, transport, electricity generation, waste management and plant decommissioning.

Fuel oil: also called heating oil is a distillate product that covers a wide range of properties; see also Nos. 1–4 Fuel oils.

Fuel treatment evaluator (FTE): a strategic assessment tool capable of aiding the identification, evaluation, and prioritization of fuel treatment opportunities.

Fuel wood: wood used for conversion to some form of energy, primarily for residential use.

Fuller's earth: a clay which has high adsorptive capacity for removing color from oils; Attapulgus clay is a widely used fuller's earth.

Functional group: the portion of a molecule that is characteristic of a family of compounds and determines the properties of these compounds.

Furfural extraction: a single-solvent process in which furfural is used to remove aromatic, naphthene, olefin, and unstable hydrocarbons from a lubricating-oil charge stock.

Furnace: an enclosed chamber or container used to burn biomass in a controlled manner to produce heat for space or process heating.

Furnace oil: a distillate fuel primarily intended for use in domestic heating equipment.

Gas cap: a part of a hydrocarbon reservoir at the top that will produce only gas.

Gaseous pollutants: gases released into the atmosphere that act as primary or secondary pollutants.

Gasification: a chemical or thermal process used to convert carbonaceous material (such as coal, petroleum, and biomass) into gaseous components such as carbon monoxide and hydrogen; a process for converting a solid or liquid fuel into a gaseous fuel useful for power generation or chemical feedstock with an oxidant and steam.

Gasifier: a device for converting solid fuel into gaseous fuel; in biomass systems, the process is referred to as pyrolytic distillation.

Gasifier cold gas efficiency (CGE): the percentage of the coal heating value that appears as chemical heating value in the gasifier product gas.

Gasohol: a mixture of 10% v/v anhydrous ethanol and 90% v/v gasoline; 7.5% v/v anhydrous ethanol and 92.5% v/v gasoline; or 5.5% v/v anhydrous ethanol and 94.5% v/v gasoline; a term for motor vehicle fuel comprising between 80% and 90% v/v unleaded gasoline and 10% and –20% v/v ethanol (see also Ethyl alcohol).

Gas oil: a petroleum distillate with a viscosity and boiling range between those of kerosene and lubricating oil.

Gas-oil ratio: ratio of the number of cubic feet of gas measured at atmospheric (standard) conditions to barrels of produced oil measured at stock tank conditions.

Gas-oil sulfonate: sulfonate made from a specific refinery stream, in this case, the gas-oil stream.

Gasoline: fuel for the internal combustion engine that is commonly, but improperly, referred to as simply gas.

Gas reversion: a combination of thermal cracking or reforming of naphtha with thermal polymerization or alkylation of hydrocarbon gases carried out in the same reaction zone.

Gas to liquids (GTL): the process of refining natural gas and other hydrocarbons into long-chain hydrocarbons, which can be used to convert gaseous waste products into fuels.

Gas turbine: a device in which fuel is combusted at pressure and the products of combustion expanded through a turbine to generate power (the Brayton Cycle); it is based on the same principle as the jet engine.

Gel point: the point at which a liquid fuel cools to the consistency of petroleum jelly.

Genetically modified organism (GMO): an organism whose genetic material has been modified through recombinant DNA technology, altering the phenotype of the organism to meet desired specifications.

Gilsonite: an asphaltite that is >90% bitumen.

Girbotol process: a continuous, regenerative process to separate hydrogen sulfide, carbon dioxide, and other acid impurities from natural gas, refinery gas, etc., using mono-, di-, or triethanolamine as the reagent.

Glance pitch: an asphaltite.

Glycol-amine gas treating: a continuous, regenerative process to simultaneously dehydrate and remove acid gases from natural gas or refinery gas.

Grahamite: an asphaltite.

Grassland pasture and range: all open land used primarily for pasture and grazing, including shrub and brush land types of pasture; grazing land with sagebrush and scattered mesquite; and all tame and native grasses, legumes, and other forage used for pasture or grazing; because of the diversity in vegetative composition, grassland pasture and range are not always clearly distinguishable from other types of pasture and range; at one extreme, permanent grassland may merge with cropland pasture or grassland may often be found in transitional areas with forested grazing land.

Gravity: see API gravity.

Gravity drainage: the movement of oil in a reservoir that results from the force of gravity.

Gravity segregation: partial separation of fluids in a reservoir caused by the gravity force acting on differences in density.

Gravity-stable displacement: the displacement of oil from a reservoir by a fluid of a different density, where the density difference is utilized to prevent gravity segregation of the injected fluid.

Gray clay treating: a fixed-bed, usually fuller's earth, vapor-phase treating process to selectively polymerize unsaturated gum-forming constituents (diolefins) in thermally cracked gasoline.

Grain alcohol: see Ethyl alcohol.

Gravimetric: gravimetric methods weigh a residue.

Gravity drainage: the movement of oil in a reservoir that results from the force of gravity.

Gravity segregation: partial separation of fluids in a reservoir caused by the gravity force acting on differences in density.

Grease car: a diesel-powered automobile rigged post-production to run on used vegetable oil.

Greenhouse effect: the effect of certain gases in the Earth's atmosphere in trapping heat from the sun.

Greenhouse gases: gases that trap the heat of the sun in the Earth's atmosphere, producing the greenhouse effect. The two major greenhouse gases are water vapor and carbon dioxide. Other greenhouse gases include methane, ozone, chlorofluorocarbons, and nitrous oxide.

Grid: an electric utility company's system for distributing power.

Growing stock: a classification of timber inventory that includes live trees of commercial species meeting specified standards of quality or vigor; cull trees are excluded.

Guard bed: a bed of an adsorbent (for example, bauxite) that protects a catalyst bed by adsorbing species detrimental to the catalyst.

Gulf HDS process: a fixed-bed process for the catalytic hydrocracking of heavy stocks to lower boiling distillates with accompanying desulfurization.

Gulfining: a catalytic hydrogen-treating process for cracked and straight-run distillates and fuel oils to reduce sulfur content; improve carbon residue, color, and general stability; and effect a slight increase in gravity.

Gum: an insoluble tacky semi-solid material formed as a result of the storage instability and/or the thermal instability of petroleum and petroleum products.

Habitat: the area where a plant or animal lives and grows under natural conditions. Habitat includes living and non-living attributes and provides all requirements for food and shelter.

HAP(s): hazardous air pollutant(s).

Hardness: the concentration of calcium and magnesium in brine.

Hardwoods: usually broad-leaved and deciduous trees.

HCPV: hydrocarbon pore volume.

Headspace: the vapor space above a sample into which volatile molecules evaporate. Certain methods sample this vapor.

Hearn method: a method used in reservoir simulation for calculating a pseudo-relative permeability curve that reflects reservoir stratification.

Heating oil: see Fuel oil.

Heating value: the maximum amount of energy that is available from burning a substance.

Heat recovery steam generator: a heat exchanger that generates steam from the hot exhaust gases from a combustion turbine.

Heavy ends: the highest boiling portion of a petroleum fraction; see also Light ends.

Heavy fuel oil: fuel oil having a high density and viscosity; generally residual fuel oil such as No. 5 and No 6 fuel oil

Heavy (crude) oil: oil that is more viscous that conventional crude oil, has a lower mobility in the reservoir but can be recovered through a well from the reservoir by the application of a secondary or enhanced recovery methods; sometimes, petroleum having an API gravity of less than 20°.

Heavy petroleum: see Heavy oil.

Hectare: common metric unit of area, equal to 2.47 acres. 100 hectares $= 1\,km^2$.

Herbaceous: non-woody type of vegetation, usually lacking permanent strong stems, such as grasses, cereals, and canola (rape).

Heteroatom compounds: chemical compounds which contain nitrogen and/or oxygen and/or sulfur and/or metals bound within their molecular structure(s).

Heterogeneity: lack of uniformity in reservoir properties such as permeability.

HF alkylation: an alkylation process whereby olefins (C_3, C_4, C_5) are combined with *iso*-butane in the presence of hydrofluoric acid catalyst.

Higgins-Leighton model: stream tube computer model used to simulate waterflood.

Hortonsphere (Horton sphere): a spherical pressure-type tank used to store a volatile liquid which prevents the excessive evaporation loss that occurs when such products are placed in conventional storage tanks.

Hot filtration test: a test for the stability of a petroleum product.

Hot spot: an area of a vessel or line wall appreciably above normal operating temperature, usually as a result of the deterioration of an internal insulating liner which exposes the line or vessel shell to the temperature of its contents.

Houdresid catalytic cracking: a continuous moving-bed process for catalytically cracking reduced crude oil to produce high-octane gasoline and light distillate fuels.

Houdriflow catalytic cracking: a continuous moving-bed catalytic cracking process employing an integrated single vessel for the reactor and regenerator kiln.

Houdriforming: a continuous catalytic reforming process for producing aromatic concentrates and high-octane gasoline from low-octane straight naphtha.

Houdry butane dehydrogenation: a catalytic process for dehydrogenating light hydrocarbons to their corresponding mono- or diolefins.

Houdry fixed-bed catalytic cracking: a cyclic regenerable process for cracking of distillates.

Houdry hydrocracking: a catalytic process combining cracking and desulfurization in the presence of hydrogen.

Huff-and-puff: a cyclic EOR method in which steam or gas is injected into a production well; after a short shut-in period, the oil and the injected fluid are produced through the same well.

Hydration: the association of molecules of water with a substance.

Hydraulic fracturing: the opening of fractures in a reservoir by high-pressure, high-volume injection of liquids through an injection well.

Hydrocarbonaceous material: a material such as bitumen that is composed of carbon and hydrogen with other elements (heteroelements) such as nitrogen, oxygen, sulfur, and metals chemically combined within the structures of the constituents; even though carbon and hydrogen may be the predominant elements, there may be very few true hydrocarbons.

Hydrocarbon compounds: chemical compounds containing only carbon and hydrogen.

Hydrocarbon-producing resource: a resource such as coal and oil shale (kerogen) which produces derived hydrocarbons by the application of conversion processes; the hydrocarbons so-produced are not naturally occurring materials.

Hydrocarbon resource: resources such as petroleum and natural gas which can produce naturally occurring hydrocarbons without the application of conversion processes.

Hydrocarbons: organic compounds containing only hydrogen and carbon.

Hydrolysis: a chemical reaction in which water reacts with another substance to form one or more new substances.

Hydroconversion: a term often applied to hydrocracking.

Hydrocracking: a catalytic high-pressure high-temperature process for the conversion of petroleum feedstocks in the presence of fresh and recycled hydrogen; carbon–carbon bonds are cleaved in addition to the removal of heteroatomic species.

Hydrocracking catalyst: a catalyst used for hydrocracking which typically contains separate hydrogenation and cracking functions.

Hydrodenitrogenation: the removal of nitrogen by hydrotreating.

Hydrodesulfurization: the removal of sulfur by hydrotreating.

Hydrofining: a fixed-bed catalytic process to desulfurize and hydrogenate a wide range of charge stocks from gases through waxes.

Hydroforming: a process in which naphtha is passed over a catalyst at elevated temperatures and moderate pressures, in the presence of added hydrogen or hydrogen-containing gases, to form high-octane motor fuel or aromatics.

Hydrogen blistering: blistering of steel caused by trapped molecular hydrogen formed as atomic hydrogen during corrosion of steel by hydrogen sulfide.

Hydrogen addition: an upgrading process in the presence of hydrogen, e.g., hydrocracking; see Hydrogenation.

Hydrogenation: the chemical addition of hydrogen to a material. In nondestructive hydrogenation, hydrogen is added to a molecule only if, and where, unsaturation with respect to hydrogen exists.

Hydrogen transfer: the transfer of inherent hydrogen within the feedstock constituents and products during processing.

Hydroprocesses: refinery processes designed to add hydrogen to various products of refining.

Hydroprocessing: a term often equally applied to hydrotreating and to hydrocracking; also often collectively applied to both.

Hydrotreating: the removal of heteroatomic (nitrogen, oxygen, and sulfur) species by treatment of a feedstock or product at relatively low temperatures in the presence of hydrogen.

Hydrovisbreaking: a non-catalytic process, conducted under similar conditions to visbreaking, which involves treatment with hydrogen to reduce the viscosity of the feedstock and produce more stable products than is possible with visbreaking.

Hydropyrolysis: a short residence time high-temperature process using hydrogen.

Hyperforming: a catalytic hydrogenation process for improving the octane number of naphtha through removal of sulfur and nitrogen compounds.

Hypochlorite sweetening: the oxidation of mercaptan derivatives (RSH) in a sour feedstock by agitation with aqueous, alkaline hypochlorite solution; used where avoidance of free-sulfur addition is desired, because of a stringent copper strip requirements and minimum expense is not the primary object.

Idle cropland: land in which no crops were planted; acreage diverted from crops to soil-conserving uses (if not eligible for and used as cropland pasture) under federal farm programs is included in this component.

Ignitability: characteristic of liquids whose vapors are likely to ignite in the presence of ignition source; also characteristic of non-liquids that may catch fire from friction or contact with water and that burn vigorously.

Illuminating oil: oil used for lighting purposes.

Immiscible: two or more fluids that do not have complete mutual solubility and co-exist as separate phases.

Immiscible carbon dioxide displacement: injection of carbon dioxide into an oil reservoir to effect oil displacement under conditions in which miscibility with reservoir oil is not obtained; see Carbon dioxide augmented waterflooding.

Immiscible displacement: a displacement of oil by a fluid (gas or water) that is conducted under conditions so that interfaces exist between the driving fluid and the oil.

Immunoassay: portable tests that take advantage of an interaction between an antibody and a specific analyte. Immunoassay tests are semi-quantitative and usually rely on color changes of varying intensities to indicate relative concentrations.

Incinerator: any device used to burn solid or liquid residues or wastes as a method of disposal.

Incompatibility: the *immiscibility* of petroleum products and also of different crude oils which is often reflected in the formation of a separate phase after mixing and/or storage; the phenomenon may involve any one or two of several possible events which are (1) phase separation, (2) precipitation of asphaltene constituents when a paraffinic crude oil is blended with a viscous crude oil, (3) when the blend is heated in pipes leading to a reactor, and also through (4) the formation of degradation products and other undesirable changes in the original properties of crude oil products. When such phenomena occur, which may not be immediately at the time of the blend but may occur after an induction period, it is often referred to as instability of the blends. See also Instability.

Incremental ultimate recovery: the difference between the quantity of oil that can be recovered by EOR methods and the quantity of oil that can be recovered by conventional recovery methods.

Inclined grate: a type of furnace in which fuel enters at the top part of a grate in a continuous ribbon, passes over the upper drying section where moisture is removed, and descends into the lower burning section. Ash is removed at the lower part of the grate.

Indirect-injection engine: an older model of diesel engine in which fuel is injected into a pre-chamber, partly combusted, and then sent to the fuel-injection chamber.

Indirect liquefaction: conversion of biomass to a liquid fuel through a synthesis gas intermediate step.

Industrial wood: all commercial round wood products except fuel wood.

Infill drilling: drilling additional wells within an established pattern.

Infrared spectroscopy: an analytical technique that quantifies the vibration (stretching and bending) that occurs when a molecule absorbs (heat) energy in the infrared region of the electromagnetic spectrum.

Inhibitor: a substance, when present in small amounts in a petroleum product, prevents or retards undesirable chemical changes from taking place in the product or in the state of the equipment in which the product is used.

Inhibitor sweetening: a treating process to sweeten gasoline of low mercaptan content, using a phenylenediamine type of inhibitor, air, and caustic.

Initial boiling point: the recorded temperature when the first drop of liquid falls from the end of the condenser.

Initial vapor pressure: the vapor pressure of a liquid of a specified temperature and zero percent evaporated.

Injection profile: the vertical flow rate distribution of fluid flowing from the wellbore into a reservoir.

Injection well: a well in an oil field used for injecting fluids into a reservoir.

Injectivity: the relative ease with which a fluid is injected into a porous rock.

In-line blending: the controlled proportioning of two or more component streams to produce a final blended product of closely defined quality from the beginning to the end of the batch which permits the blended product to be used immediately for the prescribed purpose.

In situ: in its original place; in the reservoir.

In situ combustion: an EOR process consisting of injecting air or oxygen-enriched air into a reservoir under conditions that favor burning part of the in situ petroleum, advancing this burning zone, and recovering oil heated from a nearby producing well.

Instability: the inability of a petroleum product to exist for periods of time without change to the product. See also Incompatibility.

Integrated gasification combine cycle (IGCC): a power plant in which a gasification process provides syngas to a combined cycle under an integrated control system.

Integrity: maintenance of a slug or bank at its preferred composition without too much dispersion or mixing.

Interface: the thin surface area separating two immiscible fluids that are in contact with each other.

Interfacial film: a thin layer of material at the interface between two fluids which differs in composition from the bulk fluids.

Interfacial tension: the strength of the film separating two immiscible fluids, e.g., oil and water or microemulsion and oil; measured in dynes (force) per centimeter or milli-dynes per centimeter.

Interfacial viscosity: the viscosity of the interfacial film between two immiscible liquids.

Interference testing: a type of pressure transient test in which pressure is measured over time in a closed-in well while nearby wells are produced; flow and communication between wells can sometimes be deduced from an interference test.

Interphase mass transfer: the net transfer of chemical compounds between two or more phases.

Iodine number: a measure of the iodine absorption by oil under standard conditions; used to indicate the quantity of unsaturated compounds present; also called iodine value.

Ion exchange: a means of removing cations or anions from solution onto a solid resin.

Ion exchange capacity: a measure of the capacity of a mineral to exchange ions in amount of material per unit weight of solid.

Ions: chemical substances possessing positive or negative charges in solution.

Isocracking: a hydrocracking process for conversion of hydrocarbons which operates at relatively low temperatures and pressures in the presence of hydrogen and a catalyst to produce more valuable, lower boiling products.

Isoforming: a process in which olefinic naphtha is contacted with an alumina catalyst at high temperature and low pressure to produce isomers of higher octane number.

Iso-Kel process: a fixed-bed, vapor-phase isomerization process using a precious metal catalyst and external hydrogen.

Isomate process: a continuous, non-regenerative process for isomerizing C_5-C_8 normal paraffin hydrocarbons, using aluminum chloride–hydrocarbon catalyst with anhydrous hydrochloric acid as a promoter.

Isomerate process: a fixed-bed isomerization process to convert pentane, hexane, and heptane to high-octane blending stocks.

Isomerization: the conversion of a *normal* (straight-chain) paraffin hydrocarbon into an *iso* (branched-chain) paraffin hydrocarbon having the same atomic composition.

Isopach: a line on a map designating points of equal formation thickness.

Iso-plus Houdriforming: a combination process using a conventional Houdriformer operated at moderate severity, in conjunction with one of three possible alternatives – including the use of an aromatic recovery unit or a thermal reformer; see Houdriforming.

Jet fuel: fuel meeting the required properties for use in jet engines and aircraft turbine engines.

Joule: metric unit of energy, equivalent to the work done by a force of one Newton applied over distance of one meter (= $1\,kg\,m^2/s^2$). One joule (J) = 0.239 calories (1 calorie = 4.187 J).

Kaolinite: a clay mineral formed by hydrothermal activity at the time of rock formation or by chemical weathering of rock with high feldspar content; usually associated with intrusive granite rock with high feldspar content.

Kata-condensed aromatic compounds: compounds based on linear condensed aromatic hydrocarbon systems, e.g., anthracene and naphthacene (tetracene).

Kauri butanol number: a measurement of solvent strength for hydrocarbon solvents; the higher the kauri butanol (KB) value, the stronger the solvency; the test method (ASTM D1133) is based on the principle that kauri resin is readily soluble in butyl alcohol but not in hydrocarbon solvents and the resin solution will tolerate only a certain amount of dilution and is reflected as a cloudiness when the resin starts to come out of solution; solvents such as toluene can be added in a greater amount (and thus have a higher KB value) than weaker solvents like hexane.

Kerogen: a complex carbonaceous (organic) material that occurs in sedimentary rock and shale; generally insoluble in common organic solvents.

Kerosene (kerosine): a fraction of petroleum that was initially sought as an illuminant in lamps; a precursor to diesel fuel.

K-factor: see Characterization factor.

Kilowatt (kW): a measure of electrical power equal to 1000 watts. 1 kW = 3412 Btu/hour = 1.341 horsepower.

Kilowatt hour (kWh): a measure of energy equivalent to the expenditure of one kilowatt for one hour. For example, 1 kWh will light a 100-watt light bulb for 10 hours. 1 kWh = 3412 Btu.

Kinematic viscosity: the ratio of viscosity to density, both measured at the same temperature.

Knock: the noise associated with self-ignition of a portion of the fuel–air mixture ahead of the advancing flame front.

Kriging: a technique used in reservoir description for interpolation of reservoir parameters between wells based on random field theory.

LAER: lowest achievable emission rate; the required emission rate in non-attainment permits.

Lamp burning: a test of burning oils in which the oil is burned in a standard lamp under specified conditions in order to observe the steadiness of the flame, the degree of encrustation of the wick, and the rate of consumption of the kerosene.

Lamp oil: see Kerosene.

Landfill gas: a type of biogas that is generated by decomposition of organic material at landfill disposal sites. Landfill gas is approximately 50% methane. See also Biogas.

Leaded gasoline: gasoline containing tetraethyl lead or other organometallic lead antiknock compounds.

Lean gas: the residual gas from the absorber after the condensable gasoline has been removed from the wet gas.

Lean oil: absorption oil from which gasoline fractions have been removed; oil leaving the stripper in a natural gasoline plant.

Lewis acid: a chemical species which can accept an electron pair from a base.

Lewis base: a chemical species which can donate an electron pair.

Light ends: the lower boiling components of a mixture of hydrocarbons; see also Heavy ends, Light hydrocarbons.

Light hydrocarbons: hydrocarbons with molecular weights less than that of heptane (C_7H_{16}).

Light oil: the products distilled or processed from crude oil up to, but not including, the first lubricating-oil distillate.

Light petroleum: petroleum having an API gravity greater than 20°.

Lignin: structural constituent of wood and (to a lesser extent) other plant tissues, which encrusts the walls and cements the cells together.

Ligroine (Ligroin): a saturated petroleum naphtha boiling in the range of 20°–135°C (68°F–275°F) and suitable for general use as a solvent; also called benzine or petroleum ether.

Linde copper sweetening: a process for treating gasoline and distillates with a slurry of clay and cupric chloride.

Liquid petrolatum: see White oil.

Liquefied petroleum gas: propane, butane, or mixtures thereof, gaseous at atmospheric temperature and pressure, held in the liquid state by pressure to facilitate storage, transport, and handling.

Liquid chromatography: a chromatographic technique that employs a liquid mobile phase.

Liquid/liquid extraction: an extraction technique in which one liquid is shaken with or contacted by an extraction solvent to transfer molecules of interest into the solvent phase.

Liquid sulfur dioxide-benzene process: a mixed solvent process for treating lubricating oil stocks to improve viscosity index; also used for dewaxing.

Lithology: the geological characteristics of the reservoir rock.

Live cull: a classification that includes live cull trees; when associated with volume, it is the net volume in live cull trees that are 5.0 inches in diameter and larger.

Live steam: steam coming directly from a boiler before being utilized for power or heat.

Liver: the intermediate layer of dark-colored, oily material, insoluble in weak acid and in oil, which is formed when acid sludge is hydrolyzed.

Logging residues: the unused portions of growing-stock and non-growing-stock trees cut or destroyed during logging and left in the woods.

Lorenz coefficient: a permeability heterogeneity factor.

Lower-phase microemulsion: a microemulsion phase containing a high concentration of water that, when viewed in a test tube, resides near the bottom with oil phase on top.

Lube: see Lubricating oil.

Lube cut: a fraction of crude oil of suitable boiling range and viscosity to yield lubricating oil when completely refined; also referred to as lube-oil distillates or lube stock.

Lubricating oil: a fluid lubricant used to reduce friction between bearing surfaces.

M85: an alcohol fuel mixture containing 85% methanol and 15% gasoline by volume. Methanol is typically made from natural gas, but can also be derived from the fermentation of biomass.

Macrofouling: fouling of refinery equipment and pipes by the deposition of coarse matter from either organic or biological or inorganic origin; these deposits foul the surfaces of heat exchangers and may cause deterioration of the relevant heat transfer coefficient as well as flow blockages. See also Microfouling.

MACT: maximum achievable control technology; applies to major sources of hazardous air pollutants.

Mahogany acids: oil-soluble sulfonic acids formed by the action of sulfuric acid on petroleum distillates. They may be converted to their sodium soaps (mahogany soaps) and extracted from the oil with alcohol for use in the manufacture of soluble oils, rust preventives, and special greases. The calcium and barium soaps of these acids are used as detergent additives in motor oils; see also Brown acids and Sulfonic acids.

Major source: a source that has the potential to emit a regulated pollutant that is greater than or equal to an emission threshold set by regulations.

Maltene fraction (maltenes): that fraction of petroleum that is soluble in, for example, pentane or heptane; deasphaltened oil; also the term arbitrarily assigned to the pentane-soluble portion of petroleum that is relatively high boiling (>300°C, 760 mm) (see also Petrolenes).

Marine engine oil: oil used as a crankcase oil in marine engines.

Marine gasoline: fuel for motors in marine service.

Marine sediment: the organic biomass from which petroleum is derived.

Marsh: an area of spongy waterlogged ground with large numbers of surface water pools. Marshes usually result from (1) an impermeable underlying bedrock; (2) surface deposits of glacial boulder clay; (3) a basin-like topography from which natural drainage is poor; (4) very heavy rainfall in conjunction with a correspondingly low evaporation rate; and (5) low-lying land, particularly at estuarine sites at or below sea level.

Marx–Langenheim model: mathematical equations for calculating heat transfer in a hot water or steam flood.

Mass spectrometer: an analytical technique that *fractures* organic compounds into characteristic "fragments" based on functional groups that have a specific mass-to-charge ratio.

Mayonnaise: low-temperature sludge; a black, brown, or gray deposit having a soft, mayonnaise-like consistency; not recommended as a food additive!

MCL: maximum contaminant level as dictated by regulations.

Medicinal oil: highly refined, colorless, tasteless, and odorless petroleum oil used as a medicine in the nature of an internal lubricant; sometimes called liquid paraffin.

Megawatt (MW): a measure of electrical power equal to one million watts (1000 kW).

Membrane technology: gas separation processes utilizing membranes that permit different components of a gas to diffuse through the membrane at significantly different rates.

MDL: see Method detection limit.

MEK (methyl ethyl ketone): a colorless liquid ($CH_3COCH_2CH_3$) used as a solvent; as a chemical intermediate; and in the manufacture of lacquers, celluloid, and varnish removers.

MEK deoiling: a wax-deoiling process in which the solvent is generally a mixture of methyl ethyl ketone and toluene.

MEK dewaxing: a continuous solvent dewaxing process in which the solvent is generally a mixture of methyl ethyl ketone and toluene.

MEOR: microbial-enhanced oil recovery.

Methanol: see Methyl alcohol.

Method detection limit: the smallest quantity or concentration of a substance that the instrument can measure.

Methyl t-butyl ether: an ether added to gasoline to improve its octane rating and to decrease gaseous emissions; see Oxygenate.

Mercapsol process: a regenerative process for extracting mercaptan derivatives (RSH), utilizing aqueous sodium (or potassium) hydroxide containing mixed cresols as solubility promoters.

Mercaptans: organic compounds having the general formula R-SH.

Metagenesis: the alteration of organic matter during the formation of petroleum that may involve temperatures above 200°C (390°F); see also Catagenesis and Diagenesis.

Methyl alcohol (methanol; wood alcohol): a colorless, volatile, inflammable, and poisonous alcohol (CH_3OH) traditionally formed by destructive distillation of wood or, more recently, as a result of synthetic distillation in chemical plants; a fuel typically derived from natural gas, but can also be produced from the fermentation of sugars in biomass.

Methyl ethyl ketone: see MEK.

Mica: a complex aluminum silicate mineral that is transparent, tough, flexible, and elastic.

Micellar fluid (surfactant slug): an aqueous mixture of surfactants, co-surfactants, salts, and hydrocarbons. The term micellar is derived from the word micelle, which is a submicroscopic aggregate of surfactant molecules and associated fluid.

Micelle: the structural entity by which asphaltene constituents are dispersed in petroleum.

Microcarbon residue: the carbon residue determined using a thermogravimetric method. See also Carbon residue.

Microcrystalline wax: wax extracted from certain petroleum residua and having a finer and less apparent crystalline structure than paraffin wax.

Microemulsion: a stable, finely dispersed mixture of oil, water, and chemicals (surfactants and alcohols).

Microemulsion or micellar/emulsion flooding: an augmented waterflooding technique in which a surfactant system is injected in order to enhance oil displacement toward producing wells.

Microfouling: deposition of solids in refinery equipment or pipes in which the solid is (1) particulate fouling, which is the accumulation of particles on a surface, (2) chemical reaction fouling, such as decomposition of organic matter on heating surfaces, (3) solidification fouling, which occurs when components of a flowing fluid with a high melting point freeze onto a subcooled surface, (4) corrosion fouling, which is caused by corrosion, (5) biofouling, which can often ensure after biocorrosion which is due to the action of bacteria or algae, and (6) composite fouling, whereby fouling involves more than one foulant or fouling mechanism. See also Macrofouling.

Microorganisms: animals or plants of microscopic size, such as bacteria.

Microscopic displacement efficiency: the efficiency with which an oil displacement process removes the oil from individual pores in the rock.

Mid-boiling point: the temperature at which approximately 50% of a material has distilled under specific conditions.

Middle distillate: distillate boiling between the kerosene and lubricating oil fractions.

Middle-phase microemulsion: a microemulsion phase containing a high concentration of both oil and water that, when viewed in a test tube, resides in the middle with the oil phase above it and the water phase below it.

Migration (primary): the movement of hydrocarbons (oil and natural gas) from mature, organic-rich source rocks to a point where the oil and gas can collect as droplets or as a continuous phase of liquid hydrocarbon.

Migration (secondary): the movement of the hydrocarbons as a single, continuous fluid phase through water-saturated rocks, fractures, or faults followed by accumulation of the oil and gas in sediments (traps, *q.v.*) from which further migration is prevented.

Mill residue: wood and bark residues produced in processing logs into lumber, plywood, and paper.

Mineral hydrocarbons: petroleum hydrocarbons, considered *mineral* because they come from the earth rather than from plants or animals.

Mineral oil: the older term for petroleum; the term was introduced in the nineteenth century as a means of differentiating petroleum (rock oil) from whale oil which, at the time, was the predominant illuminant for oil lamps.

Minerals: naturally occurring inorganic solids with well-defined crystalline structures.

Mineral seal oil: a distillate fraction boiling between kerosene and gas oil.

Mineral wax: yellow to dark brown, solid substances that occur naturally and are composed largely of paraffins; usually found associated with considerable mineral matter, as a filling in veins and fissures or as an interstitial material in porous rocks.

Minimum miscibility pressure (MMP): see Miscibility.

Miscibility: an equilibrium condition, achieved after mixing two or more fluids, which is characterized by the absence of interfaces between the fluids: (1) *First-contact miscibility:* miscibility in the usual sense, whereby two fluids can be mixed in all proportions without any interfaces forming. Example: At room temperature and pressure, ethyl alcohol and water are first-contact miscible. (2) *Multiple-contact miscibility (dynamic miscibility):* miscibility that is developed by repeated enrichment of one fluid phase with components from a second fluid phase with which it comes into contact. (3) *Minimum miscibility* pressure: the minimum pressure above which two fluids become miscible at a given temperature, or can become miscible, by dynamic processes.

Miscible flooding: see EOR process.

Miscible fluid displacement (miscible displacement): an oil displacement process in which an alcohol, a refined hydrocarbon, a condensed petroleum gas, carbon dioxide, liquefied natural gas, or even exhaust gas is injected into an oil reservoir at pressure levels such that the injected gas or fluid and reservoir oil are miscible; the process may include the concurrent, alternate, or subsequent injection of water.

Mitigation: identification, evaluation, and cessation of potential impacts of a process product or byproduct.

Mixed-phase cracking: the thermal decomposition of higher boiling hydrocarbons to gasoline components.

Mobility: a measure of ease with which a fluid moves through reservoir rock; the ratio of rock permeability to apparent fluid viscosity.

Mobility buffer: the bank that protects a chemical slug from water invasion and dilution and assures mobility control.

Mobility control: ensuring that the mobility of the displacing fluid or bank is equal to or less than that of the displaced fluid or bank.

Mobility ratio: the ratio of mobility of an injection fluid to mobility of fluid being displaced.

Modified alkaline flooding: the addition of a co-surfactant and/or polymer to the alkaline flooding process.

Modified/unmodified diesel engine: traditional diesel engines must be modified to heat the oil before it reaches the fuel injectors in order to handle straight vegetable oil. With modification, any diesel engine can run on veggie oil; without modification, the oil must first be converted to biodiesel.

Modified naphtha insolubles (MNI): an insoluble fraction obtained by adding naphtha to petroleum; usually, the naphtha is modified by adding paraffin constituents; the fraction might be equated to asphaltenes *if* the naphtha is equivalent to n-heptane, but usually it is not.

Moisture content (MC): the weight of the water contained in wood, usually expressed as a percentage of weight, either oven-dried or as received.

Moisture content, dry basis: moisture content expressed as a percentage of the weight of oven-wood, i.e. [(weight of wet sample − weight of dry sample)/weight of dry sample] × 100.

Moisture content, wet basis: moisture content expressed as a percentage of the weight of wood as-received, i.e. [(weight of wet sample − weight of dry sample)/weight of wet sample] × 100.

Molecular sieve: a synthetic zeolite mineral having pores of uniform size; it is capable of separating molecules on the basis of their size, structure, or both by absorption or sieving.

Motor octane method: a test for determining the knock rating of fuels for use in spark-ignition engines; see also Research octane method.

Moving-bed catalytic cracking: a cracking process in which the catalyst is continuously cycled between the reactor and the regenerator.

MSDS: material safety data sheet.

MTBE: methyl tertiary butyl ether is highly refined high-octane light distillate used in the blending of gasoline.

NAAQS: National Ambient Air Quality Standards; standards exist for the pollutants known as the criteria air pollutants: nitrogen oxides (NO_x), sulfur oxides (SO_x), lead, ozone, particulate matter less than 10 microns in diameter, and carbon monoxide (CO).

Naft: pre-Christian era (Greek) term for naphtha.

Napalm: a thickened gasoline used as an incendiary medium that adheres to the surface it strikes.

Naphtha: a generic term applied to refined, partly refined, or unrefined petroleum products and liquid products of natural gas, the majority of which distills below 240°C (464°F); the volatile fraction of petroleum which is used as a solvent or as a precursor to gasoline.

Naphthenes: cycloparaffins.

Native asphalt: see Bitumen.

Natural asphalt: see Bitumen.

Natural gas: the naturally occurring gaseous constituents that are found in many petroleum reservoirs; also there are certain reservoirs where natural gas may be the sole occupant.

Natural gas liquids (NGL): the hydrocarbon liquids that condense during the processing of hydrocarbon gases that are produced from oil or gas reservoir; see also Natural gasoline.

Natural gasoline: a mixture of liquid hydrocarbons extracted from natural gas suitable for blending with refinery gasoline.

Natural gasoline plant: a plant for the extraction of fluid hydrocarbon, such as gasoline and liquefied petroleum gas, from natural gas.

NESHAP: National Emissions Standards for Hazardous Air Pollutants; emission standards for specific source categories that emit or have the potential to emit one or more hazardous air pollutants; the standards are modeled on the best practices and most effective emission reduction methodologies in use at the affected facilities.

Neutralization: a process for reducing the acidity or alkalinity of a waste stream by mixing acids and bases to produce a neutral solution; also known as pH adjustment.

Neutral oil: a distillate lubricating oil with viscosity usually not above 200 seconds at 100°F.

Neutralization number: the weight, in milligrams, of potassium hydroxide needed to neutralize the acid in 1 g of oil; an indication of the acidity of an oil.

Nitrogen fixation: the transformation of atmospheric nitrogen into nitrogen compounds that can be used by growing plants.

Nitrogen oxides (NOx): products of combustion that contribute to the formation of smog and ozone.

Non-asphaltic road oil: any of the nonhardening petroleum distillates or residual oils used as dust layers. They have sufficiently low viscosity to be applied without heating and, together with asphaltic road oils, are sometimes referred to as dust palliatives.

Non-attainment area: any area that does not meet the national primary or secondary ambient air quality standard established (by the Environmental Protection Agency) for designated pollutants, such as carbon monoxide and ozone.

Non-forest land: land that has never supported forests and lands formerly forested where use of timber management is precluded by development for other uses; if intermingled in forest areas, unimproved roads and non-forest strips must be more than 120 feet wide, and clearings must be more than 1 acre in area to qualify as non-forest land.

Non-industrial private: an ownership class of private lands where the owner does not operate wood processing plants.

Non-ionic surfactant: a surfactant molecule containing no ionic charge.

Non-Newtonian: a fluid that exhibits a change of viscosity with flow rate.

NOx: oxides of nitrogen.

Nuclear magnetic resonance spectroscopy: an analytical procedure that permits the identification of complex molecules based on the magnetic properties of the atoms they contain.

No. 1 Fuel oil: very similar to kerosene and is used in burners where vaporization before burning is usually required and a clean flame is specified.

No. 2 Fuel oil: also called domestic heating oil; has properties similar to diesel fuel and heavy jet fuel; used in burners where complete vaporization is not required before burning.

No. 4 Fuel oil: a light industrial heating oil and is used where preheating is not required for handling or burning; there are two grades of No. 4 fuel oil, differing in safety (flash point) and flow (viscosity) properties.

No. 5 Fuel oil: a heavy industrial fuel oil which requires preheating before burning.

No. 6 Fuel oil: a heavy fuel oil and is more commonly known as Bunker C oil when it is used to fuel ocean-going vessels; preheating is always required for burning this oil.

Observation wells: wells that are completed and equipped to measure reservoir conditions and/or sample reservoir fluids, rather than to inject produced reservoir fluids.

Octane barrel yield: a measure used to evaluate fluid catalytic cracking processes; defined as (RON + MON)/2 times the gasoline yield, where RON is the research octane number and MON is the motor octane number.

Octane number: a number indicating the antiknock characteristics of gasoline.

Oil bank: see Bank.

Oil breakthrough (time): the time at which the oil–water bank arrives at the producing well.

Oil from tar sand: synthetic crude oil.

Oil mining: application of a mining method to the recovery of bitumen.

Oil originally in place (OOIP): the quantity of petroleum existing in a reservoir before oil recovery operations begin.

Oils: that portion of the maltenes which is not adsorbed by a surface active material such as clay or alumina.

Oil sand: see Tar sand.

Oil shale: a fine-grained impervious sedimentary rock which contains an organic material called kerogen.

Olefin: synonymous with *alkene*.

OOIP: see Oil originally in place.

Open-loop biomass: biomass that can be used to produce energy and bioproducts even though it was not grown specifically for this purpose; includes agricultural livestock waste, residues from forest harvesting operations, and crop harvesting.

Optimum salinity: the salinity at which a middle-phase microemulsion containing equal concentrations of oil and water results from the mixture of a micellar fluid (surfactant slug) with oil.

Organic sedimentary rocks: rocks containing organic material such as residues of plant and animal remains/decay.

Orifice-plate mixer (orifice mixer): a mixer in which two or more liquids are pumped through an orifice constriction to cause turbulence and consequent mixing action.

Overhead: that portion of the feedstock which is vaporized and removed during distillation.

Override: the gravity-induced flow of a lighter fluid in a reservoir above another heavier fluid.

Oxidation: a process which can be used for the treatment of a variety of inorganic and organic substances.

Oxidized asphalt: see Air-blown asphalt.

Oxygenate: an oxygen-containing compound that is blended into gasoline to improve its octane number and to decrease gaseous emissions; a substance which, when added to gasoline, increases the amount of oxygen in that gasoline blend; includes fuel ethanol, methanol, and methyl tertiary butyl ether (MTBE).

Oxygenated gasoline: gasoline with added ethers or alcohols, formulated according to the Federal Clean Air Act to reduce carbon monoxide emissions during winter months.

Oxygen scavenger: a chemical which reacts with oxygen in injection water, used to prevent degradation of polymer.

Ozokerite (Ozocerite): a naturally occurring wax; when refined also known as ceresin.

Pale oil: a lubricating oil or a process oil refined until its color, by transmitted light, is straw to pale yellow.

Paraffin wax: the colorless, translucent, highly crystalline material obtained from the light lubricating fractions of paraffin crude oils (wax distillates).

Paraffinum liquidum: see Liquid petrolatum.

Particle density: the density of solid particles.

Particulate: a small, discrete mass of solid or liquid matter that remains individually dispersed in gas or liquid emissions.

Particulate emissions: particles of a solid or liquid suspended in a gas, or the fine particles of carbonaceous soot and other organic molecules discharged into the air during combustion.

Particulate matter (particulates): particles in the atmosphere or on a gas stream that may be organic or inorganic and originate from a wide variety of sources and processes.

Particle size distribution: the particle size distribution (of a catalyst sample) expressed as a percent of the whole.

Partitioning: in chromatography, the physical act of a solute having different affinities for the stationary and mobile phases.

Partition ratios, K: the ratio of total analytical concentration of a solute in the stationary phase, CS, to its concentration in the mobile phase, CM.

Pattern: the areal pattern of injection and producing wells selected for a secondary or enhanced recovery project.

Pattern life: the length of time a flood pattern participates in oil recovery.

Pay zone thickness: the depth of a tar sand deposit from which bitumen (or a product) can be recovered.

Penex process: a continuous, non-regenerative process for isomerization of C_5 and/or C_6 fractions in the presence of hydrogen (from reforming) and a platinum catalyst.

Pentafining: a pentane isomerization process using a regenerable platinum catalyst on a silica–alumina support and requiring outside hydrogen.

Pepper sludge: the fine particles of sludge produced after acid treatment which may remain in suspension.

Peri-condensed aromatic compounds: compounds based on angular condensed aromatic hydrocarbon systems, e.g., phenanthrene, chrysene, and picene.

Permeability: the ease of flow of the water through the rock.

Petrol: a term commonly used in some countries for gasoline.

Petrolatum: a semi-solid product, ranging from white to yellow in color, produced during refining of residual stocks; see Petroleum jelly.

Petrolenes: the term applied to that part of the pentane-soluble or heptane-soluble material that is low boiling ($<300°C$, $<570°F$, 760 mm) and can be distilled without thermal decomposition (see also Maltenes).

Petroleum (crude oil): a naturally occurring mixture of gaseous, liquid, and solid hydrocarbon compounds usually found trapped deep underground beneath impermeable cap rock and above a lower dome of sedimentary rock such as shale; most petroleum reservoirs occur in sedimentary rocks of marine, deltaic, or estuarine origin.

Petroleum asphalt: see Asphalt.

Petroleum ether: see Ligroine.

Petroleum jelly: a translucent, yellowish to amber or white, hydrocarbon substance (melting point: $38°C–54°C$) having almost no odor or taste, derived from petroleum and used principally in medicine and pharmacy as a protective dressing and as a substitute for fats in ointments and cosmetics; also used in many types of polishes and in lubricating greases, rust preventives, and modeling clay; obtained by dewaxing heavy lubricating oil stocks.

Petroleum refinery: see Refinery.

Petroleum refining: a complex sequence of events that result in the production of a variety of products.

Petroleum sulfonate: a surfactant used in chemical flooding prepared by sulfonating selected crude oil fractions.

Petroporphyrins: see Porphyrins.

Phase: a separate fluid that co-exists with other fluids; gas, oil, water, and other stable fluids such as microemulsions are all called phases in EOR research.

Phase behavior: the tendency of a fluid system to form phases as a result of changing temperature, pressure, or the bulk composition of the fluids or of individual fluid phases.

Phase diagram: a graph of phase behavior. In chemical flooding, a graph showing the relative volume of oil, brine, and sometimes, one or more microemulsion phases. In carbon dioxide flooding, a graph showing conditions for formation of various liquid, vapor, and solid phases.

Phase properties: types of fluids, compositions, densities, viscosities, and relative amounts of oil, microemulsion, or solvent, and water formed when a micellar fluid (surfactant slug) or miscible solvent (e.g., carbon dioxide) is mixed with oil.

Phase separation: the formation of a separate phase that is usually the prelude to coke formation during a thermal process; the formation of a separate phase as a result of the instability/incompatibility of petroleum and petroleum products; the precipitation of asphaltene constituents when a paraffinic crude oil is blended with a viscous crude oil or when the blend is heated in pipes leading to a reactor leading to the formation of degradation products and other undesirable changes in the original properties of crude oil products. When such phenomena occur, which may not be immediately at the time of the blend but can occur after an induction period, it is often referred to as instability of the blends. See also Incompatibility.

pH adjustment: neutralization.

Phosphoric acid polymerization: a process using a phosphoric acid catalyst to convert propene, butene, or both to gasoline or petrochemical polymers.

Photoionization: a gas chromatographic detection system that utilizes a photoionization *detector* (*PID*) ultraviolet lamp as an ionization source for analyte detection. It is usually used as a selective detector by changing the photon energy of the ionization source.

Photosynthesis: process by which chlorophyll-containing cells in green plants concert incident light to chemical energy, capturing carbon dioxide in the form of carbohydrates.

PINA analysis: a method of analysis for paraffins, *iso*-paraffins, naphthenes, and aromatics.

PIONA analysis: a method of analysis for paraffins, *iso*-paraffins, olefins, naphthenes, and aromatics.

Pipe still: a still in which heat is applied to the oil while being pumped through a coil or pipe arranged in a suitable firebox.

Pipestill gas: the most volatile fraction that contains most of the gases that are generally dissolved in the crude. Also known as pipestill light ends.

Pipestill light ends: see *Pipestill gas.*

Pitch: the non-volatile, brown to black, semi-solid to solid viscous product from the destructive distillation of many bituminous or other organic materials, especially coal.

Platforming: a reforming process using a platinum-containing catalyst on an alumina base.

PNA: a polynuclear aromatic compound.

PNA analysis: a method of analysis for paraffins, naphthenes, and aromatics.

Polar aromatics: resins; the constituents of petroleum that are predominantly aromatic in character and contain polar (nitrogen, oxygen, and sulfur) functions in their molecular structure(s).

Pollutant: a chemical (or chemicals) introduced into the land, water, and air systems that are not indigenous to these systems; also an indigenous chemical (or chemicals) introduced into the land water and air systems in amounts greater than the natural abundance.

Pollution: the introduction into the land water and air systems of a chemical or chemicals that are not indigenous to these systems or the introduction into the land water and air systems of indigenous chemicals in greater than natural amounts.

Polyacrylamide: very high molecular weight material used in polymer flooding.

Polycyclic aromatic hydrocarbons (PAHs): polycyclic aromatic hydrocarbons are a suite of compounds comprised of two or more condensed aromatic rings. They are found in many petroleum mixtures, and they are predominantly introduced to the environment through natural and anthropogenic combustion processes.

Polyforming: a process charging both C_3 and C_4 gases with naphtha or gas oil under thermal conditions to produce gasoline.

Polymer: in EOR, any very high molecular weight material that is added to water to increase viscosity for polymer flooding.

Polymer augmented waterflooding: waterflooding in which organic polymers are injected with the water to improve areal and vertical sweep efficiency.

Polymer gasoline: the product of polymerization of gaseous hydrocarbons to hydrocarbons boiling in the gasoline range.

Polymerization: the combination of two olefin molecules to form a higher molecular weight paraffin.

Polymer stability: the ability of a polymer to resist degradation and maintain its original properties.

Polynuclear aromatic compound: an aromatic compound having two or more fused benzene rings, e.g., naphthalene and phenanthrene.

Polysulfide treating: a chemical treatment used to remove elemental sulfur from refinery liquids by contacting them with a non-regenerable solution of sodium polysulfide.

PONA analysis: a method of analysis for paraffins (P), olefins (O), naphthenes (N), and aromatics (A).

Pore diameter: the average pore size of a solid material, e.g., catalyst.

Pore space: a small hole in reservoir rock that contains fluid or fluids; a four-inch cube of reservoir rock may contain millions of inter-connected pore spaces.

Pore volume: total volume of all pores and fractures in a reservoir or part of a reservoir; also applied to catalyst samples.

Porosity: the percentage of rock volume available to contain water or other fluid.

Porphyrins: organometallic constituents of petroleum that contain vanadium or nickel; the degradation products of chlorophyll that became included in the protopetroleum.

Positive bias: a result that is incorrect and too high.

Possible reserves: reserves where there is an even greater degree of uncertainty but about which there is some information.

Potential reserves: reserves based upon geological information about the types of sediments where such resources are likely to occur and they are considered to represent an educated guess.

Pour point: the lowest temperature at which oil will pour or flow when it is chilled without disturbance under definite conditions.

Powerforming: a fixed-bed naphtha-reforming process using a regenerable platinum catalyst.

Power-law exponent: an exponent used to model the degree of viscosity change of some non-Newtonian liquids.

Precipitation number: the number of milliliters of precipitate formed when 10 mL of lubricating oil is mixed with 90 mL of petroleum naphtha of a definite quality and centrifuged under definitely prescribed conditions.

Preflush: a conditioning slug injected into a reservoir as the first step of an EOR process.

Pressure cores: cores cut into a special coring barrel that maintains reservoir pressure when brought to the surface; this prevents the loss of reservoir fluids that usually accompanies a drop in pressure from reservoir to atmospheric conditions.

Pressure gradient: rate of change of pressure with distance.

Pressure maintenance: augmenting the pressure (and energy) in a reservoir by injecting gas and/ or water through one or more wells.

Pressure pulse test: a technique for determining reservoir characteristics by injecting a sharp pulse of pressure in one well and detecting it in surrounding wells.

Pressure transient testing: measuring the effect of changes in pressure at one well on other wells in a field.

Primary oil recovery: oil recovery utilizing only naturally occurring forces.

Primary wood-using mill: a mill that converts round wood products into other wood products; common examples are sawmills that convert saw logs into lumber and pulp mills that convert pulpwood round wood into wood pulp.

Primary structure: the chemical sequence of atoms in a molecule.

Primary structure: a chemical that, when injected into a test well, reacts with reservoir fluids form a detectable chemical compound.

Probable reserves: mineral reserves that are nearly certain but about which a slight doubt exists.

Process heat: heat used in an industrial process rather than for space heating or other housekeeping purposes.

Producer gas: fuel gas high in carbon monoxide (CO) and hydrogen (H_2), produced by burning a solid fuel with insufficient air or by passing a mixture of air and steam through a burning bed of solid fuel.

Producibility: the rate at which oil or gas can be produced from a reservoir through a wellbore.

Producing well: a well in an oil field used for removing fluids from a reservoir.

Propane asphalt: see Solvent asphalt.

Propane deasphalting: solvent deasphalting using propane as the solvent.

Propane decarbonizing: a solvent extraction process used to recover catalytic cracking feed from heavy fuel residues.

Propane dewaxing: a process for dewaxing lubricating oils in which propane serves as solvent.

Propane fractionation: a continuous extraction process employing liquid propane as the solvent; a variant of propane deasphalting.

Protopetroleum: a generic term used to indicate the initial petroleum product formed after its precursors have undergone some kind of changes.

Proved reserves: mineral reserves that have been positively identified as recoverable with current technology.

PSD: prevention of significant deterioration.

PTE: potential to emit; the maximum capacity of a source to emit a pollutant, given its physical or operation design, and considering certain controls and limitations.

Pulpwood: round wood, whole-tree chips, or wood residues that are used for the production of wood pulp.

Pulse-echo ultrasonic borehole televiewer: well-logging system wherein a pulsed, narrow acoustic beam scans the well as the tool is pulled up the borehole; the amplitude of the reflecting beam is displayed on a cathode-ray tube resulting in a pictorial representation of wellbore.

Purge and trap: a chromatographic sample introduction technique in volatile components that are purged from a liquid medium by bubbling gas through it. The components are then concentrated by "trapping" them on a short intermediate column, which is subsequently heated to drive the components onto the analytical column for separation.

Purge gas: typically helium or nitrogen, used to remove analytes from the sample matrix in purge/trap extractions.

Pyrobitumen: see Asphaltoid.

Pyrolysis: the thermal decomposition of biomass at high temperatures (greater than 400°F or 200°C) in the absence of air; the end product of pyrolysis is a mixture of solids (char), liquids (oxygenated oils), and gases (methane, carbon monoxide, and carbon dioxide) with proportions determined by operating temperature, pressure, oxygen content, and other conditions; exposure of a feedstock to high temperatures in an oxygen-poor environment.

Pyrophoric: substances that catch fire spontaneously in air without an ignition source.

Quad: one quadrillion Btu (10^{15} Btu) = 1.055 exajoules (EJ) or approximately 172 million barrels of oil equivalent.

Quadrillion: 1×10^{15}

Quench: the sudden cooling of hot material discharging from a thermal reactor.

RACT: reasonably Available Control Technology standards; implemented in areas of non-attainment to reduce emissions of volatile organic compounds and nitrogen oxides.

Raffinate: that portion of the oil which remains undissolved in a solvent refining process.

Ramsbottom carbon residue: see Carbon residue.

Raw crude oil: crude oil direct from the wellbore, before it is treated in a gas separation plant; typically contains nonhydrocarbon contaminants.

Raw materials: minerals extracted from the earth prior to any refining or treating.

Recovery boiler: a pulp mill boiler in which lignin and spent cooking liquor (black liquor) are burned to generate steam.

Rectification zone: the zone in a distillation tower in which the more volatile component(s) is (are) removed through contacting the rising vapor with the down-flowing liquid; at the stripping section, the down-flowing liquid is stripped of the more volatile component by the rising vapor; the distillation tower can be equipped with trays or packings or both; also known as the rectification section or the rectifying section.

Recycle ratio: the volume of recycle stock per volume of fresh feed; often expressed as the volume of recycle divided by the total charge.

Recycle stock: the portion of a feedstock which has passed through a refining process and is recirculated through the process.

Recycling: the use or reuse of chemical waste as an effective substitute for a commercial products or as an ingredient or feedstock in an industrial process.

Reduced crude: a residual product remaining after the removal, by distillation or other means, of an appreciable quantity of the more volatile components of crude oil.

Refinery: a series of integrated unit processes by which petroleum can be converted to a slate of useful (saleable) products.

Refinery gas: a gas (or a gaseous mixture) produced as a result of refining operations.

Refining: the processes by which petroleum is distilled and/or converted by application of a physical and chemical process to form a variety of products are generated.

Reformate: the liquid product of a reforming process.

Reformed gasoline: gasoline made by a reforming process.

Reforming: the conversion of hydrocarbons with low octane numbers into hydrocarbons having higher octane numbers, e.g., the conversion of a n-paraffin into a iso-paraffin.

Reformulated gasoline (RFG): gasoline designed to mitigate smog production and to improve air quality by limiting the emission levels of certain chemical compounds such as benzene and other aromatic derivatives; often contains oxygenates.

Refractory lining: a lining, usually made of ceramic, capable of resisting and maintaining high temperatures.

Refuse-derived fuel (RDF): fuel prepared from municipal solid waste; non-combustible materials such as rocks, glass, and metals are removed, and the remaining combustible portion of the solid waste is chopped or shredded; the combustible portion of municipal solid waste after removal of glass and metals.

Reid vapor pressure: a measure of the volatility of liquid fuels, especially gasoline.

Regeneration: the activation or reactivation of a catalyst by burning off the coke deposits.

Regenerator: a reactor for catalyst reactivation.

Relative permeability: the permeability of rock to gas, oil, or water, when any two or more are present, expressed as a fraction of the permeability of the rock.

Renewable energy sources: solar, wind, and other non-fossil fuel energy sources.

Rerunning: the distillation of an oil which has already been distilled.

Research octane method: a test for determining the knock rating, in terms of octane numbers, of fuels for use in spark-ignition engines; see also Motor octane method.

Reserves: well-identified resources that can be profitably extracted and utilized with the existing technology.

Reservoir: a rock formation below the earth's surface containing petroleum or natural gas; a domain where a pollutant may reside for an indeterminate time.

Reservoir simulation: analysis and prediction of reservoir performance with a computer model.

Residual asphalt: see Straight-run asphalt.

Residual fuel oil: obtained by blending the residual product(s) from various refining processes with suitable diluent(s) (usually middle distillates) to obtain the required fuel oil grades.

Residual oil: see Residuum; petroleum remaining in situ after oil recovery.

Residual resistance factor: the reduction in permeability of rock to water caused by the adsorption of polymer.

Residues: bark and woody materials that are generated in primary wood-using mills when round wood products are converted to other products.

Residuum (resid; plural: residua): the residue obtained from petroleum after nondestructive distillation has removed all the volatile materials from crude oil, e.g., an atmospheric (345°C, 650°F+) residuum.

Resins: that portion of the maltenes that is adsorbed by a surface active material such as clay or alumina; the fraction of deasphaltened oil that is insoluble in liquid propane but soluble in n-heptane.

Resistance factor: a measure of resistance to flow of a polymer solution relative to the resistance to flow of water.

Resource: the total amount of a commodity (usually a mineral but can include non-minerals such as water and petroleum) that has been estimated to be ultimately available.

Retention: the loss of chemical components due to adsorption onto the rock's surface, precipitation, or to trapping within the reservoir.

Retention time: the time it takes for an eluate to move through a chromatographic system and reach the detector. Retention times are reproducible and can therefore be compared to a standard for analyte identification.

Rexforming: a process combining platforming with aromatics extraction, wherein low-octane raffinate is recycled to the Platformer.

Rich oil: absorption oil containing dissolved natural gasoline fractions.

Riser: the part of the bubble-plate assembly which channels the vapor and causes it to flow downward to escape through the liquid; also the vertical pipe where fluid catalytic cracking reactions occur.

Rock asphalt: bitumen which occurs in formations that have a limiting ratio of bitumen-to-rock matrix.

Rock matrix: the granular structure of a rock or porous medium.

Rotation: period of years between establishment of a stand of timber and the time when it is considered ready for final harvest and regeneration.

Round wood products: logs and other round timber generated from harvesting trees for industrial or consumer use.

Run-of-the-river reservoirs: reservoirs with a large rate of flow-through compared to their volume.

Salinity: the concentration of salt in water.

Sand: a course granular mineral mainly comprising quartz grains that are derived from the chemical and physical weathering of rocks rich in quartz, notably sandstone and granite.

Sand face: the cylindrical wall of the wellbore through which the fluids must flow to or from the reservoir.

Sandstone: a sedimentary rock formed by compaction and cementation of sand grains; can be classified according to the mineral composition of the sand and cement.

SARA analysis: a method of fractionation by which petroleum is separated into saturates, aromatics, resins, and asphaltene fractions.

SARA separation: see SARA analysis.

Saturated steam: steam at boiling temperature for a given pressure.

Saturates: paraffins and cycloparaffins (naphthenes).

Saturation: the ratio of the volume of a single fluid in the pores to pore volume, expressed as a percent and applied to water, oil, or gas separately; the sum of the saturations of each fluid in a pore volume is 100%.

Saybolt Furol viscosity: the time, in seconds (Saybolt Furol Seconds, SFS), for 60 mL of fluid to flow through a capillary tube in a Saybolt Furol viscometer at specified temperatures between 70°F and 210°F; the method is appropriate for high-viscosity oils such as transmission, gear, and heavy fuel oils.

Saybolt Universal viscosity: the time, in seconds (Saybolt Universal Seconds, SUS), for 60 mL of fluid to flow through a capillary tube in a Saybolt Universal viscometer at a given temperature.

Scale wax: the paraffin derived by removing the greater part of the oil from slack wax by sweating or solvent deoiling.

Screen factor: a simple measure of the viscoelastic properties of polymer solutions.

Screening guide: a list of reservoir rock and fluid properties critical to an EOR process.

Scrubber: a device that uses water and chemicals to clean air pollutants from combustion exhaust.

Scrubbing: purifying a gas by washing with water or chemical; less frequently, the removal of entrained materials.

Secondary oil recovery: application of energy (e.g., water flooding) to recovery of crude oil from a reservoir after the yield of crude oil from primary recovery diminishes.

Secondary pollutants: a pollutant (chemical species) produced by interaction of a primary pollutant with another chemical or by dissociation of a primary pollutant or by other effects within a particular ecosystem.

Secondary recovery: oil recovery resulting from injection of water, or an immiscible gas at moderate pressure, into a petroleum reservoir after primary depletion.

Secondary structure: the ordering of the atoms of a molecule in space relative to each other.

Secondary tracer: the product of the chemical reaction between reservoir fluids and an injected primary tracer.

Secondary wood processing mills: a mill that uses primary wood products in the manufacture of finished wood products, such as cabinets, moldings, and furniture.

Sediment: an insoluble solid formed as a result of the storage instability and/or the thermal instability of petroleum and petroleum products.

Sedimentary: formed by or from deposits of sediments, especially from sand grains or silts transported from their source and deposited in water, such as sandstone and shale, or from calcareous remains of organisms, such as limestone.

Sedimentary strata: typically consist of mixtures of clay, silt, sand, organic matter, and various minerals; formed by or from deposits of sediments, especially from sand grains or silts transported from their source and deposited in water, such as sandstone and shale, or from calcareous remains of organisms, such as limestone.

Selective solvent: a solvent which, at certain temperatures and ratios, will preferentially dissolve more of one component of a mixture than of another and thereby permit partial separation.

Separation process: an upgrading process in which the constituents of petroleum are separated, usually without thermal decomposition, e.g., distillation and deasphalting.

Separator-nobel dewaxing: a solvent (trichloroethylene) dewaxing process.

Separatory funnel: glassware shaped like a funnel with a stoppered rounded top and a valve at the tapered bottom, used for liquid/liquid separations.

Shear: mechanical deformation or distortion, or partial destruction of a polymer molecule as it flows at a high rate.

Shear rate: a measure of the rate of deformation of a liquid under mechanical stress.

Shear-thinning: the characteristic of a fluid whose viscosity decreases as the shear rate increases.

Shell fluid catalytic cracking: a two-stage fluid catalytic cracking process in which the catalyst is regenerated.

Shell still: a still formerly used in which the oil was charged into a closed, cylindrical shell and the heat required for distillation was applied to the outside of the bottom from a firebox.

Side-stream: a liquid stream taken from any one of the intermediate plates of a bubble tower.

Side-stream stripper: a device used to perform further distillation on a liquid stream from any one of the plates of a bubble tower, usually by the use of steam.

Single well tracer: a technique for determining residual oil saturation by injecting an ester, allowing it to hydrolyze; following dissolution of some of the reaction products in residual oil, the injected solutions are produced back and are then analyzed.

Slack wax: the soft, oily crude wax obtained from the pressing of paraffin distillate or wax distillate.

Slime: a name used for petroleum in ancient texts.

Slim tube testing: laboratory procedure for the determination of minimum miscibility pressure using long, small-diameter, sand-packed, oil-saturated, stainless steel tube.

Sludge: a semi-solid to solid product which results from the storage instability and/or the thermal instability of petroleum and petroleum products.

Slug: a quantity of fluid injected into a reservoir during enhanced oil recovery.

Slurry hydroconversion process: a process in which the feedstock is contacted with hydrogen under pressure in the presence of a catalytic coke-inhibiting additive.

Slurry phase reactors: tanks into which wastes, nutrients, and microorganisms are placed.

Smoke point: a measure of the burning cleanliness of jet fuel and kerosine.

Sodium hydroxide treatment: see Caustic wash.

Sodium plumbite: a solution prepared from a mixture of sodium hydroxide, lead oxide, and distilled water; used in making the doctor test for light oils such as gasoline and kerosine.

Solubility parameter: a measure of the solvent power and polarity of a solvent.

Solutizer-steam regenerative process: a chemical treating process for extracting mercaptan derivatives (RSH) from gasoline or naphtha, using solutizers (potassium isobutyrate, potassium alkyl phenolate) in strong potassium hydroxide solution.

Solvent: a liquid in which certain kinds of molecules dissolve. While they typically are liquids with low-boiling points, they may include high-boiling liquids, supercritical fluids, or gases.

Solvent asphalt: the asphalt produced by solvent extraction of residua or by light hydrocarbon (propane) treatment of a residuum or an asphaltic crude oil.

Solvent deasphalting: a process for removing asphaltic and resinous materials from reduced crude oils, lubricating oil stocks, gas oils, or middle distillates through the extraction or precipitant action of low molecular weight hydrocarbon solvents; see also Propane deasphalting.

Solvent decarbonizing: see Propane decarbonizing.

Solvent deresining: see Solvent deasphalting.

Solvent dewaxing: a process for removing wax from oils by means of solvents usually by chilling a mixture of solvent and waxy oil, filtration or by centrifuging the wax which precipitates, and solvent recovery.

Solvent extraction: a process for separating liquids by mixing the stream with a solvent that is immiscible with part of the waste but that will extract certain components of the waste stream.

Solvent gas: an injected gaseous fluid that becomes miscible with oil under reservoir conditions and improves oil displacement.

Solvent naphtha: a refined naphtha of restricted boiling range used as a solvent; also called petroleum naphtha; petroleum spirits.

Solvent refining: see Solvent extraction.

Sonication: a physical technique employing ultrasound to intensely vibrate a sample media in extracting solvent and to maximize solvent/analyte interactions.

Sonic log: a well log based on the time required for sound to travel through rock, useful in determining porosity.

Sour crude oil: crude oil containing an abnormally large amount of sulfur compounds; see also Sweet crude oil.

SOx: oxides of sulfur.

Soxhlet extraction: an extraction technique for solids in which the sample is repeatedly contacted with solvent over several hours, increasing extraction efficiency.

Specific gravity: the mass (or weight) of a unit volume of any substance at a specified temperature compared to the mass of an equal volume of pure water at a standard temperature; see also Density.

Spent catalyst: catalyst that has lost much of its activity due to the deposition of coke and metals.

Spontaneous ignition: ignition of a fuel, such as coal, under normal atmospheric conditions; usually induced by climatic conditions.

Stabilization: the removal of volatile constituents from a higher boiling fraction or product (q.v. stripping); the production of a product which, to all intents and purposes, does not undergo any further reaction when exposed to the air.

Stabilizer: a fractionating tower for removing light hydrocarbons from an oil to reduce vapor pressure particularly applied to gasoline.

Stand (of trees): a tree community that possesses sufficient uniformity in composition, constitution, age, spatial arrangement, or condition to be distinguishable from adjacent communities.

Standpipe: the pipe by which catalyst is transported between the reactor and the regenerator.

Stationary phase: in chromatography, the porous solid or liquid phase through which an introduced sample passes. Different affinities of the stationary phase for a sample allow the components in the sample to be separated or resolved.

Steam cracking: a conversion process in which the feedstock is treated with superheated steam.

Steam distillation: distillation in which vaporization of the volatile constituents is effected at a lower temperature by introduction of steam (open steam) directly into the charge.

Steam drive injection (steam injection): EOR process in which steam is continuously injected into one set of wells (injection wells) or other injection source to effect oil displacement toward and production from a second set of wells (production wells); steam stimulation of production wells is *direct steam stimulation*, whereas steam drive by steam injection to increase production from other wells is *indirect steam stimulation*.

Steam stimulation: injection of steam into a well and the subsequent production of oil from the same well.

Steam turbine: a device for converting energy of high-pressure steam (produced in a boiler) into mechanical power which can then be used to generate electricity.

Stiles method: a simple approximate method for calculating oil recovery by waterflood that assumes separate layers (stratified reservoirs) for the permeability distribution.

Storage stability (or storage instability): the ability (inability) of a liquid to remain in storage over extended periods of time without appreciable deterioration as measured by gum formation and the depositions of insoluble material (sediment).

Straight-run asphalt: the asphalt produced by the distillation of asphaltic crude oil.

Straight-run products: obtained from a distillation unit and used without further treatment.

Straight vegetable oil (SVO): any vegetable oil that has not been optimized through the process of transesterification.

Strata: layers including the solid iron-rich inner core, molten outer core, mantle, and crust of the earth.

Straw oil: pale paraffin oil of straw color used for many process applications.

Stripper well: a well that produces (strips from the reservoir) oil or gas.

Stripping: a means of separating volatile components from less volatile ones in a liquid mixture by the partitioning of the more volatile materials to a gas phase of air or steam (q.v. stabilization).

Sulfonic acids: acids obtained by petroleum or a petroleum product with strong sulfuric acid.

Sulfuric acid alkylation: an alkylation process in which olefins (C_3, C_4, and C_5) combine with *iso*-butane in the presence of a catalyst (sulfuric acid) to form branched chain hydrocarbons used especially in gasoline blending stock.

Supercritical fluid: an extraction method where the extraction fluid is present at a pressure and temperature above its critical point.

Superheated steam: steam which is hotter than boiling temperature for a given pressure.

Surface active material: a chemical compound, molecule, or aggregate of molecules with physical properties that cause it to adsorb at the interface between *two* immiscible liquids, resulting in a reduction of interfacial tension or the formation of a microemulsion.

Surfactant: a type of chemical, characterized as one that reduces interfacial resistance to mixing between oil and water or changes the degree to which water wets reservoir rock.

Suspensoid catalytic cracking: a non-regenerative cracking process in which cracking stock is mixed with slurry of catalyst (usually clay) and cycle oil and passed through the coils of a heater.

Sustainable: an ecosystem condition in which biodiversity, renewability, and resource productivity are maintained over time.

SW-846: an EPA multi-volume publication entitled *Test Methods for Evaluating Solid Waste, Physical/Chemical Methods*; the official compendium of analytical and sampling methods that have been evaluated and approved for use in complying with the RCRA regulations and that functions primarily as a guidance document setting forth acceptable, although not required, methods for the regulated and regulatory communities to use in response to RCRA-related sampling and analysis requirements. SW-846 changes over time as new information and data are developed.

Sweated wax: a crude petroleum-based wax that has been freed from oil by having been passed through a sweater.

Sweating: the separation of paraffin oil and low-melting wax from paraffin wax.

Sweep efficiency: the ratio of the pore volume of reservoir rock contacted by injected fluids to the total pore volume of reservoir rock in the project area. (See also Areal sweep efficiency and Vertical sweep efficiency.)

Sweet crude oil: crude oil containing little sulfur; see also Sour crude oil.

Sweetening: the process by which petroleum products are improved in odor and color by oxidizing or removing the sulfur-containing and unsaturated compounds.

Swelling: increase in the volume of crude oil caused by absorption of EOR fluids, especially carbon dioxide. Also, increase in volume of clays when exposed to brine.

Swept zone: the volume of rock that is effectively swept by injected fluids.

Synthesis gas (syngas): a gas produced by the gasification of a solid or liquid fuel that consists primarily of carbon monoxide and hydrogen.

Synthetic crude oil (syncrude): a hydrocarbon product produced by the conversion of coal, oil shale, or tar sand bitumen that resembles conventional crude oil; can be refined in a petroleum refinery.

Synthetic ethanol: ethanol produced from ethylene, a petroleum byproduct.

Tar: the volatile, brown to black, oily, viscous product from the destructive distillation of many bituminous or other organic materials, especially coal; a name used arbitrarily for petroleum in ancient texts.

Target analyte: target analytes are compounds that are required analytes in U.S. EPA analytical methods. BTEX and PAHs are examples of petroleum-related compounds that are target analytes in U.S. EPA Methods.

Tar sand (bituminous sand): a formation in which the bituminous material (bitumen) is found as a filling in veins and fissures in fractured rocks or impregnating relatively shallow sand, sandstone, and limestone strata; a sandstone reservoir that is impregnated with a heavy, extremely viscous, black hydrocarbonaceous, petroleum-like material that cannot be retrieved through a well by conventional or enhanced oil recovery techniques.

(FE 76-4): the several rock types that contain an extremely viscous hydrocarbon which is not recoverable in its natural state by conventional oil well production methods including currently used enhanced recovery techniques; see Bituminous sand.

Tertiary structure: the three-dimensional structure of a molecule.

Tetraethyl lead (TEL): an organic compound of lead, $Pb(CH_3)_4$, which, when added in small amounts, increases the antiknock quality of gasoline.

Thermal coke: the carbonaceous residue formed as a result of a non-catalytic thermal process: the Conradson carbon residue and the Ramsbottom carbon residue.

Thermal cracking: a process which decomposes, rearranges, or combines hydrocarbon molecules by the application of heat without the aid of catalysts.

Thermal polymerization: a thermal process to convert light hydrocarbon gases into liquid fuels.

Thermal process: any refining process which utilizes heat without the aid of a catalyst.

Thermal recovery: see EOR process.

Thermal reforming: a process using heat (but no catalyst) to effect molecular rearrangement of low-octane naphtha into gasoline of higher antiknock quality.

Thermal stability (thermal instability): the ability (inability) of a liquid to withstand relatively high temperatures for short periods of time without the formation of carbonaceous deposits (sediment or coke).

Thermochemical conversion: use of heat to chemically change substances from one state to another, e.g., to make useful energy products.

Thermofor catalytic cracking: a continuous, moving-bed catalytic cracking process.

Thermofor catalytic reforming: a reforming process in which the synthetic, bead-type catalyst of coprecipitated chromia (Cr_2O_3) and alumina (Al_2O_3) flows down through the reactor concurrent with the feedstock.

Thermofor continuous percolation: a continuous clay treating process to stabilize and decolorize lubricants or waxes.

Thief zone: any geologic stratum not intended to receive injected fluids in which significant amounts of injected fluids are lost; fluids may reach the thief zone due to an improper completion or a faulty cement job.

Thin layer chromatography (TLC): a chromatographic technique employing a porous medium of glass coated with a stationary phase. An extract is spotted near the bottom of the medium and placed in a chamber with solvent (mobile phase). The solvent moves up the medium and separates the components of the extract based on affinities for the medium and solvent.

Throttling device: the generic name of any device or process that dissipates pressure energy by irreversibly converting it into thermal energy. See also Throttling valve.

Throttling valve: a type of valve that can be used to start, stop, and regulate the flow of fluid through a rotodynamic pump. When the flow of a pump is regulated using a throttling valve, the system curve is changed. The operating point moves to the left on the pump curve when the flow is decreased. See also Throttling device valve.

Timberland: forest land that is producing or is capable of producing crops of industrial wood, and that is not withdrawn from timber utilization by statute or administrative regulation.

Time-lapse logging: the repeated use of calibrated well logs to quantitatively observe changes in measurable reservoir properties over time.

Tipping fee: a fee for disposal of waste.

Ton (short ton): 2000 pounds.

Ton (Imperial ton, long ton, shipping ton): 2240 pounds; equivalent to 1000 kg or in crude oil terms approximately 7.5 barrels of oil.

Topped crude: petroleum that has had volatile constituents removed up to a certain temperature, e.g., 250°C+ (480°F+) topped crude; not always the same as a residuum.

Topping: the distillation of crude oil to remove light fractions only

Topping and back pressure turbines: turbines which operate at exhaust pressure considerably higher than atmospheric (non-condensing turbines); often multistage with relatively high efficiency.

Topping cycle: a cogeneration system in which electric power is produced first. The reject heat from power production is then used to produce useful process heat.

Total petroleum hydrocarbons (TPH): the family of several hundred chemical compounds that originally come from petroleum.

Tower: equipment for increasing the degree of separation obtained during the distillation of oil in a still.

TPH E: gas chromatographic test for TPH extractable organic compounds.

TPH V: gas chromatographic test for TPH volatile organic compounds.

TPH-D(DRO): gas chromatographic test for TPH diesel-range organics.

TPH-G(GRO): gas chromatographic test for TPH gasoline-range organics.

Trace element: those elements that occur at very low levels in a given system.

Tracer test: a technique for determining fluid flow paths in a reservoir by adding small quantities of easily detected material (often radioactive) to the flowing fluid and monitoring their appearance at production wells. Also used in cyclic injection to appraise oil saturation.

Transesterification: the chemical process in which an alcohol reacts with the triglycerides in vegetable oil or animal fats, separating the glycerin and producing biodiesel.

Transmissibility (transmissivity): an index of producibility of a reservoir or zone, the product of permeability and layer thickness.

Traps: sediments in which oil and gas accumulate from which further migration is prevented.

Traveling grate: a type of furnace in which assembled links of grates are joined together in a perpetual belt arrangement. Fuel is fed in at one end and ash is discharged at the other.

Treatment: any method, technique, or process that changes the physical and/or chemical character of petroleum.

Triaxial borehole seismic survey: a technique for detecting the orientation of hydraulically induced fractures, wherein a tool holding three mutually seismic detectors is clamped in the borehole during fracturing; fracture orientation is deduced through analysis of the detected microseismic perpendicular events that are generated by the fracturing process.

Trickle hydrodesulfurization: a fixed-bed process for desulfurizing middle distillates.

Trillion: 1×10^{12}.

True boiling point (True boiling range): the boiling point (boiling range) of a crude oil fraction or a crude oil product under standard conditions of temperature and pressure.

True boiling point curve (TBP curve): the composition of any crude oil sample is approximated by a true boiling point curve. The method used is a batch distillation operation, using a large number of stages – typically in excess of 60 and high reflux to distillate ratio (in excess of 5). The temperature at any point on the temperature-volumetric yield curve represents the true boiling point of the hydrocarbon material present at the given volume percent point distilled. True boiling point distillation curves are generally determined for the whole crude oil and not for any of the crude oil products.

Tube-and-tank cracking: an older liquid-phase thermal cracking process.

Turbine: a machine for converting the heat energy in steam or high temperature gas into mechanical energy. In a turbine, a high velocity flow of steam or gas passes through successive rows of radial blades fastened to a central shaft.

Turn down ratio: the lowest load at which a boiler will operate efficiently as compared to the boiler's maximum design load.

Ultimate analysis: elemental composition.

Ultimate recovery: the cumulative quantity of oil that will be recovered when revenues from further production no longer justify the costs of the additional production.

Ultrafining: a fixed-bed catalytic hydrogenation process to desulfurize naphtha and upgrade distillates by essentially removing sulfur, nitrogen, and other materials.

Ultraforming: a low-pressure naphtha-reforming process employing onstream regeneration of a platinum-on-alumina catalyst and producing high yields of hydrogen and high-octane number reformate.

Unassociated molecular weight: the molecular weight of asphaltenes in a non-associating (polar) solvent, such as dichlorobenzene, pyridine, or nitrobenzene.

Unconformity: a surface of erosion that separates younger strata from older rocks.

Unifining: a fixed-bed catalytic process to desulfurize and hydrogenate refinery distillates.

Unisol process: a chemical process for extracting mercaptan sulfur and certain nitrogen compounds from sour gasoline or distillates using regenerable aqueous solutions of sodium or potassium hydroxide containing methanol.

Unit process: one of a grouped operation in a refinery system that can be defined and separated from others.

Universal viscosity: see Saybolt Universal viscosity.

Unresolved complex: the thousands of compounds that a gas chromatograph *mixture (UCM)* is unable to fully separate.

Unstable: usually refers to a petroleum product that has more volatile constituents present or refers to the presence of olefin and other unsaturated constituents.

UOP alkylation: a process using hydrofluoric acid (which can be regenerated) as a catalyst to unite olefins with *iso*-butane.

UOP copper sweetening: a fixed-bed process for sweetening gasoline by converting mercaptan derivatives (RSH) to disulfide derivatives (RSSR) by contact with ammonium chloride and copper sulfate in a bed.

UOP fluid catalytic cracking: a fluid process of using a reactor-over-regenerator design.

Upgrading: the conversion of petroleum to value-added saleable products.

Upper-phase microemulsion: a microemulsion phase containing a high concentration of oil that, when viewed in a test tube, resides on top of a water phase.

Urea dewaxing: a continuous dewaxing process for producing low pour point oils and using urea which forms a solid complex (adduct) with the straight-chain wax paraffins in the stock; the complex is readily separated by filtration.

Vacuum distillation: a secondary distillation process which uses a partial vacuum to lower the boiling point of residues from primary distillation and extract further blending components; distillation under reduced pressure.

Vacuum residuum: a residuum obtained by distillation of a crude oil under vacuum (reduced pressure); that portion of petroleum which boils above a selected temperature such as 510°C (950°F) or 565°C (1050°F).

Vapor-phase cracking: a high-temperature, low-pressure conversion process.

Vapor-phase hydrodesulfurization: a fixed-bed process for desulfurization and hydrogenation of naphtha.

Vertical sweep efficiency: the fraction of the layers or vertically distributed zones of a reservoir that are effectively contacted by displacing fluids.

Visbreaking: a process for reducing the viscosity of heavy feedstocks by controlled thermal decomposition.

Viscosity: a measure of the ability of a liquid to flow or a measure of its resistance to flow; the force required to move a plane surface of area 1 square meter over another parallel plane surface 1 m away at a rate of 1 m per second when both surfaces are immersed in the fluid; the higher the viscosity, the slower the liquid flows.

VGC (viscosity-gravity constant): an index of the chemical composition of crude oil defined by the general relation between specific gravity, sg, at 60°F and Saybolt Universal viscosity, SUV, at 100°F:

$$a = 10sg - 1.0752\log(SUV - 38) / 10sg - \log(SUV - 38)$$

The constant, a, is low for the paraffin crude oils and high for the naphthenic crude oils.

VI (Viscosity index): an arbitrary scale used to show the magnitude of viscosity changes in lubricating oils with changes in temperature.

Viscosity-gravity constant: see VGC.

Viscosity index: see VI.

VOC (VOCs): volatile organic compound(s), compounds that are regulated because they are precursors to ozone; carbon-containing gases and vapors from incomplete gasoline combustion and from the evaporation of solvents.

Volatile compounds: a relative term that may mean (1) any compound that will purge, (2) any compound that will elute before the solvent peak (usually those < C6), or (3) any compound that will not evaporate during a solvent removal step.

Volatile Organic Compounds (VOCs): name given to light organic hydrocarbons which escape as vapor from fuel tanks or other sources, and during the filling of tanks. VOCs contribute to smog.

Volumetric sweep: the fraction of the total reservoir volume within a flood pattern that is effectively contacted by injected fluids.

VSP: vertical seismic profiling, a method of conducting seismic surveys in the borehole for detailed subsurface information.

Waste streams: unused solid or liquid byproducts of a process.

Waste vegetable oil (WVO): grease from the nearest fryer which is filtered and used in modified diesel engines, or converted to biodiesel through the process of transesterification and used in any diesel-fueled vehicle.

Water-cooled vibrating grate: a boiler grate made up of a tuyere grate surface mounted on a grid of water tubes interconnected with the boiler circulation system for positive cooling; the structure is supported by flexing plates allowing the grid and grate to move in a vibrating action; ash is automatically discharged.

Waterflood: injection of water to displace oil from a reservoir (usually a secondary recovery process).

Waterflood mobility ratio: mobility ratio of water displacing oil during waterflooding. (See also Mobility ratio.)

Waterflood residual: the waterflood residual oil saturation; the saturation of oil remaining after waterflooding in those regions of the reservoir that have been thoroughly contacted by water.

Watershed: the drainage basin contributing water, organic matter, dissolved nutrients, and sediments to a stream or lake.

Watson characterization factor: see Characterization factor.

Watt: the common base unit of power in the metric system; one watt equals one joule per second or the power developed in a circuit by a current of one ampere flowing through a potential difference of one volt. One Watt = 3.412 Btu/hour.

Wax: see Mineral wax and Paraffin wax.

Wax distillate: a neutral distillate containing a high percentage of crystallizable paraffin wax, obtained on distillation of paraffin or mixed-base crude and on reducing neutral lubricating stocks.

Wax fractionation: a continuous process for producing waxes of low oil content from wax concentrates; see also MEK deoiling.

Wax manufacturing: a process for producing oil-free waxes.

Weathered crude oil: crude oil which, due to natural causes during storage and handling, has lost an appreciable quantity of its more volatile components; also indicates uptake of oxygen.

Wellbore: the hole in the earth comprising a well.

Well completion: the complete outfitting of an oil well for either oil production or fluid injection; also the technique used to control fluid communication with the reservoir.

Wellhead: that portion of an oil well above the surface of the ground.

Wet gas: gas containing a relatively high proportion of hydrocarbons which are recoverable as liquids; see also Lean gas.

Wet scrubbers: devices in which a countercurrent spray liquid is used to remove impurities and particulate matter from a gas stream.

Wettability: the relative degree to which a fluid will spread on (or coat) a solid surface in the presence of other immiscible fluids.

Wettability number: a measure of the degree to which a reservoir rock is water-wet or oil-wet based on capillary pressure curves.

Wettability reversal: the reversal of the preferred fluid wettability of a rock, e.g., from water-wet to oil-wet, or vice versa.

Wheeling: the process of transferring electrical energy between buyer and seller by way of an intermediate utility or utilities.

White oil: a generic name applied to highly refined, colorless, low-volatility hydrocarbon oils that cover a wide range of viscosity.

Whole-tree harvesting: a harvesting method in which the whole tree (above the stump) is removed.

Wobbe Index (or Wobbe Number): the calorific value of a gas divided by the specific gravity.

Wood alcohol: see Methyl alcohol.

Yarding: the initial movement of logs from the point of felling to a central loading area or landing.

Zeolite: a crystalline aluminosilicate used as a catalyst and having a particular chemical and physical structure.

CONVERSION FACTORS

1 acre = 43,560 sq ft
1 acre foot = 7758.0 bbl
1 atmosphere = 760 mm Hg = 14.696 psi = 29.91 in. Hg
1 atmosphere = 1.0133 bars = 33.899 ft. H_2O
1 barrel (oil) = 42 gal = 5.6146 cu ft
1 barrel (water) = 350 lb. at 60°F
1 barrel per day = 1.84 cu cm per second
1 Btu = 778.26 ft-lb.
1 centipoise × 2.42 = lb. mass/(ft) (hour), viscosity
1 centipoise × 0.000672 = lb. mass/(ft) (seconds), viscosity
1 cubic foot = 28,317 cu cm = 7.4805 gal
Density of water at 60°F = 0.999 g/cu cm = 62.367 lb./cu ft = 8.337 lb./gal
1 gallon = 231 cu in. = 3,785.4 cu cm = 0.13368 cu ft
1 horsepower-hour = 0.7457 kWh = 2544.5 Btu
1 horsepower = 550 ft-lb./second = 745.7 watts
1 inch = 2.54 cm
1 meter = 100 cm = 1000 mm = 10^{6}µm = 10^{10} Å (Δ)
1 ounce = 28.35 g
1 pound = 453.59 g = 7,000 grains
1 square mile = 640 acres

SI METRIC CONVERSION FACTORS

(E = exponent; i.e. E + 03 = 10^3 and E − 03 = 10^{-3}
acre-foot × 1.233482: E + 03 = meters cubed
barrels × 1.589873: E − 01 = meters cubed
centipoise × 1.000000: E − 03 = pascal seconds
darcy × 9.869233: E − 01 = micrometers squared
feet × 3.048000: E − 01 = meters
pounds/acre-foot × 3.677332: E − 04 = kg/m^3
pounds/square inch × 6.894757: E + 00 = kilo pascals
dyne/cm × 1.000000: E + 00 = mN/m
parts per million × 1.000000: E + 00 = mg/kg

Index